FUNDAMENTALS OF STRUCTURAL INTEGRITY

FUNDAMENTALS OF STRUCTURAL INTEGRITY
Damage Tolerant Design and Nondestructive Evaluation

ALTEN F. GRANDT, JR.

WILEY

JOHN WILEY & SONS, INC.

Library of Congress Cataloging-in-Publication Data

Grandt, A. F. (Alten F.), 1945–
    Fundamentals of structural integrity : damage tolerant design and nondestructive
evaluation / by Alten F. Grandt, Jr.
        p. cm.
    Includes bibliographical references and index.
    ISBN 0-471-21459-0 (cloth)
        1. Airplanes—Materials—Testing.   2. Airframes—Design and construction.   3.
Non-destructive testing.   4. Fault tolerance (Engineering).   5. Structural analysis
(Engineering).   I. Title.

TL674.G68   2003
629.134'1—dc21   2002192407

To Barbara and our children, Jennifer, Steven, and Scott

CONTENTS

PREFACE

The objective of this book is to provide a comprehensive description of damage tolerant design and nondestructive evaluation methods for preventing structural failure from manufacturing or service-induced damage. While it is widely recognized that these two technologies are key to achieving structural integrity and there are excellent books on either topic, they are rarely presented together. Indeed, damage tolerance analysis (i.e., fracture mechanics) and nondestructive inspection are usually studied independently, with the result that responsibility for their implementation is often relegated to different individuals within a given organization. It is the author's contention, however, that one should be knowledgeable in *both* damage tolerance and inspection procedures in order to appreciate the many interconnected details that must function in consort for these two critical technologies to prevent structural failures. Moreover, there is a valuable synergy from studying these two subjects together that leads to a much better understanding of their subtle interactions and complementary action.

The text is divided into four major sections: Part I, Introduction; Part II, Damage Tolerance Analysis; Part III, Nondestructive Evaluation; and Part IV, Applications. The objective of the four chapters in Part I is to introduce the concept of structural integrity and how it may be achieved through damage tolerant design and nondestructive inspection. The first chapter provides examples of threats to structural integrity from various sources of preexistent and service-induced damage and sets the overall context for the book. The second chapter reviews basic strength of materials concepts, while the third and fourth chapters overview damage tolerance analysis and nondestructive inspection methods.

The five chapters in Part II provide further details of the capabilities and limitations of the crack growth methodology introduced in Chapter 3. Chapter 5 discusses crack

xix

tip plasticity issues that restrict applications of fracture mechanics concepts and result in peculiar crack growth behavior. Chapter 6 expands the presentation of fracture caused by preexistent cracks, while Chapter 7 discusses the many intricacies of fatigue crack growth. Chapter 8 summarizes methods for obtaining the stress intensity factor solutions needed for crack analyses, and Chapter 9 overviews procedures for assessing the formation of service-induced damage by fatigue or corrosion.

The six chapters in Part III extend the discussion of nondestructive evaluation introduced in Chapter 4. Chapters 10–14 provide in-depth presentation of the major inspection techniques used to locate structural damage: visual, dye penetrant, radiography, ultrasonic, eddy current, and magnetic particle. Other emerging and specialized inspection methods are also discussed in Chapter 15.

Part IV concludes with various applications of structural integrity concepts. Chapter 16 discussed practical aspects of designing a new structure to be resistant to initial manufacturing and service-induced damage. Chapter 17 focuses on issues related to continued or extended service of older structures that near or exceed their original design life goals. Chapter 18 provides final summarizing and concluding comments regarding the structural integrity concepts developed here.

The Appendix gives a brief summary of the capabilities of the AFGROW fatigue crack growth analysis computer program. (This appendix was prepared by the author of AFGROW, Mr. J. A. Harter.) Procedures for obtaining this public domain software are described along with a short description of its use. Some homework problems in Chapters 7, 16, and 17 also employ this software.

This book is intended for senior undergraduate and graduate students in aerospace, materials, mechnical, or civil engineering but will also be of interest to practicing engineers who desire a single reference that combines treatment of both damage tolerant design procedures and nondestructive inspection methods. While emphasis is on aerospace applications, the concepts and principles apply to other types of high-performance structures. The book is envisioned as a text for a one-semester course that overviews both inspection and damage tolerance as a coherent subject, although it could also be used, with supplemental material, for a two-semester sequence on damage tolerance and inspection.

Many people and organizations have helped prepare the author with the background and motivation for this book. He is particularly indebted to the opportunity afforded by his first professional position as a materials research engineer with the Air Force Materials Laboratory (AFML) at Wright-Patterson Air Force Base from 1971 to 1979, where as a freshly minted Ph.D. trained in theoretical fracture mechanics he quickly became engaged with those involved in the practical aspects of designing and maintaining damage tolerant aircraft. This was an exciting decade at AFML as it was during the 1970s that the U.S. Air Force pioneered many of the damage tolerant design criteria that are used today. Conducting basic fracture mechanics research to help implement these new requirements brought the author in contact with many of the nation's top engineers and scientists who are engaged in related research and development. This period also introduced the author to the nondestructive inspection community and provided the opportunity to learn first hand how nondestructive evaluation must work in consort with damage tolerance analysis to ensure structural integrity. The author's subsequent move to Purdue University in 1979 provided the additional opportunity for organizing and presenting this industrial exposure in a teaching context and to continue basic research in the structural integrity area.

The author is obliged to many individuals for making this book possible and, although all cannot be named, would like to recognize a few with special mention. T. Swift, C. F. Tiffany, and the late Dr. J. W. Lincoln reviewed the orginal conceptual outline and provided valuable suggestions and encouragement. These gentlemen are truly icons in the damage tolerance community, and students would do well to strive for careers that emulate the impact they have made on structural integrity. The author's first exposure to fracture mechanics concepts came as a graduate student at the University of Illinois, and he is indebted to his advisor, the late Professor G. M. Sinclair, and to Dr. J. P. Gallagher for providing rigorous instruction in the fundamentals of crack growth analysis. Dr. R. L. Crane reviewed the nondestructive evaluation sections and provided many valuable comments and suggestions. The review of other sections by Dr. T. Nicholas is also gratefully acknowledged. C. Annis, R. Bucci, A. Gunderson, D. Sexton, D. Tritsch, and J. Wildey were helpful sources for technical information reported here. Special thanks are due to J. A. Harter for preparing the description of the AFGROW software included in the Appendix. The author also wishes to acknowledge his current and former graduate students who obtained the original research results described in the text and to thank them and the many other students who reviewed various draft sections of this book. Any errors in the text remain, of course, the responsibility of the author. The financial support provided by the Raisbeck Engineering Distinguished Professorship at Purdue University is also most appreciated. Finally, the author wishes to thank his parents, his wife, Barbara, and our children, Jennifer, Steven, and Scott, for their inspiration and encouragement.

PART I

INTRODUCTION

The objective of the four chapters in Part I is to introduce the concept of structural integrity and how it may be achieved through damage tolerant design and nondestructive inspection. The first chapter provides examples of threats to structural integrity from various sources of preexistent and service-induced damage and sets the overall context for the book. The second chapter reviews elementary topics of mechanics of materials, and the third and fourth chapters overview damage tolerance analysis and nondestructive inspection methods. The damage tolerance analysis concepts introduced in Chapter 3 are then developed more extensively in Part II, and Chapter 4's introduction to nondestructive evaluation is further expanded in Part III. Part IV describes applications of these structural integrity concepts to various types of structures.

CHAPTER 1

INTRODUCTION

1.1 OVERVIEW

Damage tolerant design and nondestructive inspection are key tasks for ensuring that structures operate safely for extended periods of service. Damage tolerance is the ability to resist fracture from preexistent cracks for a given period of time and is an essential attribute of components whose failure could result in catastrophic loss of life or property. Nondestructive inspections define the maximum size of life-limiting defects that could be present at a given time. Inspection requirements are determined by the anticipated service loads and by the damage tolerance designed into the structure. This chapter demonstrates the importance of damage tolerance and nondestructive inspection by discussion of several service mishaps that resulted from undetected manufacturing and/or service-induced damage. Various structural failure mechanisms are reviewed, followed by an introduction to inspection techniques for locating cracks and other forms of structural damage. The chapter concludes with a brief description of several design criteria for achieving long-life, damage tolerant structures.

Although structural failures are rare, those that do occur are often related to preexistent manufacturing or service-induced damage (e.g., material anomalies or fatigue cracking). The two key elements for preventing such failures are nondestructive inspection and damage tolerant design. Nondestructive inspection provides the first step toward structural integrity by identifying damaged components that must be repaired or discarded. Since all inspection techniques have limitations, however, some components will contain undetected cracks. Damage tolerant design then provides the second line of defense against premature fracture by incorporating structural configurations and materials that are resistant to crack growth and fracture.

Damage tolerance is the ability of a structure to resist a specified amount of damage for a given period of service. The damage in question may result from initial manufacturing errors or service-induced degradation, whereas the time period of interest could represent the desired service life, the time between maintenance actions, or the period required to safely cease operation and remove all personnel from danger. Damage tolerance is the protection provided to unanticipated sources of impairment and is a key attribute for structures and machines whose failure could result in catastrophic loss of life or property.

A dramatic example of a damage tolerant structure is the B-17 Flying Fortress shown in Figure 1.1. This photograph shows a gaping hole in the bomber's fuselage incurred during a midair collision with a German fighter during World War II. The ultimate goal of a damage tolerant design is vividly demonstrated by the ability of this aircraft to resist such extensive trauma and land safely. Although it is not possible to foresee any particular mishap, the designer must provide redundant load paths, select "tough" materials, and employ other design and maintenance strategies that lead to such "forgiving" structures.

An example of poor damage tolerance, combined with inadequate inspection capability, is represented by the April 11, 1965, rocket motor fracture shown in Figure 1.2. This 75-foot-long by 22-foot-diameter (23 × 7-m) high-strength steel cylinder exploded during a hydrostatic proof test at approximately one-half of the desired pressure load [1]. The fracture origin was traced to an undetected 1.5-in-long by 0.1-in-wide (3.8 × 0.25-cm) flaw in one of the welds that joined the 3/4-in-(2-cm-) thick steel plates to form the rocket motor case. Although the weld region had been inspected

B-17F/Bf-109 midair collision on February 1, 1943 over Tunisia.

B-17 flew 90 minutes and landed safely.

(USAF Museum Photographs)

Figure 1.1. Example of a damage tolerant aircraft. This Boeing B-17 Flying Fortress collided with another aircraft during World War II and, although sustaining large amounts of structural damage, landed safely (Photographs courtesy of the U.S. Air Force Museum.)

Figure 1.2. Example of a structure with poor damage tolerance. The 1965 fracture of this 260-in-diameter solid-fuel rocket case originated at the 1.5-in-long by 0.1-in-wide welding flaw shown in the lower photograph. Failure occurred during a hydrostatic proof test at an internal pressure of 542 psi, approximately 56% of the desired proof pressure. (From *ASTM Materials Research Standards,* Vol. 7, No. 3, March 1967. Copyright ASTM. Reprinted with permission.)

several times by dye penetrant, radiographic, and ultrasonic test methods, the critical flaw was not found until after fracture occurred. The fact that such a 1.5-in-long (3.8-cm) manufacturing flaw could cause catastrophic failure of a lightly loaded structure indicated significant shortcomings in the design and manufacturing procedures of that time and highlighted the need for increased emphasis on damage tolerance and improved inspection methods.

The objectives of this book are to summarize the nondestructive inspection and damage tolerant design concepts that guarantee structural integrity for long periods of service. The first goal entails determining the maximum crack size that could be present following various types of nondestructive inspection, whereas the second involves evaluating the structural resistance to undetected damage. This latter objective includes determining both the failure load for a given crack size (residual strength) and the length of time for small, subcritical cracks to grow to a size that will cause failure in service. Explicit criteria are developed for analyzing crack-induced fracture and subcritical crack growth mechanisms (e.g., fatigue or stress corrosion cracking). These criteria are then incorporated into design procedures for selecting materials and structural configurations that resist crack growth for specified periods of service.

1.2 MANUFACTURING AND SERVICE-INDUCED DAMAGE

Although it is rare to operate a structure with known damage, failure can occur from preexistent flaws that escape detection. Many structural components, for example, exhibit the typical "bathtub" service life curve shown in Figure 1.3. Following initial "shakedown" failures caused by latent defects introduced during manufacturing or materials processing, the failure rate decreases for a period, until service-induced damage becomes the life-limiting feature. Typical examples of initial manufacturing and service degradation are discussed below.

1.2.1 Manufacturing Damage

Manufacturing errors that occur during material processing, machining, and structural assembly can lead to premature component failure. Examples of material flaws include porosity, constituent particles, inclusions, forging or casting defects, and

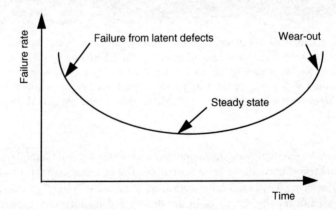

Figure 1.3. Schematic representation of "bathtub" failure rate curve showing high incidence of initial failures due to latent defects and high failure rate late in life caused by service-induced damage.

improper thermal/mechanical treatment of the basic alloy. Machining problems include gouges and tears, rough surfaces, burrs and scratches at fastener holes or fillet locations, and other local material trauma caused by improper tool usage. Material surface condition plays an especially critical role in preventing damage formation. Final component assembly offers additional opportunity for damage in the form of "nicks and dings" from rough handling, improper welds, permanent deformations or cracking from force fits, as well as missing or damaged subcomponents. These machining and assembly errors are often accompanied by harmful residual stresses introduced during the manufacturing process.

Figure 1.4 shows an infamous example of a manufacturing flaw that caused the loss of a new F-111 aircraft and its crew on December 22, 1969. In this case, a relatively

Figure 1.4. This 1969 failure of a new F-111 aircraft was caused by an undetected forging defect (lower photograph) that quickly grew to failure by fatigue in the high-strength steel wing pivot fitting structure (top photo).

small forging defect in the high-strength D6-ac steel wing attachment structure grew by fatigue and led to catastrophic fracture after approximately 100 hours of routine flight. Although this component was inspected at several stages during manufacturing and postfailure examination showed residue from a dye penetrant inspection, the 1-in- (2.5-cm-) long forging defect was not recognized and wound up in a critical component where it quickly grew to failure by cyclic loading [2].

Again, the fact that such a small defect could escape detection and lead to premature loss of the aircraft under normal operating conditions caused great concern in the technical community. One of the key results of this F-111 incident was the reassessment of how U.S. Air Force (USAF) aircraft are designed and inspected. It became widely recognized that nondestructive inspection alone could not prevent premature failure and that it was necessary to employ additional means to protect the structure from unanticipated damage. This accident, along with several other incidents such as the rocket failure shown in Figure 1.2, led to design criteria that *assume* the structure contains *initial* cracks based on the size that could go undetected [3, 4]. The designer's goal then is to select materials and structural configurations that resist the growth of these potential undetected cracks for specified periods of time or until they can be readily detected and repaired. The success of such an approach is, of course, dependent on reliable inspection procedures that result in small initial crack size assumptions.

1.2.2 Service-Induced Damage

In addition to normal wear and tear, service-induced damage includes several forms of corrosion, creep, or fatigue crack formation following cyclic loading. Overloads, thermal degradation, hydrogen embrittlement, fretting, or other forms of abuse suffered during service may also bring about permanent deformations and other detrimental changes in material properties. These damage mechanisms may also be aggravated by improper maintenance or repair during prior service. Finally, the possibility for foreign object damage (FOD) in the form of hail damage, bird strikes, accidental impacts, uncontained engine bursts, or battle damage in the case of military vehicles must also be considered. Again, these various forms of service-induced trauma are often accompanied by detrimental residual stresses that result from component abuse.

Examples of service-induced damage are given by the mid–air collision between the World War II bomber and fighter discussed previously and by the two commercial airline incidents shown in Figures 1.5 and 1.6. Figure 1.5 actually represents two examples of service-induced damage (engine and airframe) and demonstrates a typical chain of failure events that must be considered by the designer. In this 1997 incident, a turbine engine exploded on takeoff, throwing high-speed engine fragments through the fuselage, killing two passengers [5]. The initial engine fracture was caused by fatigue cracking at tie-rod holes in the fan disk. This "engine burst" then precipitated a second event when metal shrapnel penetrated the adjacent aircraft structure. Since this accident occurred on takeoff, the fuselage was not pressurized, and penetration by the high-speed engine fragments did not fracture the airframe. If the engine had

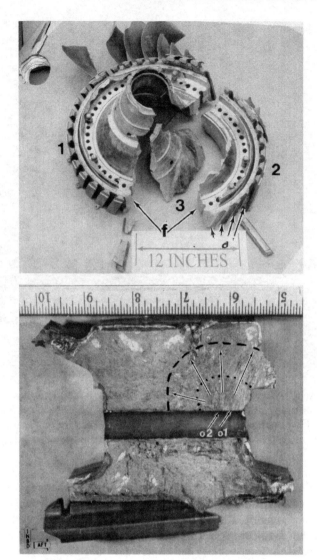

Figure 1.5. Example of 1997 fatigue failure of turbine engine component that led to FOD incident when fan disk fragments penetrated aircraft fuselage. Disk failure resulted from a fatigue crack that originated along the bore of a bolt hole. (Photographs courtesy of the National Transportation Safety Board.)

exploded during flight, however, when the fuselage was pressurized, one must ensure that this FOD would not cause catastrophic loss of the entire aircraft.

Figure 1.6 shows another dramatic example of service-induced damage that led to partial defeat of the crack arrestment features built into modern airframes. On April 28, 1988, a large segment of fuselage skin suddenly peeled away during flight of a

Figure 1.6. April 1988 example of service-induced damage that led to explosive decompression and loss of large portion of fuselage skin when small fatigue cracks suddenly linked together. The subsequent fracture was eventually arrested by fuselage frame structure, and the aircraft landed safely (Photograph courtesy of the National Transportation Safety Board.)

Boeing 737 aircraft that had seen many hours of service [6]. Although one flight attendant was lost upon the initial explosive decompression, the aircraft flew for another 20 minutes and landed safely. The failure originated along a row of rivets that joined two pieces of the thin aluminum skin. This particular aircraft had a large number of flights (i.e., many pressurization cycles) and experienced degradation of some adhesive bonds, resulting in formation of small fatigue cracks at many of the rivet holes. These small cracks suddenly linked together, forming a large lead crack that rapidly propagated through the pressurized skin, before eventually arresting at fuselage frames.

As mentioned in the previous example, aircraft designers do anticipate that large cracks can develop in service (caused, perhaps, by FOD from an engine burst), and provision is provided to arrest these skin cracks by fuselage frames and stiffeners before they can cause catastrophic fracture. It is routine, for example, to demonstrate by test and analysis that a crack that forms in the skin between two stiffeners cannot propagate beyond a third stringer without arresting, thus allowing the internal pressure to escape, relieving the load in the fuselage skin. When the lead crack formed in this case, however, it spread across several frames before arresting and led to much more structural damage than anticipated. The problem in this instance was that the lead crack did not propagate into new, undamaged material, but instead grew along a row of holes that already contained numerous small cracks developed during the many hours of prior service. This multiple-site-damage (MSD) problem was a new failure mode unique to "aging aircraft" and led to much research on the MSD issue. Note, however, that although the MSD defeated the primary crack arrestment mechanism, the lead crack was eventually stopped before the entire aircraft was lost.

1.3   FAILURE MODES

The goal of a structural engineer is to design components that effectively perform a specified function for a given period of service. The design process includes determining member sizes and shapes, selection of materials, and in many cases, specifying manufacturing processes. Implicit in the design goal is the requirement that the particular item operate in a safe manner for a given period (service life) without endangering human life or other property. To meet this objective, it is essential to understand and anticipate all possible failure modes that could occur in service and to provide the structure with the ability to resist these failure mechanisms.

Typical structural failure modes are outlined in the following subsections. When examining these failure mechanisms, it is important to remember several points. First, one needs to distinguish between pristine structure that has been manufactured to "ideal" standards and components that have been subjected to wear and tear in service or those which could contain initial manufacturing damage. Indeed, the initial structural and material condition has a tremendous influence on which failure modes will limit structural performance. Second, one must expect damage to develop and grow in service (i.e., by fatigue and/or corrosion) and that an initially "perfect" structure will likely develop fatigue cracks, corrosion, or some other form of damage that could ultimately lead to failure. Thus, the limiting failure modes often change with use. Finally, it must be recognized that structural damage will eventually form and that all structures will have a finite life.

The paramount operational goal is to ensure that the structure is retired or repaired before its structural life is exceeded. This objective requires that the designer explicitly specify life goals for the object of interest (e.g., number of aircraft flights, years a bridge is in service, number of revolutions of a rotating component). The designer and operator must also anticipate worst case scenarios for the service loads and environment and must determine which forms of damage are likely to develop with use. As background to this challenge, it is helpful at this time to provide a brief review of typical structural failure modes. This short summary is intended to set the context for damage-related failure mechanisms that are discussed in more depth later.

1.3.1   Elastic Deformations

Since all structures are made from deformable materials, an important analysis task is to determine changes in component dimensions caused by applied loads. As discussed more thoroughly in Chapter 2, many materials behave in an elastic manner when forces are small and result in deflections that are readily predicted by strength-of-materials concepts (see Section 2.6.1). These elastic deformations are controlled by the stiffness of the material and are recoverable when the component is unloaded (i.e., it returns to its original dimensions).

Although elastic deformations are often not fatal by themselves, there are many instances when they can lead to structural "failure." Excessive deformations can, for example, eliminate tight clearances, resulting in interference between moving components that prevent a machine from performing properly. Another failure example comes from the study of airfoils, where it is possible for aerodynamic forces to distort the cross section of a wing so much that it loses lift. The unloaded wing then quickly returns to its original shape (i.e., elastic recovery), where it once again develops the original aerodynamic loading and repeats the load–deform–unload process over and over again. This aerodynamic "flutter" can result in resonant vibrations that deteriorate to unstable conditions or lead to fatigue cracking.

1.3.2   Inelastic Deformations

If the applied forces exceed a given value for a particular material, the loaded member may continue to deform without fracturing but will experience additional "inelastic" deformations. Although some of the total changes in dimensions may be recovered upon unloading (i.e., the elastic component), the member has permanently changed shape. The maximum load that can be applied without causing inelastic deformation is related to the material yield stress and is discussed in more detail in Section 2.6.2. Although most structures are designed to prevent yielding, it should be noted that this failure mode may not lead to immediate fracture, and the component may continue to carry additional load. Thus, inelastic deformations often provide an additional measure of safety that is important for damage tolerant structures.

1.3.3   Buckling

Buckling is a failure mode unique to slender members loaded in compression or shear. In this case, an instability develops when the compression loading results in lateral deflections that, in turn, cause an additional bending moment that leads to further deflections and increased bending. Buckling can occur at very small elastic loads and is controlled by the material stiffness, the unsupported length, and the component cross-sectional moment of inertia. Buckling is an important failure mode for thin, compression-dominated structure (e.g., upper wing skins in aircraft, building support columns) and is discussed in more detail in Section 2.6.3.

1.3.4   Fatigue

Cyclic fatigue is the failure mode associated with *repeated* loading and is one of the main factors that limits the life of mechanical devices. Consider, for example, a structural member that it is subjected to repeated application and removal of remotely

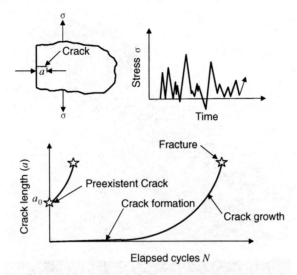

Figure 1.7. Schematic representation of fatigue failure mechanism showing crack formation and propagation following a period of cyclic loading.

applied load cycles, as shown in Figure 1.7. The peak values of the maximum and minimum load per cycle may be tension or compression and may change during the component life (i.e., variable-amplitude loading). No single load application is large enough, however, to fracture new components.

After repeated load cycles, small cracks will form, often at multiple locations in the structure. (This initial period of cycling is known as the crack nucleation or initiation life.) At the outset these cracks are too small to cause fracture, but they do extend slowly after repeated cycling. Eventually some coalesce, leading to a dominant crack(s) that continues to grow in a stable manner. Finally, the dominant crack reaches a size that causes fracture, and the member fails in a sudden, catastrophic manner. As shown in Figure 1.8, the final fracture surface often exhibits a "brittle" appearance and may demonstrate characteristic markings that are remnants of the crack tip position at various stages in life. These "beach marks" are visible to the naked eye and are one of the distinguishing characteristics of a fatigue fracture. More detailed examination of the fracture surface under a microscope may reveal additional patterns of closely spaced lines, or "striations," that record the crack tip position after each cycle of loading (see Figure 1.9). The striation spacing is the fatigue crack growth rate (i.e., increase in crack length per cycle), and if present, striations offer conclusive proof of a fatigue failure.

The fatigue failure mode is one of the main topics of this text and is discussed at great length in subsequent chapters. For the present time, however, note several points about this important failure mechanism. First, fatigue requires cyclic loading to occur. The number of cycles that can be applied before fracture takes place (i.e., the fatigue life) depends on the amplitude and mean value of the cyclic load, the

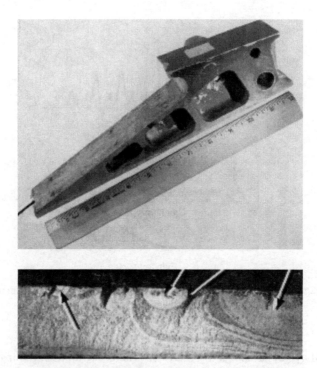

Fracture Surface

Figure 1.8. Fatigue failure of aircraft wing outboard flap track wing fitting (top photograph). Note brittle appearance of fracture surface (lower photograph) and beach markings that indicate growth and coalescence of multiple fatigue cracks.

sequence of applied loads (when variable-amplitude load histories are involved), and the component shape and material. The condition of the specimen surface is extremely important, as rough surfaces, notches, or material impurities can rapidly lead to cracks, eliminating the "crack formation" portion of the fatigue life. (Recall, e.g., the F-111 component shown in Figure 1.4, where an undetected forging flaw led to a fatigue crack that quickly grew to failure.)

Residual stresses can also have a significant influence on the fatigue life. Intentional introduction of compressive residual stresses by shotpeening or coldworking can be quite beneficial, for example, and is a common technique for extending the fatigue life of a component. On the other hand, residual tensile stresses, such as those introduced by welding or other manufacturing processes, can be extremely detrimental to fatigue.

Since fatigue can occur at relatively small cyclic loads and is so dependent on quality of construction, fatigue is a major consideration in the design and manufacture of many mechanical devices. Nondestructive inspection is essential to locate manufacturing anomalies that could lead to premature fatigue cracking and to detect cracks that do form before they grow to failure.

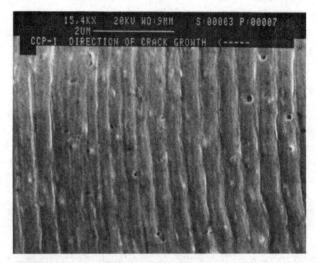

Figure 1.9. Scanning electron microscope photograph of fatigue fracture surface showing fatigue striations that are remnants of crack tip position after each cycle of loading. The length of the scale bar in the photo legend is 20μm (7.9 × 10^{-4} in). (From Reference 7.)

1.3.5 Creep

Creep is a time dependent failure mode that occurs when a member continues to deform under extended application of a *static* load. When the load is removed, some of the elongations may be recovered (elastic), whereas some deformations may remain at zero load (inelastic). These latter changes in dimension may be permanent or the member may gradually return to its original state after an additional period of time as these deformations continue to "relax." Creep is an important failure mode for metal components that operate for long periods at elevated temperatures and high loads (e.g., turbine engine blades). In addition, many polymers creep at room temperature.

1.3.6 Corrosion

Corrosion is material degradation due to chemical attack and can occur in several forms—galvanic (dissimilar metal), pitting, exfoliation, intergranular attack, filiform, or, when combined with the presence of a tensile stress, stress corrosion cracking. This time dependent failure mechanism is highly dependent on the structural material and environment combination in question and is often accelerated by increasing temperature. Corrosion is a complex chemical phenomenon that can cause general thickness loss (and a corresponding increase in stress) as well as stress concentrations (i.e., pits and other localized areas of attack) that lead to fatigue cracking or fracture. Corrosion is difficult to predict, and its prevention requires careful materials selection, protective coatings, and periodic maintenance. Although corrosion can occur independent of applied loading (i.e., aircraft can continue to corrode without being flown), it frequently acts in conjunction with static or cyclic loading (e.g., stress corrosion or corrosion fatigue) to represent a particularly dangerous failure mode. Various aspects of corrosion are discussed in further detail in later chapters.

1.3.7 Fracture

Final catastrophic failure results when the member cleaves into two or more parts. Although fracture may occur in pristine structure that is subjected to overloads, it is frequently initiated at smaller loads by preexistent flaws (e.g., the rocket motor of Figure 1.2) or by service-induced damage (e.g., the fatigue failures of Figures 1.4, 1.5, 1.6, and 1.8). Fracture can occur in a ductile manner that requires considerable expenditure of energy but can also happen suddenly with little warning. This latter result may exhibit "brittle" behavior with little evidence of plasticity and is a particularly dangerous failure mode. The fracture resistance of a material is characterized by its "toughness," a quantity that, as reviewed in Section 2.5, may be measured in smooth tension specimens. Toughness can be a function of temperature and loading rate as well as geometric constraint to yielding (e.g., specimen thickness).

The main focus of this book, however, deals with crack-induced fracture and is concerned with determining the fracture load as a function of crack size. As shown schematically in Figure 1.10, for example, the stress required to fracture a tension member decreases significantly with the presence of a crack. Since cracks can grow in a subcritical manner by fatigue or stress corrosion cracking, the maximum load-carrying ability of the structure can decrease with time, as shown by the schematic

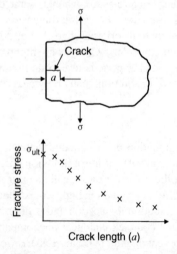

Figure 1.10. Schematic representation of the reduction in fracture stress in a tension member caused by the presence of a crack. Note that the material ultimate stress, measured in an uncracked member, no longer represents the maximum load that can be sustained when a crack is present.

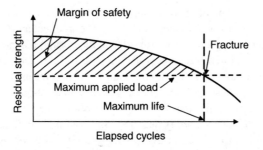

Figure 1.11. Schematic representation of the reduction in residual strength as a function of the number of applied load cycles. Note that fatigue crack growth decreases the maximum load-carrying ability until fracture occurs at the applied service load.

"residual strength" diagram in Figure 1.11. Note here that there is an initial margin of safety provided by the fact that the fracture stress exceeds the maximum applied load. This residual strength, however, decays with time (as small cracks form and grow by cyclic loading or as corrosion develops over time), until fracture occurs at the normal service load. At this time the component is no longer safe and must be retired or repaired. Chapter 3 introduces the concept of "crack toughness" and overviews the linear elastic fracture mechanics approach for determining residual strength. Additional details for predicting and preventing crack-induced fracture are then discussed in subsequent chapters.

1.4 NONDESTRUCTIVE EVALUATION

This section provides a brief overview to various nondestructive evaluation (NDE) techniques that are discussed in more detail in Chapter 4 and subsequent chapters. Since fracture loads and component life are extremely sensitive to preexistent damage, NDE plays a key role in preventing structural failures. Inspections of new structures are essential to detect manufacturing flaws that can cause immediate fracture or serve as sources for early fatigue cracking. Repeated inspections following extended periods of service are also needed to locate fatigue cracks, corrosion, FOD, or other forms of accidental damage before they can grow to a size that compromises structural integrity.

The main NDE techniques employed in practice are visual, radiography, ultrasonic, eddy current, magnetic particle, and dye penetrant methods. (These methods are discussed in detail in Chapters 10–14.) Although limited to relatively large anomalies, visual inspection is the most common method for locating surface damage, especially when employed with magnification and lighting aids. Internal flaws not visible to the human eye can often be found by X-rays or gamma rays (radiography) or by use of high-frequency sound waves (ultrasonics) that penetrate deep into the interior of the structure. Small surface or near-surface cracks can be readily found in electrically conductive materials by examining changes in small eddy currents generated near the component surface by an alternating magnetic field. Surface cracks in ferromagnetic materials may also be located by magnetizing the component and then allowing small magnetic particles to migrate to interruptions in the normal magnetic field caused by cracking. Brightly colored fluids may also be used to seep between crack faces and locate surface-breaking cracks by the popular dye penetrant inspection method.

When employing these and other nondestructive inspection methods, however, it is important to keep in mind that all NDE techniques have limits regarding the size of the anomaly that can be *reliably* detected. Successful NDE depends on many factors, including selection of the correct technique for the given application, proper calibration and operation of test equipment, surface preparation, flaw shape and orientation, and a host of human factors (e.g., inspector training, experience, alertness, confidence, expectations). While extremely small cracks can often be located under ideal conditions, it is unfortunately possible to miss much larger cracks during field-level inspections (e.g., the F-111 accident). Thus, when evaluating the results of an inspection, it is important to focus on determining the *largest flaw* that can be *missed* rather than emphasizing the smallest crack that can be found. As discussed later, various statistical methods are available to establish probability-of-detection limits for a given inspection application.

1.5 FATIGUE DESIGN CRITERIA

Having noted that limitations in inspection methods result in the possibility that structures contain small, undetected cracks that can grow by fatigue, now consider a "big picture" view of various design strategies to prevent component failure. Whereas there are several approaches to structural integrity, the general keys to long-life designs are proper materials selection, setting low stress levels, providing multiple load path and crack-arresting features, and implementation of rigorous inspection programs.

As engineers pushed design envelopes in the pursuit of efficiency with advanced materials and construction techniques, shortcomings began to appear in design criteria, resulting in unanticipated failures (e.g., the rocket motor case and the F-111 fractures mentioned earlier). These failures resulted in a flurry of research that advanced understanding of the particular problem at hand and led to improved design methodology and inspection methods. Development of fracture mechanics techniques during the 1960s and 1970s, for example, led to sophisticated methods to explicitly treat initial damage as a design variable when evaluating residual strength and fatigue life. Whereas specific design requirements for commercial and military aircraft are detailed in References 8–9, the present discussion is an overview intended to help set the context for subsequent development of technical methodology to accomplish structural integrity objectives.

1.5.1 Infinite-Life Design

Early designers treated fatigue by attempting to keep component stresses below the endurance limit for the material of interest. The endurance limit is the maximum constant-amplitude stress that can be applied to a *pristine* specimen without causing a fatigue failure (i.e., specimens do not fracture after application of more than 10^7 cycles of loading). The endurance limits for steels are typically on the order of half the ultimate stress, but other structural materials such as aluminum do not have a distinct endurance limit and may eventually fail at even very small cyclic loads.

Although the endurance limit concept was useful for solving fatigue problems that originally plagued railroad equipment and other machines that made the industrial revolution possible, this approach has many serious limitations. The endurance limit, if it exists at all for the material of interest, is extremely sensitive to the condition of the test specimen. Notches, small scratches, or other "nicks and dings" serve as stress concentrations that quickly cause localized fatigue cracking and greatly reduce the endurance limit so that little, if any, damage tolerance can be achieved by this approach. Moreover, residual stresses introduced by manufacturing or by localized yielding during variable-amplitude loading often have deleterious effects. Since it is impractical, if not impossible, to design high-performance structures for infinite lives by the simple endurance limit methodology, engineers have recognized that infinite life may not be an achievable goal for many applications. It is now accepted that most structures will have a finite fatigue life. The objective of the designer, then, is to determine what that service life *should be* and then to ensure that the actual component *exceeds* that goal.

1.5.2 Safe-Life Design

The safe-life approach treats fatigue as a crack nucleation process and does not explicitly consider the possibility for crack growth (failure is assumed when cracks are first formed). Since fatigue tests often demonstrate considerable "scatter," as shown schematically in Figure 1.12, extensive component testing is needed to determine the expected mean life for the desired service loading. The mean life is then divided by a safety factor (often 4) to determine the maximum allowable service life that provides a low probability for fatigue failure. The safe-life approach led to several inadequate aircraft designs in the 1960s, however, and its current use is discouraged,

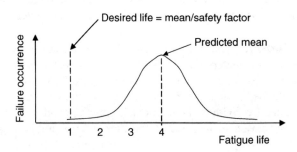

Figure 1.12. Schematic representation of the safe-life design approach showing the variation in fatigue life obtained by testing components to the desired load spectrum. The maximum allowed service life is obtained by dividing the mean test life by a factor of safety (often equal to 4).

although aircraft landing gear and many helicopter components are still designed to safe-life criteria. Some failures of U.S. Air Force safe-life designed aircraft are described below [4]:

- *KC-135*   Although the safe-life for this transport/tanker aircraft was determined to be 13,000 flight hours, there were 14 cases of unstable cracking in lower wing skins that occurred between 1800 and 5000 hours. These premature failures resulted in a costly wing modification for the entire fleet at 8500 service hours.
- *F-5*   One of these fighters failed by fatigue cracking in the lower wing skin after 1900 flight hours although the safe-life for this area had been shown to be 4000 hours.
- *F-111*   The safe-life for this aircraft was 4000 flight hours, but as discussed previously (recall Figure 1.4), one aircraft was lost to fatigue after only 105 hours of service.

The problem in these cases, and the Achilles heel for the safe-life approach, was the presence of *unanticipated* structural or material *damage* in specific aircraft that greatly reduced the crack nucleation portion of the fatigue process (the safe-life is established by testing initially *undamaged* components). Although the mean life was divided by 4, this safety factor was not large enough to account for the reduction in fatigue life caused by the undetected initial damage in these aircraft.

1.5.3   Damage Tolerant Design

In order to overcome this shortcoming of the safe-life approach, damage tolerant design methods were developed that *assume* the structure contains *initial cracks*. As mentioned previously, damage tolerance is the ability of a structure to resist fracture from cracks of a given size for a specified time period. The initial crack size is usually based on inspection limits and is expected to be a conservative assumption (aircraft are not allowed to fly with known cracks). There are two general approaches, with variations, that may be followed to guarantee that the structure (with its assumed cracks) does not fail in service: slow crack growth and fail-safe design.

Slow Crack Growth   The slow crack growth (or safe crack growth) design criterion selects component materials and sets stress levels so that the assumed preexistent crack will not grow to failure during service (see Figure 1.13) and is the normal approach for single load path structure. For increased safety, the allowed service life is usually obtained by dividing the total crack growth period by a factor of 2. The component would have to be inspected at this time before continued operation would be permitted.

Fail-Safe Design   The fail-safe design approach shown schematically in Figure 1.14 is another technique for achieving damage tolerance. The goal here is to employ

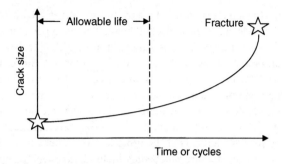

Figure 1.13. Schematic representation of the slow crack growth approach to damage tolerant design. The component is assumed to contain a preexistent crack that is not allowed to grow to failure during the life of the structure.

multiple load paths and/or crack arrest features so that a single component failure does not lead to immediate loss of the entire structure. The load carried by the broken member is immediately picked up by adjacent structure and total fracture is avoided. It is essential, however, that the original failure be detected and promptly repaired, because the extra load they carry will shorten the fatigue lives of remaining components. The 1977 loss of a Boeing 707 transport aircraft after 16,723 flights, for

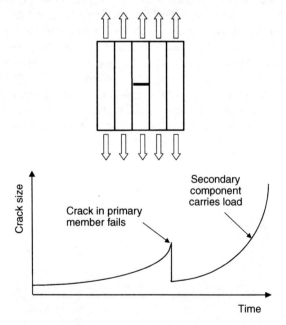

Figure 1.14. Schematic representation of the fail-safe approach to damage tolerant design. Multiple load paths and/or crack arrestment features are designed to contain fracture of an individual component without leading to catastrophic loss of the entire structure.

example, occurred when its right-hand horizontal stabilizer separated shortly before landing [4]. In this case, a fatigue crack had formed in the upper spar cap and grew for 7200 flights before fracturing the spar. The aircraft then flew for another 100 flights as redundant components took up the extra load as intended. Since the original spar fracture was not detected during its postfailure period, however, cracks eventually developed in the remaining fail-safe structure, leading to final catastrophic loss of the aircraft. This accident vividly demonstrated that the fail-safe approach must be accompanied by *thorough* inspections that locate the original member's failure. It, along with other similar incidents at the time, led to changes in certification procedures that require use of fracture mechanics methods to develop supplementary inspection documents (SIDs) for continued operation of aging aircraft. (Further aspects of the fail-safe design approach and its relationship to nondestructive inspection are discussed in Chapter 16.)

1.5.4 Retirement for Cause

As nondestructive inspection and crack growth analysis methods have matured, it has become possible to employ a "retirement for cause" strategy to extend the lives of existing hardware that have reached their theoretical life limits. Although the initial design life may have been based on crack formation concepts (i.e., safe-life design), the actual current damage state is established through rigorous nondestructive evaluation. Fracture mechanics concepts then determine the remaining additional service life (if any) that can be safely exploited for individual components.

The retirement-for-cause approach is shown schematically in Figure 1.15, where periodic inspections locate damaged components that are then repaired or replaced. The structures are subsequently returned to service for another specified period, when the inspection/repair process is repeated. The inspection intervals are based on the time for an *undetected* crack to grow to fracture (divided by an appropriate safety factor). This process can be repeated indefinitely until the costs of inspection and repair become prohibitive (many cracks would be expected to develop after long lifetimes). The basis for this procedure is to *reliably* determine the maximum crack

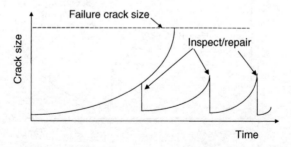

Figure 1.15. Schematic representation of a maintenance plan where periodic inspections and repairs are designed to keep preexistent cracks from growing to a size that causes failure of the structure.

size that could exist following an inspection and repair cycle and then determine the remaining operational life of a component with this potential crack.

1.6   CONCLUDING REMARKS

The objective of this chapter has been to introduce the challenges facing the engineer who seeks to design, build, and maintain safe structures. Ensuring structural integrity requires anticipation of all possible failure mechanisms and then providing adequate resistance to these various threats. Whereas every effort is made to ensure high-quality construction, experience has shown that it is impossible to build complex mechanical devices that are free of initial manufacturing or material flaws. Moreover, accidental damage may occur during use, along with fatigue, corrosion, or other forms of material degradation. Thus, nondestructive evaluation, both before and during service, is an essential step for achieving structural reliability. Damage tolerant designs provide additional protection against such trauma by keeping stress levels small, employing damage resistant materials, and providing crack-arresting structural features.

The foundations for structural integrity have been compared with the stability provided by a "three-legged stool." One leg is a rigorous inspection program that locates manufacturing or service-induced damage before it can lead to failure. Another leg is superior residual strength that ensures the catastrophic fracture load always exceeds the largest applied service load. The third leg of the damage tolerance stool is a long subcritical crack growth life that gives many opportunities to locate and repair any damage that does develop before it grows to a size that causes final fracture. These three legs work together to ensure structural integrity throughout the service life. Subsequent chapters of this book overview various NDE techniques and the fracture mechanics methodology needed to develop and implement comprehensive fracture control plans that guarantee success of this three-pronged approach to structural integrity.

PROBLEMS

1.1   Briefly discuss the "bathtub" failure rate curve that has been observed for many types of structures.

1.2   Briefly describe the fatigue failure mode. Give some examples from your personal experience.

1.3   Discuss the impact of preexistent damage of the fracture behavior of a component. Give some examples from your personal experience.

1.4   Describe the characteristic appearance of a fatigue failure. What microscopic and macroscopic evidence is there for the fatigue failure mode?

1.5   Briefly describe the corrosion failure mode. Give some examples from your personal experience.

1.6 Briefly describe and compare the elastic, inelastic, and creep deformation failure modes. How are these failure modes influenced by initial structural damage? Give some examples from your personal experience.

1.7 Briefly describe the buckling failure mode. How is buckling influenced by initial structural damage? Give some examples from your personal experience.

1.8 Briefly describe and contrast the failure modes that you would expect to control the design of upper and lower portions of aircraft wing structures. What material properties would you expect to be important in these two areas?

1.9 Briefly discuss the considerations required to select a material(s) for a new structural design. How would these deliberations influence the research needed to successfully transition a "new" material from the laboratory to engineering applications?

1.10 Discuss how "new" components can become damaged during manufacturing. What are possible consequences of this initial damage? Give some examples from your personal experience.

1.11 Discuss how "old" components can become damaged during use. What are possible consequences of this service-induced damage? Give some examples from your personal experience.

1.12 Briefly compare and contrast the fail-safe and safe-life design approaches to fatigue life.

1.13 Briefly define damage tolerance. Give some examples from your personal experience.

1.14 Briefly describe retirement for cause. Give some examples from your personal experience.

1.15 Briefly discuss infinite life in the concept of fatigue analysis. Is infinite life a realistic goal for the structural designer? Give some examples to justify your answer.

1.16 Briefly describe some human factors that could influence the results of a nondestructive inspection.

1.17 Briefly discuss the three-legged stool approach to damage tolerance. Indicate how each leg works in conjunction with the other two to achieve structural integrity.

1.18 Consult recent media reports of a structural failure (e.g., search the internet). Indicate the failure mode responsible for the mishap and discuss the role of initial structural damage in the incident. What was the role of nondestructive inspection in the event? What actions were taken to correct the deficiencies in the design, construction, or maintenance of the structure in question?

1.19 Briefly discuss the types of loads and service environment that you would expect the following components to see in service. What failure modes would you expect to govern the design of these components? What service life (i.e., number of cycles, length of time) would be required for these structures?

(a) Aircraft wing (compare and contrast upper and lower surfaces)

(b) Transport aircraft fuselage

(c) Aircraft landing gear

(d) Rocket motor case

(e) Turbine engine fan blades

(f) Turbine engine fan disk

(g) Automobile suspension structure

(h) Automobile drive shaft

(i) Support structure for an automobile airbag

(j) Support column on a highway bridge

(k) Bracket that connects a basketball hoop to the backboard

REFERENCES

1. J. E. Srawley and J. B. Esgar, *Investigation of Hydrotest Failure of Thiokol Chemical Corporation 260-Inch-Diameter SL-1 Motor Case,* NASA Technical Memorandum TM X-1195, Lewis Research Center, Cleveland, Ohio, January 1966.

2. U. A. Hinders, "F-111 Design Experience—Use of High Strength Steel," paper presented at the AIAA 2nd Aircraft Design and Operations Meeting, 1970.

3. M. P. Kaplan and J. L. Lincoln, "The U. S. Air Force Approach to Aircraft Damage Tolerant Design," in *ASM Handbook, Fatigue and Fracture,* Vol. 19: ASM International, Materials Park, Ohio, 1996, pp. 577–588.

4. T. Swift, "Damage Tolerance Certification of Commercial Aircraft," in *ASM Handbook, Fatigue and Fracture,* Vol. 19: ASM International, Materials Park, Ohio, 1996, pp. 566–576.

5. J. Ott, "JT8D Hub Failure Sparks Intense Inquiry," *Aviation Week & Space Technology,* July 15, 1996, pp. 29–30.

6. "Safety of Aging Aircraft Undergoes Reassessment, *Aviation Week & Space Technology,* May 16, 1988, pp. 16–18.

7. D. S. Dawicke, "Fatigue Crack Closure," Ph.D. Thesis, Purdue University School of Aeronautics and Astronautics, West Lafayette, Indiana, 1989.

8. "Fatigue Evaluation, Damage Tolerance and Fatigue Evaluation of Structure," Section 25.571, *Airworthiness Standards: Transport Category Airplanes,* DOT Part 25, Federal Aviation Administration, Washington, D.C., March 31, 1998.

9. *Aircraft Structural Integrity Program, General Guidelines For,* MIL-HDBK-1530B, U.S. Air Force, July 3, 2002.

CHAPTER 2

REVIEW OF PRELIMINARY CONCEPTS

2.1 OVERVIEW

The goal of this chapter is to review preliminary concepts needed for more detailed discussions of damage tolerance analysis that follow in later chapters. Introductory strength-of-materials topics dealing with stress and strain, material properties, and common failure modes are briefly discussed. Additional details on these subjects may be found in References 1–8.

2.2 STRESS

Consider the member shown in Figure 2.1 which is in static equilibrium under the applied external forces $\mathbf{F}_i$ and couples $\mathbf{M}_i$. (Here bold symbols represent vectors, and the subscript i takes the values 1, 2, ...) Cutting the object with plane 1–1 exposes new surfaces that were *internal* to the original object. For the member to remain in equilibrium, there must be a resultant force $\mathbf{R}$ and couple $\mathbf{C}$ acting on this cross section. The values of $\mathbf{R}$ and $\mathbf{C}$ are readily found by the equations of equilibrium and are typically expressed in terms of their orthogonal x–y–z components $(R_x, R_y, R_z, C_x, C_y,$ and $C_z)$. Here x is the direction perpendicular to the plane of the cross section, and the y and z axes lie in plane 1–1. Since the resultant force–couple system depends on the cross section location, $\mathbf{R}$ and $\mathbf{C}$ will, in general, vary with x. (Shear and moment diagrams are frequently constructed to determine the variation in these quantities along the length of the member and to locate planes where maximum and minimum values occur.)

Since failure often initiates at a particular *point* in a body, it is useful to have a *local* measure of the manner in which $\mathbf{R}$ and $\mathbf{C}$ are distributed across the section of

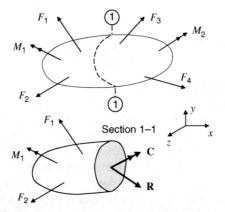

Figure 2.1. Schematic representation of resultant force and couple developed on a plane that sections a solid body subjected to an equilibrated system of forces and couples.

interest. Consider, for example, the small area ΔA centered at point (y, z) on the cross section, as shown in Figure 2.2. The local force acting over area ΔA consists of a normal component ΔN_i perpendicular to the cross section (in the x direction) and a resultant in-plane shear force ΔV_i. (This latter shear force may also be resolved into y and z components.) Summation of all the ΔN_i and ΔV_i over the entire cross section leads to the normal and in-plane components of **R**.

It is useful to examine the limiting values of these normal and shear forces as the area ΔA shrinks to a point. Define, for example, the normal and shear stress quantities σ and τ given below:

Normal Stress

$$\sigma = \lim_{\Delta A \to 0} \frac{\Delta N_i}{\Delta A} \qquad (2.1)$$

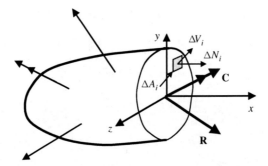

Figure 2.2. View of sectioned plane showing how resultant force **R** and couple **C** are composed of a series of normal ΔN and shear ΔV forces acting on small area ΔA.

Shear Stress

$$\tau = \lim_{\Delta A \to 0} \frac{\Delta V_i}{\Delta A} \tag{2.2}$$

Expanding these definitions to include stresses that act on the perpendicular faces of a small cube of material removed from point (x, y, z) in the loaded body as shown in Figure 2.3 leads to the three normal stresses σ_{xx}, σ_{yy}, and σ_{zz} and the six shear stresses τ_{xy}, τ_{yx}, τ_{xz}, τ_{zx}, τ_{yz}, and τ_{zy}. Here the first subscript refers to the direction normal to the plane of interest, while the second subscript indicates the direction in that plane. It is also readily shown by conditions of equilibrium that $\tau_{xy} = \tau_{yx}$, $\tau_{xz} = \tau_{zx}$, and $\tau_{yz} = \tau_{zy}$, so that there are only three independent normal stresses and three independent shear stresses acting at a point. Equilibrium considerations also result in the following differential equations relating these stresses [5–7]:

$$\frac{\partial \sigma_{xx}}{\partial x} + \frac{\partial \tau_{xy}}{\partial y} + \frac{\partial \tau_{xz}}{\partial z} = 0$$

$$\frac{\partial \tau_{xy}}{\partial x} + \frac{\partial \sigma_{yy}}{\partial y} + \frac{\partial \tau_{yz}}{\partial z} = 0 \tag{2.3}$$

$$\frac{\partial \tau_{xy}}{\partial x} + \frac{\partial \tau_{yz}}{\partial y} + \frac{\partial \sigma_{zz}}{\partial z} = 0$$

The general goal of a stress analysis is to determine how these six stresses vary throughout the body and how they are related to the remotely applied forces, moments, and cross-sectional dimensions of the member of interest. Plane orientations that give maximum and minimum values of stress are readily determined by transformation equations or by graphical means (e.g., Mohr's circle). There are, for example, a uniquely defined set of directions that yield orthogonal planes where the shear stresses are zero and the normal stresses take on maximum or minimum values,

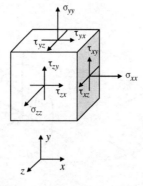

Figure 2.3. Small cube of material removed from loaded member showing normal σ and shear stresses τ that act on perpendicular faces.

known as principal stresses. In addition, there is another unique element orientation where the shear stress achieves a maximum value. (As discussed later in Section 2.6.2, yielding often occurs along this maximum shear stress plane in ductile materials.)

Recalling the fact that **R** and **C** are the *resultant* force and couple acting on the cross section leads to the following relations between the stresses and resultant force–couple system. (See Figure 2.4 and note the sign convention shown there for positive directions of the components of **R** and **C**.)

$$F \equiv R_x = \int_A \sigma_{xx} \, dA$$

$$V_y \equiv R_y = \int_A \tau_{xy} \, dA \qquad V_z \equiv R_z = \int_A \tau_{xz} \, dA$$

(2.4)

$$T \equiv C_x = \int_A y\tau_{xz} \, dA - \int_A z\tau_{xy} \, dA$$

$$M_y \equiv C_y = \int_A z\sigma_{xx} \, dA \qquad M_z \equiv C_z = -\int_A y\sigma_{xx} \, dA$$

These relations, along with observations regarding deformations and assumptions of material behavior, may be exploited to develop stress analysis formulas for various structural configurations. Some well-known results obtained from strength of materials are given below. (See Figures 2.5–2.9 below for definition of terms.)

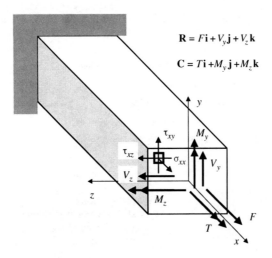

$$\mathbf{R} = F\mathbf{i} + V_y\mathbf{j} + V_z\mathbf{k}$$

$$\mathbf{C} = T\mathbf{i} + M_y\mathbf{j} + M_z\mathbf{k}$$

Figure 2.4. Schematic drawing relating stresses at a point with resultant force and couple acting on a cross section.

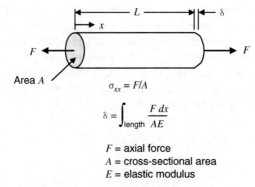

$$\sigma_{xx} = F/A$$

$$\delta = \int_{length} \frac{F\,dx}{AE}$$

F = axial force
A = cross-sectional area
E = elastic modulus

Figure 2.5. Straight rod subjected to axial force F showing formation of normal stress σ and elongation δ.

Straight Axial Member Loaded with Remote Force F *(Figure 2.5)*

$$\sigma_{xx} = \frac{F}{A} \tag{2.5}$$

Note that this formula requires that the axial force F be applied through the centroid of the cross-sectional area A.

Circular Rod Loaded with Twisting Torque T *(Figure 2.6)*

$$\tau = \frac{T\rho}{J} \tag{2.6}$$

Equation 2.6 is restricted to elastic behavior and circular cross sections.

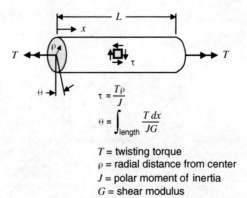

$$\tau = \frac{T\rho}{J}$$

$$\theta = \int_{length} \frac{T\,dx}{JG}$$

T = twisting torque
ρ = radial distance from center
J = polar moment of inertia
G = shear modulus

Figure 2.6. Circular rod subjected to twisting torque T showing formation of shear stresses τ and angular rotation θ.

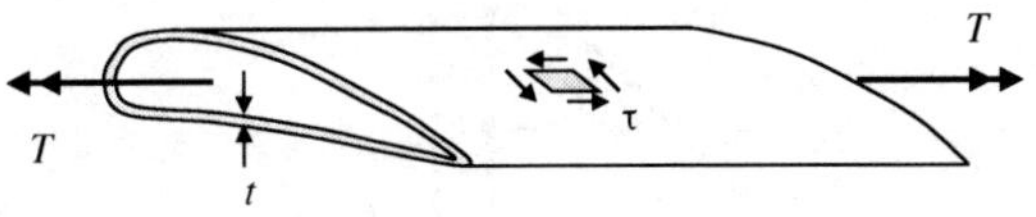

$$\tau = T/2At$$

$$\theta_L = \oint \frac{T \, ds}{4A^2 \, Gt}$$

A = area enclosed by mean cell wall
t = wall thickness
G = shear modulus
T = twisting torque

Figure 2.7. Thin-walled tube (closed section) subjected to twisting torque T showing formation of shear stress τ and angular rotation per unit length θ_L.

Closed Thin-Walled Tube Subjected to Twisting Torque T (Figure 2.7)

$$\tau = \frac{T}{2At} \tag{2.7}$$

Beam Loaded with Bending Moment M and Shear Force V (Figure 2.8)

$$\sigma_{xx} = -\left(k_2 M_z + k_1 M_y\right) y + \left(k_3 M_y + k_1 M_z\right) z \tag{2.8}$$

$$\tau t = \left(k_1 V_z - k_2 V_y\right) \int_{A'} y \, dA + \left(k_1 V_y - k_3 V_z\right) \int_{A'} z \, dA \tag{2.9}$$

where

$$k_1 = \frac{I_{yz}}{D} \qquad k_2 = \frac{I_y}{D} \qquad k_3 = \frac{I_z}{D}$$

$$D = I_y I_z - I_{yz}^2$$

In Equations 2.8 and 2.9, I_z, I_y, and I_{yz} are the cross-sectional moments of inertia computed about the centroidal y–z axis. Note that both formulas are restricted to elastic behavior and assume the positive bending moment and shear force directions shown in Figure 2.8. In addition, the shear forces V_y and V_z are assumed to go through the shear center of the cross section for application of Equation 2.9. The area A' is defined by a "cut" along the cross section. The location of the cut specifies where the shear stress τ (or the shear flow $q = \tau t$) is computed. When the y component of the applied bending moment M_y and the z component of the shear force V_z are both zero and the y or z axis is a line of symmetry ($I_{yz} = 0$), Equations 2.8 and 2.9 simplify the familiar *symmetric* cross section results below:

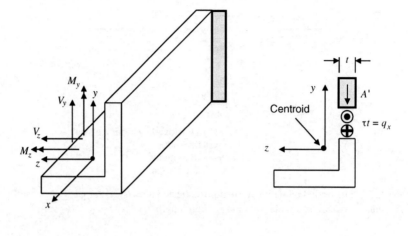

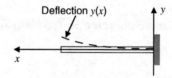

Figure 2.8. Beam loaded with bending moment **M** and shear force **V** which cause formation of shear τ, normal stresses $\sigma_x \equiv \sigma_b$, and beam *deflection* $y(x)$.

$$\sigma_{xx} = -\frac{M_z y}{I_z}$$

$$\tau = \frac{VQ}{I_z t} \quad \text{where} \quad Q = \int_{A'} y \, dA$$

Thin-Walled Cylindrical Pressure Vessel (Figure 2.9)

$$\sigma_{\text{hoop}} = \frac{PR}{t} \quad \sigma_{\text{long}} = \frac{PR}{2t} \tag{2.10}$$

Here P is the internal pressure, R is the radius of the cylinder, and t is the wall thickness that is assumed to be much less than the radius.

Thin-Walled Spherical Pressure Vessel

$$\sigma_{\text{hoop}} = \sigma_{\text{long}} = \frac{PR}{2t} \tag{2.11}$$

Again, P is the internal pressure and the radius R is much greater than the wall thickness t.

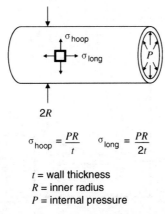

$$\sigma_{hoop} = \frac{PR}{t} \qquad \sigma_{long} = \frac{PR}{2t}$$

t = wall thickness
R = inner radius
P = internal pressure

Figure 2.9. Thin-walled cylindrical pressure vessel showing formation of normal stresses in hoop and longitudinal directions.

2.3 STRAIN

Since all structures are made from materials that deform under applied loads, it is important to have a means for quantifying these changes in dimensions. There are, in general, four types of displacements that can occur in a loaded member:

- translation of the body as a whole,
- rotation of the body as a whole,
- changes in length, and
- changes in angles.

These latter two "deformations" are of interest here and lead to the following definitions for normal and shear strains ε and γ, respectively:

$$\varepsilon = \overbrace{\text{unit length} \to 0}^{\text{limit}} \left(\frac{\text{change in length}}{\text{unit length}} \right) \tag{2.12}$$

$$\gamma = \text{change in angle between two perpendicular lines}$$

These strains are dimensionless quantities and are measured with respect to a particular coordinate system. Consider, for example, two initially perpendicular line segments AB and AC (lengths dx and dy) which, as shown in Figure 2.10, are "deformed" to the positions A', B', and C'. Let the movement of point A be given by functions u and v, where

$$u = u(x, y) = x \text{ displacement of point } A \text{ due to loads}$$
$$v = v(x, y) = y \text{ displacement of point } A \text{ due to loads} \tag{2.13}$$

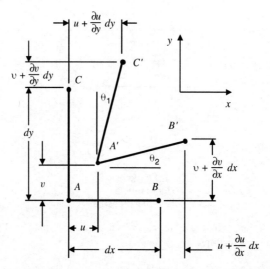

Figure 2.10. Schematic representation of the deformation of two perpendicular line segments *AB* and *AC*.

Now, assuming small deformations, the *x* and *y* displacements of point *B* to position *B'* are given by

$$u + du = u + \frac{\partial u}{\partial x}\, dx = x \text{ displacement of point } B$$

$$v + dv = v + \frac{\partial v}{\partial x}\, dx = y \text{ displacement of point } B$$

Likewise, the movement of point *C* to location *C'* is given by

$$u + \frac{\partial u}{\partial y}\, dy = x \text{ displacement of point } C$$

$$v + \frac{\partial v}{\partial y}\, dy = y \text{ displacement of point } C$$

Applying the definitions for strain (Equations 2.12) gives the following normal strains:

$$\varepsilon_{xx} = \frac{\text{change in length } AB}{\text{length } AB} = \frac{u + (\partial u/\partial x)\, dx - u}{dx} = \frac{\partial u}{\partial x} \qquad (2.14)$$

$$\varepsilon_{yy} = \frac{\text{change in length } AC}{\text{length } AC} = \frac{v + (\partial v/\partial y)\, dy - v}{dy} = \frac{\partial v}{\partial y} \qquad (2.15)$$

The shear strain γ_{xy} is given by the change in the perpendicular angle formed by the original line segments AB and AC. Assuming small angle changes gives

$$\gamma_{xy} = \text{change in perpendicular} = \theta_1 + \theta_2 = \tan(\theta_1 + \theta_2) = \tan\theta_1 + \tan\theta_2$$

Now

$$\tan\theta_1 = \frac{u + (\partial u/\partial y)\, dy - u}{dy} = \frac{\partial u}{\partial y}$$

$$\tan\theta_2 = \frac{v + (\partial v/\partial x)\, dx - v}{dx} = \frac{\partial v}{\partial x}$$

which leads to the final result for the shear strain:

$$\gamma_{xy} = \frac{\partial u}{\partial y} + \frac{\partial v}{\partial x} \tag{2.16}$$

Similar arguments for the general three-dimensional case lead to the following strain–displacement relations:

$$\varepsilon_{xx} = \frac{\partial u}{\partial x} \qquad \varepsilon_{yy} = \frac{\partial v}{\partial y} \qquad \varepsilon_{zz} = \frac{\partial w}{\partial z}$$

$$\gamma_{xy} = \frac{\partial u}{\partial y} + \frac{\partial v}{\partial x} \qquad \gamma_{yz} = \frac{\partial v}{\partial z} + \frac{\partial w}{\partial y} \qquad \gamma_{xz} = \frac{\partial u}{\partial z} + \frac{\partial w}{\partial x} \tag{2.17}$$

Here

$$u = u(x, y, z) = x \text{ displacement of point } A$$

$$v = v(x, y, z) = y \text{ displacement of point } A$$

$$w = w(x, y, z) = z \text{ displacement of point } A$$

Although beyond the scope of the present discussion, compatibility conditions provide additional relationships between the six strains defined above that involve partial derivatives of the strains with respect to position (x, y, z). It should be noted that the strains given in Equations 2.17 are dimensionless quantities, although normal strains are often expressed in units of length per length and shear strains are measured in radians.

It is important to recognize that the strains will, in general, vary from point to point in the loaded member and depend on the choice of reference axes (i.e., they are functions of x, y, and z). Orientations of maximum and minimum normal strains (principal strains) again correspond to directions of zero shear strain and may be found by transformation equations or by Mohr's circle.

There is also a sign convention employed here that assumes positive normal strains increase lengths, whereas positive shear strains decrease angles. Moreover, these expressions assume *small* displacements, although similar arguments can be used to develop corresponding (albeit more complex) "large displacement" values for the strains [5–7]. Finally, it should be pointed out that some authors use a slightly different definition of shear strain that is defined as one-half of the change in perpendicular angles [7].

2.4 STRESS–STRAIN RELATIONS

The prior discussions of stress and strain have been conducted without reference to any particular material. If one, however, wants to relate stress and strain, experiments must be conducted to determine the "constitutive" relation for the material of interest. The simplest experiment in this regard is the standard tension test shown in Figure 2.11 and discussed in Section 2.5. Here a straight bar with a uniform diameter is loaded with a uniaxial force, and the elongation δ_x is measured over a specified gage length, along with the change in diameter δ_d. Dividing the applied force by the cross-sectional area A gives the normal stress σ_{xx}, and dividing the elongation δ_x by the gage length gives the normal strain ε_{xx}.

Plotting σ_{xx} versus ε_{xx} for various values of the applied force F gives the familiar engineering stress–strain curve shown in Figure 2.12. As discussed further in the following section, many materials initially exhibit a linear stress–strain relationship with slope E. Although the experimental curve shown in Figure 2.12 relates stress and strain for the tension test, it must be emphasized that this relationship only applies for the *special case* of uniaxial loading (i.e., the simple tension

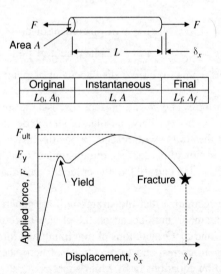

Figure 2.11. Schematic representation of tension test used to define standard material properties.

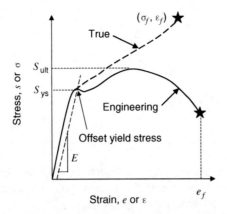

Figure 2.12. Comparison of engineering and true stress–strain curves.

test involves a state of stress where $\sigma_{xx} = F/A$ and all of the other five stresses, $\sigma_{yy} = \sigma_{zz} = \tau_{xy} = \tau_{xz} = \tau_{yz} = 0$). Since most structural applications will involve more complex stress configurations, a general goal is to relate the six components of stress ($\sigma_{xx}, \sigma_{yy}, \sigma_{zz}, \tau_{xy}, \tau_{xz}, \tau_{yz}$) with the six independent components of strain ($\varepsilon_{xx}, \varepsilon_{yy}, \varepsilon_{zz}, \gamma_{xy}, \gamma_{xz}, \gamma_{yz}$).

Guided by the special case of the tension test, where a linear relation was initially obtained between stress and strain, one may assume the following linear elastic relation between the six components of stress and strain:

$$
\begin{aligned}
\varepsilon_{xx} &= C_{11}\sigma_x + C_{12}\sigma_y + C_{13}\sigma_z + \sigma_{14}\tau_{xy} + C_{15}\tau_{xz} + C_{16}\tau_{yz} \\
\varepsilon_{yy} &= C_{21}\sigma_x + \cdots \qquad\qquad\qquad\qquad\qquad + C_{26}\tau_{yz} \\
\varepsilon_{zz} &= C_{31}\sigma_x + \cdots \qquad\qquad\qquad\qquad\qquad + C_{36}\tau_{yz} \\
\gamma_{xy} &= C_{41}\sigma_x + \cdots \qquad\qquad\qquad\qquad\qquad + C_{46}\tau_{yz} \\
\gamma_{xz} &= C_{51}\sigma_x + \cdots \qquad\qquad\qquad\qquad\qquad + C_{56}\tau_{yz} \\
\gamma_{yz} &= C_{61}\sigma_x + C_{62}\sigma_y + C_{63}\sigma_z + C_{64}\sigma_{xy} + C_{65}\tau_{xz} + C_{66}\tau_{yz}
\end{aligned}
\tag{2.18}
$$

Here the 36 constants C_{ij} are properties of the material of interest and may depend on location or orientation. For a general linear elastic *anisotropic* material, only 21 of the 36 constants are independent. (An anisotropic material is one whose properties change with specimen orientation.) If further restrictive assumptions are made on the material behavior, it is possible to show that many of the remaining C_{ij} are equal or zero. In particular, for a linear isotropic material (one whose properties are the same in all directions), there are only two independent constants, and the generalized Hooke's law (Equations 2.18) reduces to the following:

$$\varepsilon_{xx} = \frac{\sigma_{xx}}{E} - \frac{v}{E}(\sigma_{yy} + \sigma_{zz})$$

$$\varepsilon_{yy} = \frac{\sigma_{yy}}{E} - \frac{v}{E}(\sigma_{xx} + \sigma_{zz})$$

$$\varepsilon_{zz} = \frac{\sigma_{zz}}{E} - \frac{v}{E}(\sigma_{xx} + \sigma_{yy})$$

(2.19)

$$\gamma_{xy} = \frac{\tau_{xy}}{G} \qquad \gamma_{xz} = \frac{\tau_{xz}}{G} \qquad \gamma_{yz} = \frac{\tau_{yz}}{G}$$

The three material constants given in Equations 2.19 are the elastic modulus E, shear modulus G, and Poisson's ratio v defined in the following section. These three quantities are further related by the expression

$$E = 2G(1 + v)$$

(2.20)

2.5 TENSILE PROPERTIES

Reconsider the standard tension test shown in Figure 2.11, where a straight specimen with original length l_0 and cross-sectional area A_0 is subjected to an axial force F. The resulting axial displacement δ_x is measured and plotted versus the applied force as shown. This force–displacement relation is initially linear for many common structural materials. At a particular value of applied force F_y, however, the force may drop slightly with increased applied displacement but then continues to increase nonlinearly to a maximum value F_{ult}. At this maximum load point, ductile materials often exhibit a rapid decrease in local cross section (commonly referred to as a "neck"), and the applied force decreases while the total specimen length continues to increase until fracture occurs. When fracture occurs, the final applied force is F_f, the final displacement is δ_f, and the final cross-sectional area is A_f.

The results of this tension test are used to define several common material properties. Before discussing these quantities, however, it is useful to distinguish between "engineering" stress and strain (s and e) and "true" stress and strain (σ and ε). As described below, the difference between these quantities depends on whether stress and strain are based on the original specimen length and cross-sectional area (L_0 and A_0) or on the instantaneous dimensions (L and A):

Engineering Stress and Strain

$$s = \frac{F}{A_0} \quad \text{(stress)}$$

(2.21)

$$e = \frac{\delta_x}{L_0} \quad \text{(strain)}$$

True Stress and Strain

$$\sigma = \frac{F}{A} \quad \text{(stress)}$$

$$\varepsilon = \int_{L_0}^{L} \frac{dL}{L} = \ln\left(\frac{L}{L_0}\right) \quad \text{(strain)}$$

(2.22)

The engineering and true strains are related as follows:

$$L = L_0 + \delta_x$$

$$\varepsilon = \ln\left(\frac{L}{L_0}\right) = \ln\left(\frac{L_0 + \delta_x}{L_0}\right) = \ln\left(1 + \frac{\delta_x}{L_0}\right)$$

which gives

$$\varepsilon = \ln(1 + e) \tag{2.23}$$

Note that ε and e are nearly equal for small strains but that $\varepsilon < e$ when deformations (and strains) are large. It should also be pointed out that Equation 2.23 is only valid until necking occurs in the test specimen.

Similarly, one can relate the engineering and true stress quantities. At any value of the applied load, one can say

$$F = sA_0 = \sigma A$$

so that

$$\sigma = s\left(\frac{A_0}{A}\right)$$

Now, assume that the volume V of the test specimen remains constant throughout the test:

$$V = AL = A_0 L_0$$

$$\frac{A_0}{A} = \frac{L}{L_0} = \frac{L_0 + \delta_x}{L_0} = 1 + e$$

Thus, the relation between true stress σ and engineering stress s is given by

$$\sigma = s(1 + e) \tag{2.24}$$

Note here that the true stress is larger than the engineering stress, although this difference is negligible when the engineering strain is small (i.e., $\sigma \approx s$ when $e \ll 1$).

Now return to the measured force–displacement curve (F versus δ_x) shown in Figure 2.11. This result is put in a more general form when converted to the engineering and true stress–strain curves shown in Figure 2.12. Note that the "engineering" stress–strain curve is identical in shape to the original force–displacement plot in Figure 2.11, since the axes are simply scaled by constant values (i.e., $s = F/A_0$ and $e = \delta_x/L_0$). The "true" stress–strain curve shown in Figure 2.12 is obtained from the engineering curve by using Equations 2.23 and 2.24. These curves may now be used to define several material properties.

The *engineering* stress, where inelastic deformation begins, is known as the material yield stress S_{ys}, and the tensile stress computed at maximum load is defined as the ultimate stress S_{ult}. Since the point of initial yield is not always obvious, various "offset" yield stresses are often defined by intersecting the stress–strain curve with a line drawn parallel to the original linear portion of the curve, as shown in Figure 2.12. This line is drawn offset by some predetermined value of engineering strain (e.g., 0.1 or 0.2% strain). The slope of the initial linear portion of the stress–strain curve is known as the elastic modulus E, and the local slope after yield is defined as the tangent modulus E_t. Note that all of these aforementioned quantities have units of stress (e.g., 1000 lb/in^2 = 1 ksi = 1000 psi = 6.894×10^6 N/m^2 = 6.894 MPa). The change in diameter, δ_d, measured during the elastic portion of the stress–strain curve divided by the original diameter defines the engineering strain e_t perpendicular to the loading direction. This transverse strain caused by the applied normal stress s_{xx} is used to determine the dimensionless Poisson's ratio $v = -e_t E/s_{xx}$.

Two common measures of material "toughness" are given by the reduction in area, RA, and the engineering strain at fracture (also known as the percent elongation):

$$\text{Reduction in area:} \quad RA = \frac{A_0 - A_f}{A_0}$$

$$\% \, RA = RA \times 100\%$$

$$\text{Elongation:} \quad e_f = \frac{L_f - L_0}{L_f} = \frac{\delta_f}{L_f}$$

$$(2.25)$$

$$\text{Percent elongation} = e_f \times 100\%$$

The true stress and strain at fracture define the true fracture strength σ_f and the fracture ductility ε_f. These quantities may be computed as follows from posttest measurements of the failed specimen:

$$\sigma_f = \frac{F_f}{A_f}$$

$$\varepsilon_f = \ln\left(\frac{L_f}{L_0}\right) = \ln\left(\frac{A_0}{A_f}\right) = \ln\left(\frac{1}{1 - RA}\right)$$

$$(2.26)$$

Note that the true fracture stress should be "corrected for necking" by multiplication by a stress concentration factor to account for the local reduction in area when necking occurs [9]. Finally, the area under the true stress–strain curve is often defined as the true fracture toughness U_p and is another measure of material ductility. It should be emphasized at this time that these mechanical properties are based on the results of tests conducted on *smooth* specimens and do not necessarily characterize the behavior of "damaged" material. In particular, an alternate measure of fracture toughness obtained from *cracked* test specimens is described in Chapter 3.

2.6 FAILURE MODES

As discussed in Chapter 1, a key element of any structural design is to anticipate the various failure mechanisms that can occur in service. This section expands on that earlier description of the elastic deflection, inelastic deformation, buckling, and the creep failure modes. Subsequent chapters will develop the fatigue, fracture, and corrosion failure mechanisms in more depth.

2.6.1 Elastic (Recoverable) Deformations

Recall that Section 1.3.1 discussed how elastic deflections can limit the operation of structures in some cases and can be considered a mode of failure. Elastic deflections are readily obtained from finite element models or may be computed by energy methods (e.g., Castigliano's theorem; see References 1–3). Although a discussion of these displacement analysis methods is beyond the scope of the current review, some typical deflection formulas are summarized below. These solutions correspond to the stress analysis results given in Section 2.2 for the common structural members shown in Figures 2.5–2.8.

Straight Axial Member Loaded with Remote Force F ***(Figure 2.5)***

$$\delta = \int_{\text{length}} \frac{F \, dx}{AE} \tag{2.27}$$

Here F is the axial force, A is the cross-sectional area, and E is the elastic modulus at location x in the member of interest. Note that, in general, all three quantities must be expressed in terms of x before integrating to determine the deflection δ. If these values are constant along the length L of the member, one obtains the familiar "FLEA" formula for the elastic deflection:

$$\delta = \frac{FL}{EA}$$

Circular Rod Loaded with Twisting Torque T ***(Figure 2.6)***

$$\theta = \int_{\text{length}} \frac{T \, dx}{JG} \tag{2.28}$$

Here θ is the angular rotation (measured in radians), T is the applied torque, J is the polar moment of inertia for the *circular* cross section, and G is the shear modulus.

Again, these latter three quantities must be expressed in terms of the axial coordinate x in order to perform the integration over length L.

Thin-Walled Tube Subjected to Twisting Torque T (Figure 2.7)

$$\theta_L = \oint \frac{T\,ds}{4A^2Gt} \tag{2.29}$$

Here θ_L is the rotation of the tube per unit length (measured in radians per length), t is the skin thickness, T is the torque, and A is the area enclosed by the cell wall. The torque T and area A are specified for the particular cross section of interest (i.e., at location x), whereas t must be expressed in terms of the cross-sectional position parameter s in order to compute the line integral around the cell perimeter for θ_L. Also note that Equation 2.29 is only valid for thin-walled, *closed* cross sections (although the cross section may be noncircular).

Beam Loaded with Bending Moment M (Figure 2.8)

$$y''(x) \equiv \frac{d^2y}{dx^2} = \frac{1}{E}\left(k_2 M_z + k_1 M_y\right)$$

$$z''(x) \equiv \frac{d^2z}{dx^2} = -\frac{1}{E}\left(k_3 M_y + k_1 M_z\right) \tag{2.30}$$

These second-order differential equations may be solved for the beam deflections $y(x)$ and $z(x)$ once the bending moments M_z and M_y, cross-sectional moment of inertia I_y, I_z, and I_{yz}, elastic modulus E, and appropriate boundary equations are specified. (The k's in Equations 2.30 are the same as in Equations 2.8 and 2.9.) Note that, in general, M_y, M_z, the I's that define the k's, and E must again be expressed in terms of the axial variable x. If the y or z axes are lines of symmetry for the cross section and one also assumes $M_y = 0$, Equation 2.30 reduces to the familiar symmetric cross-sectional result:

$$y'' \equiv \frac{d^2y}{dx^2} = \frac{M_z}{EI_z}$$

As a final comment on these deflection formulas, note that all are controlled by the material stiffness (elastic modulus E or shear modulus G). Also note that all deflections obtained here assume elastic behavior.

2.6.2   Inelastic Deformations (Yielding)

As discussed in Section 1.3.2, most structural components are designed to resist permanent deformations when subjected to service loads. For simple uniaxial tension members, this goal is achieved by simply keeping the applied normal stress below the material yield stress. The issue is not as simple, however, when more complex states of stress are involved (i.e., multiaxial combinations of normal and shear stresses).

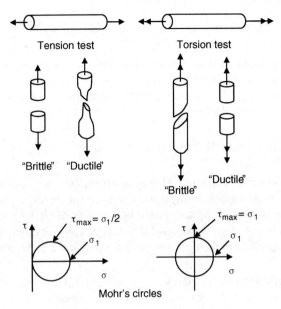

Figure 2.13. Schematic representation of tension and torsion test results for brittle and ductile materials showing failure planes and Mohr's circles of stress. Note that brittle materials fail along planes of maximum normal stress, whereas ductile failure occurs along maximum shear stress planes.

Consider, for example, the results of circular rods that are loaded to failure in either simple tension or torsion, as shown in Figure 2.13. Note that if the rods are made from a "brittle" material, the tension specimen fractures along a plane perpendicular to the direction of the applied force F, whereas the torsion specimen fails along a helical plane that makes a 45° angle with the vector of the applied twisting torque T. Specimens made from a "ductile" material, however, fail along a 45° plane to the loading axis for the tension test and along a plane perpendicular to the loading axis in the torsion test.

In order to correlate the results of these tests, we must seek the common denominator that led to failure in these various cases. Upon further examination of these results, it will be noted that the ductile specimens failed along planes of maximum shearing stresses in both the tension and torsion tests, whereas the brittle materials failed along planes of maximum normal stress (consider the Mohr's circles given in Figure 2.13 for these two cases). These considerations suggest that failure in ductile materials is based on shearing stresses, whereas failure in the brittle materials is expected to be controlled by normal stresses.

Maximum Shear Stress (Tresca) Criterion Assume, for example, that yield occurs in a ductile material when the maximum shear stress at a point in a loaded member exceeds a critical value for the material of interest, as stated by the equation

$$\text{Yield occurs when } \tau_{max} \geq \text{material constant} = \frac{\sigma_{ys}}{2} \qquad (2.31)$$

Here the appropriate material constant is determined by the results of a particular test. If, for example, we consider the special case of the tension test (and its Mohr's circle given in Figure 2.13), we note that yield begins when the maximum shear stress equals the half tensile yield stress σ_{ys}:

$$\tau_{max} = \text{one-half of the principal stress} = \frac{\sigma_1}{2} = \frac{F}{2A} = \frac{\sigma_{ys}}{2}$$

We now assume that Equation 2.31 may be generalized to all states of stress, once the value of the maximum shear stress is determined for the problem of interest. The maximum shear stress may be determined by a variety of methods (such as Mohr's circle). It is readily shown, for example, that $\tau_{max} = \frac{1}{2}|\sigma_1 - \sigma_3|$, where σ_1 and σ_3 are the largest and smallest of the three principal stresses at the given point.

Octahedral Shear Stress Criterion   A similar yield condition is obtained by assuming that the octahedral shear stress τ_{oct} given by Equation 2.32 controls yielding:

$$\begin{aligned}
\tau_{oct} &= \tfrac{1}{3}\sqrt{(\sigma_x - \sigma_y)^2 + (\sigma_y - \sigma_z)^2 + (\sigma_x - \sigma_z)^2 + 6\left(\tau_{xy}^2 + \tau_{xz}^2 + \tau_{yz}^2\right)} \\
&= \tfrac{1}{3}\sqrt{(\sigma_1 - \sigma_2)^2 + (\sigma_1 - \sigma_3)^2 + (\sigma_2 - \sigma_3)^2}
\end{aligned} \qquad (2.32)$$

Here, τ_{oct} is the shear stress that occurs on the octahedral plane, which is a plane that makes equal angles with the three principal stress axes, and is a uniquely defined quantity for a given state of stress. Note from Equation 2.32 that the octahedral shear stress may be expressed in terms of either the general state of stress (i.e., the three normal and three shear stresses acting at a point) or the three principal stresses (which occur along planes of zero shear stress). Assuming that yield occurs when the octahedral shear stress reaches a constant value for a particular material gives the following equation:

$$\text{Yield occurs when } \tau_{oct} \geq \text{material constant} = \tfrac{1}{3}\sqrt{2}\sigma_{ys} \qquad (2.33)$$

As before, the material constant is found from the special case of the tension test, where yielding begins when $\sigma_x = F/A = $ material yield stress σ_{ys} and the other two normal stresses and three shear stresses are all zero. Now, the results of the tension case can be extended to the following general statement that yielding occurs by the octahedral shear stress criterion when Equation 2.33 is satisfied:

$$\tfrac{1}{3}\sqrt{2}\sigma_{ys} = \tfrac{1}{3}\sqrt{(\sigma_x - \sigma_y)^2 + (\sigma_y - \sigma_z)^2 + (\sigma_x - \sigma_z)^2 + 6\left(\tau_{xy}^2 + \tau_{xz}^2 + \tau_{yz}^2\right)} \qquad (2.34)$$

Equation 2.34 can also be derived by an alternate approach employing energy methods, and when determined in this manner, Equation 2.34 is known as the von

Mises yield condition. While the maximum shear stress (Tresca) and the octahedral shear stress (von Mises) criteria are often employed to determine yielding in ductile structural metals, it should be noted that they will give different results for various states of stress. Although they give the same result for uniaxial tension, they differ by as much as 12% in their predictions for the maximum twisting torque T that can be applied in a torsion test.

Returning to the "brittle" tension and torsion specimens shown in Figure 2.13, recall that failure in these cases occurred along planes of maximum normal stress (again consider the Mohr's circles of Figure 2.13). This consideration leads to the following two other theories of failure commonly employed for "brittle" materials.

Maximum Normal Stress Criterion   This criterion assumes failure is controlled by the value of the maximum principal stress σ_1 at the given point of interest. Again, in the special case of the tension test, failure occurs when the maximum principal stress σ_1 equals the material ultimate stress, leading to the equation

$$\text{Failure occurs when } \sigma_1 \geq \text{material constant} = \text{ultimate stress } \sigma_{\text{ult}} \quad (2.35)$$

Note that to apply this result to more complex states of stress, the maximum principal stress must first be determined (e.g., from Mohr's circle).

Maximum Principal Strain (St. Venant's) Criterion   Another theory of failure commonly employed for brittle materials is to assume that fracture is controlled by the maximum normal strain ε_1 that occurs at a point. This condition may be expressed by the equation

$$\varepsilon_1 = \frac{\sigma_1}{E} - \frac{\nu}{E}(\sigma_2 + \sigma_3) \geq \text{material constant} \quad (2.36)$$

Note that when applying this expression to failure of the special case of the tension test, $\sigma_1 = F/A = $ material ultimate stress $ = \sigma_{\text{ult}}$, and $\sigma_2 = \sigma_3 = 0$, leading to the following general expression that can be applied to other states of stress as well:

$$\text{Fails when } \sigma_1 - \nu(\sigma_2 + \sigma_3) \geq \sigma_{\text{ult}} \quad (2.37)$$

Sometimes the limiting material constants of Equations 2.35 and 2.37 are given in terms of the material yield stress rather than the ultimate stress, implying yield conditions similar to the maximum shear stress (Equation 2.31) and octahedral shear stress (Equation 2.33) yield criteria. Since the yield stress and ultimate stress are often similar for "brittle" materials, however, this change has little practical effect for those materials.

A final point involving the four theories of failure discussed in this section is to note that all give the same failure loads for uniaxial loading situations. They will, however, provide different results for components subjected to more complex states of stress (i.e., situations involving both normal and shear stresses). Selection of the best theory of failure for a given material in these more general situations should be based on the results of a test program involving specimens subjected to several states of stress.

2.6.3 Buckling

Slender members loaded in compression or shear can fail in an unstable manner at very small elastic stresses. An intuitive appreciation for the significance of this problem can be gained by considering the example shown in Figure 2.14. Here an ordinary wooden yardstick is loaded with a compressive force P. The cross section of the yard stick is 1.25×0.125 in (3.18×0.32 cm), whereas typical material properties for wood include an ultimate strength of 7000 psi (48.3 MPa) and an elastic modulus of 2,000,000 psi (13,800 MPa). Using the maximum principal stress condition discussed in the preceding section gives the following value for the maximum force P that could be applied to the yardstick:

$$P = \sigma_{ult} A = (7000)(1.25)(0.125) = 1094 \text{ lb} = 4.87 \text{ N}$$

While a wooden yardstick may support 1000 lb (4.5 N) when loaded in *tension*, it would clearly fail at a much smaller force when subjected to a *compressive* force. Intuition tells us that the yardstick will "bow out" under the compression load, as shown by the dotted line in Figure 2.14, leading to a large unstable bending moment that quickly causes failure. This unstable out of-plane deformation is the buckling failure mechanism that is so dangerous to slender compression (or shear) members.

Elastic Column Buckling Strength-of-materials theory states that *elastic* column buckling is controlled by the Euler buckling equation given by

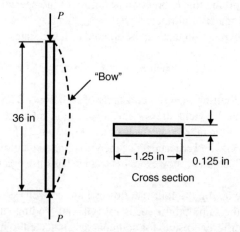

Figure 2.14. Schematic view of wooden yardstick loaded with compressive force P that results in out-of-plane bending that leads to unstable buckling failure.

$$P_{cr} = \frac{\pi^2 EI}{(kL)^2} \qquad (2.38)$$

Here P_{cr} is the critical buckling load, E is the elastic modulus, I is the moment of intertia for the column cross section (computed about the axis that gives the smallest value), L is the unsupported column length, and k is a dimensionless coefficient that depends on the constraint applied to the ends of the specimen. In general, k decreases as more constraint is provided at the column ends (see Table 2.1).

Note from Equation 2.38 that the buckling load increases linearly with the elastic modulus E and the cross-sectional moment of inertia I but is inversely related to the square of kL (kL is known as the effective column length). Thus, the buckling resistance is greatly increased by shorting the column length and/or increasing the end constraint.

Now, assuming pinned end conditions ($k = 1$), the buckling load for the yardstick example considered previously may be found from Equation 2.38:

$$P_{cr} = \frac{\pi^2 (2,000,000)(1.25)(0.125)^3}{(36)^2 (12)} = 3.1 \text{ lb}$$

Note that this answer is much more realistic than the 1094-lb (4.87-N) value determined previously and demonstrates the drastic consequences of the buckling failure mode for long, slender members. *Clearly the engineer must anticipate the appropriate failure mode for a given application in order to design a "failure resistant" component.*

The Euler column buckling formula may be given in terms of stress by dividing both sides of Equation 2.38 by the cross-sectional area A. If one also expresses the moment of inertia I in terms of the radius of gyration r, where $I = Ar^2$, the column buckling *stress* σ_{cr} is given as

$$\sigma_{cr} = \frac{\pi^2 E}{(kL/r)^2} \qquad (2.39)$$

The quantity kL/r is known as the *slenderness ratio* and is an important measure of the column resistance to buckling. Plotting the primary column buckling stress σ_{cr} versus the slenderness ratio as shown in Figure 2.15 demonstrates three regions of column behavior:

TABLE 2.1   Euler Buckling Coefficients for Various Column End Support Conditions

End Condition	Both Ends Pinned	One End Pinned, Other End Clamped	Both Ends Clamped	One End Clamped, Other End Free
Buckling Coefficient k (dimensionless)	1.0	0.7	0.5	2.0

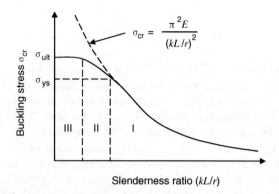

Figure 2.15. Schematic plot of primary column buckling stress versus slenderness ratio (kL/r) showing three regions of buckling behavior.

- Region I—for large kL/r buckling occurs at elastic stresses and is controlled by Equation 2.39 (which is limited to elastic behavior).
- Region II—for intermediate kL/r buckling occurs in an inelastic manner and cannot be predicted by Equation 2.39. Note that use of the Euler buckling formula would give unconservative buckling loads in this regime (the actual buckling stress is shown schematically in Figure 2.15).
- Region III—for very short columns (small kL/r) buckling is not an issue and failure results from reaching the material ultimate stress.

Inelastic Buckling   Inelastic column buckling (region II) may be estimated by replacing the elastic modulus in Equation 2.39 with the tangent modulus E_t (i.e., the instantaneous slope of the stress–strain curve after yielding has occurred). Note, however, that since E_t is not constant after yielding but decreases with additional loading, an iterative solution is required to obtain the buckling stress.

Sheet Buckling   Thin sheets can also be extremely sensitive to buckling failure. Consider, for example, the sheets shown in Figure 2.16 which are loaded in compression, shear, or bending (or combinations of these loads). The elastic buckling stress for these sheets are given by the equation

$$\sigma_{cr}, \ \tau_{cr}, \ \text{or} \ \sigma_{b,cr} = \frac{\pi^2 k_c E}{12(1 - v^2)}\left(\frac{t}{b}\right)^2 \tag{2.40}$$

Here σ_{cr} is the buckling stress due to compression loading, τ_{cr} is the shear stress that causes buckling, and $\sigma_{b,cr}$ is the buckling stress for bending (i.e., the outer fiber bending stress My/I). Other terms in this *elastic* result include the elastic modulus E, Poisson's ratio v, the sheet thickness t, the short dimension b of the sheet, and the buckling coefficient k_c. In this case k_c depends on both the long and short sheet

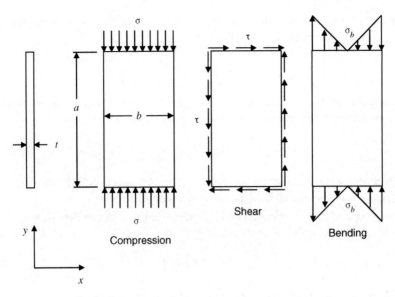

Figure 2.16. Schematic representation of thin sheets subjected to compression, shear, or bending loads.

dimensions a and b, the type of load (i.e., compression, shear, or bend), and the edge constraint along the sheet edges. Some typical values for the sheet-buckling coefficients for long sheets ($a/b \geq 3$) are given in Table 2.2 [8].

Sheet-buckling conditions for *combinations* of compression, shear, and bending may be estimated by the following interaction equations [8] that involve load ratios R_i:

TABLE 2.2   Dimensionless Sheet Buckling Coefficients ($a/b \geq 3$)

Type of Loading	Edge Support	Buckling Coefficient, k_c
Compression	All edges simply supported	4.00
	All edges clamped	6.98
	Three edges simply supported, One unloaded edge free	0.43
	Three edges clamped, One unsupported edge free	1.28
Shear	All edges simply supported	5.35
	All edges clamped	8.98
Bending	All edges simply supported	23.9
	All edges clamped	41.8

Biaxial Compression   Here the sheet is assumed to be loaded in simultaneous compression in the x and y directions:

$$R_{cx} + R_{cy} = 1 \qquad (2.41)$$

Compression and Shear   In this case the sheet is loaded in compression in one direction, whereas all four edges are simultaneously loaded in shear:

$$R_c + R_s^2 = 1 \qquad (2.42)$$

(Note that while Reference 8 squares the R_s term in Equation 2.42, other references suggest it be raised to the 1.5 power rather than squared.)

Bending and Shear   Here a bending stress is applied in one direction, and shear is applied along all four sheet edges:

$$R_b^2 + R_s^2 = 1 \qquad (2.43)$$

The load ratios R_i given in Equations 2.41–2.43 are defined as

$$R_s = \frac{\text{applied shear stress}}{\text{buckling stress for shear loading only}}$$

$$R_c = \frac{\text{applied compressive stress}}{\text{buckling stress for compression loading only}}$$

$$R_b = \frac{\text{applied bending stress}}{\text{buckling stress for bending loading only}}$$

2.6.4   Creep

Creep is the time dependent failure mode that occurs when a member continues to deform under a *fixed* static load. The creep phenomenon is shown schematically in Figure 2.17 where a straight member is subjected to a uniaxial force F at time t_0. The force is held constant until time t_1 and then removed. A corresponding axial displacement occurs immediately when the force is first applied, but the bar continues to stretch during the period when the load is held constant. When the force is removed at time t_1, the elastic component of the displacement is recovered but there remains a residual deformation at zero load. Depending on specific conditions, the bar may continue to relax to its original length with further time at zero load.

Although creep is possible at any temperature, it is usually only an issue when temperatures exceed 40% of the melting point T_m. At temperatures above $0.4T_m$, materials often demonstrate three distinct creep behaviors, as shown schematically in Figure 2.18. The first, or primary, stage of creep is characterized by an immediate elastic strain ε_0 followed by an increasing rate of deformation under the fixed load. The strain rate then levels out to a constant rate (i.e., the steady state or "minimum

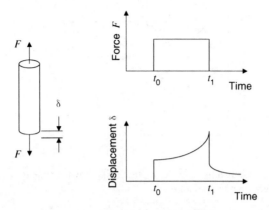

Figure 2.17. Schematic representation of force displacement behavior for creep failure mode.

creep rate") during the secondary stage of creep. Finally, the third or tertiary stage of creep sees the strain rate continue to increase until the specimen ruptures when it develops a "neck" and incurs a rapid increase in local stress. The presence of all three creep regimes depends on the applied stress and temperature, and the second and third stages may not occur at temperatures below $0.4T_m$.

A common method for determining the creep resistance of a material is the stress–rupture test where a uniaxial member is subjected to a constant stress σ and temperature T and allowed to creep to failure. The time t_r to specimen rupture is measured and plotted as a function of applied stress for various temperatures, as shown schematically in Figure 2.19. The test results for individual temperatures may then be fit with a series of exponential or power law functions such as those given by [9]

$$t_r = K_1 e^{a\sigma} \tag{2.44}$$

$$t_r = K_2 \sigma^m \tag{2.45}$$

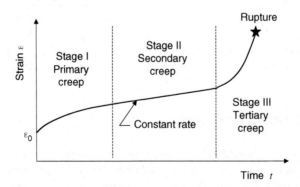

Figure 2.18. Schematic representation of creep–rupture test showing three stages of creep.

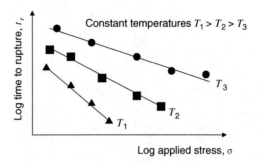

Figure 2.19. Schematic results of creep–rupture tests showing time to failure versus applied stress for various constant temperatures.

Here K_1, a, K_2, and m are empirical constants determined by fitting Equations 2.44 and 2.45 to the test data for individual temperatures.

Other models may be used to find a single three-parameter characterization of the time to rupture, applied stress, and test temperature. Equation 2.46 is one such expression, and other approaches can be found in References 9–11:

$$t_r = K_3 \sigma^m e^{[Q_r/RT]} \tag{2.46}$$

Here σ is the applied stress and T is the absolute temperature measured in kelvins (K). The constants in Equation 2.46 include the empirical term K_3, the universal gas constant R (8.314 kJ/mol), the stress exponent for rupture m, and the activation energy for rupture Q_r.

Creep is a potential problem for structures that operate for long periods of time at elevated temperatures. Typical examples include turbine engine components (e.g., turbine blades) that see large centrifugal loads combined with the high operating temperatures. Creep is a consequence of new deformation mechanisms that can occur at elevated temperatures (e.g., grain boundary deformation, changes in slip systems or metallurgical stability, high-temperature oxidation). Creep resistance is a key parameter for high-temperature components and has led to the development of many specialized alloys for these critical applications. Since creep frequently occurs at material grain boundaries, some high-temperature materials employ directionally solidified or single-crystal microstructures to control or eliminate grain boundaries.

2.7   CONCLUDING REMARKS

This chapter has provided a brief summary of the concepts of stress and strain used to provide local measures of force and displacement in structural components. These quantities are related through the material constitutive laws and are used to define various material properties. Stresses and strains may also be correlated with the elastic and inelastic, buckling, and creep "failure modes" discussed previously.

The remainder of this text returns to the original goal of preventing failure of "damaged" structures and materials. Chapter 3 introduces the linear elastic fracture mechanics approach for characterizing the resistance of *cracked* members to fracture, fatigue, and stress corrosion. These damage-related failure modes are then developed more in Part II.

PROBLEMS

2.1 Rigid bar *ABC* is pinned at point *A* and held by a cable at points *B* and *C*. The cable passes over a massless, frictionless pulley that is pinned at point *D*. The bar is weightless and is loaded with the distributed force shown in Figure P2.1.

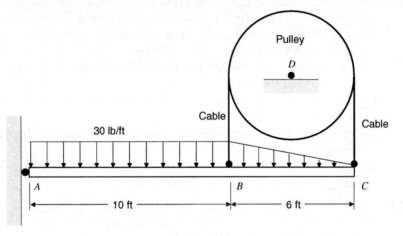

Figure P2.1

(a) Find the reactions at point *A* and the force in the cable at points *B* and *C*. Show all free-body diagrams used for your solution and specify the magnitude and direction of all forces.

(b) Draw shear and moment diagrams for member *ABC* and specify the magnitude and location of the maximum shear force and moment in member *ABC*. Indicate the positive directions of shear and moment.

2.2 The state of plane stress shown in Figure P2.2 exists at a point in a material with an elastic modulus $E = 10,000$ ksi and a Poisson's ratio $v = \frac{1}{3}$.

(a) Find the magnitude and orientation of the maximum shear stress at the point.

(b) Find the magnitude and orientations of the principal stresses.

(c) Find the magnitude and directions of the principal strains.

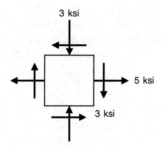

Figure P2.2

2.3 The situation where all three principal stresses have the same values ($\sigma_1 = \sigma_2 = \sigma_3 = \sigma$) is known as the case of hydrostatic stress. Assume this case of stress exists in a material with a yield stress $\sigma_{ys} = 50$ ksi, Poisson's ratio $v = \frac{1}{3}$, and an elastic modulus $E = 10,000$ ksi.

   (a) Find the value of σ that initiates yielding by the maximum shear stress (Tresca) yield criterion.

   (b) Find the value of σ that initiates failure by the maximum principal stress criterion.

   (c) Find the value of σ that initiates failure by the maximum principal strain criterion.

   (d) Find the value of σ that initiates failure by the octahedral shear stress criterion (equivalent to the von Mises criterion).

2.4 Plastic flow initiates in the solid circular rod shown in Figure P2.4 when the applied torque $T = 10,000\pi$ in-lb. Assume failure is controlled by the octahedral shear stress (von Mises) criterion.

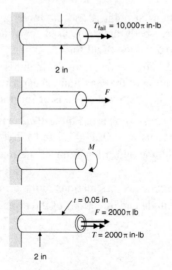

Figure P2.4

(a) What axial force F initiates failure in an identical rod (axial load only)?

(b) What bending moment M initiates failure in an identical rod (bending load only)?

(c) Will the combined action of a twisting torque $T = 2000\pi$ in-lb and $F = 2000\pi$ lb axial force fail the thin-walled tube made from the same material?

2.5 The tensile yield stress σ_{ys} and the torsion yield stress τ_{ys} are material properties measured from axial and torsion members, as shown in Figure P2.5.

(a) Determine the relation between σ_{ys} and τ_{ys} by the maximum shear stress criterion (i.e., compute σ_{ys}/τ_{ys}).

(b) Determine σ_{ys}/τ_{ys} by the octahedral shear stress criterion.

(c) Determine σ_{ys}/τ_{ys} by the maximum principal stress criterion.

(d) Determine σ_{ys}/τ_{ys} by the maximum principal strain criterion.

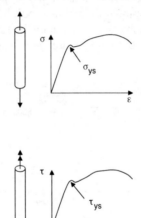

Figure P2.5

2.6 A thin-walled circular tube is twisted and pulled with 5000 lb axial force F and twisting couple T shown in Figure P2.6. The wall thickness t of the tube is 0.05 in, and the average tube diameter D_{avg} is 1.0 in. The tube is made from a material with an elastic modulus $E = 10,000$ ksi, Poisson's ratio $= \frac{1}{3}$, an ultimate stress of 55 ksi, and a tensile yield stress of 50 ksi.

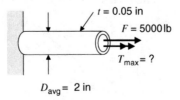

Figure P2.6

    (a) Find the torque T that will initiate failure by the maximum principal stress criterion.

    (b) Find the torque T that will initiate failure by the maximum principal strain criterion.

    (c) Find the torque T that will initiate failure by the maximum shear stress criterion.

    (d) Find the torque T that will initiate failure by the octahedral shear stress (von Mises) criterion.

2.7 A long cylindrical pressure vessel is 4 ft in diameter and has a $\frac{3}{8}$ in wall thickness. It is made from a ductile alloy that has 35 ksi yield stress, 58 ksi ultimate stress, and 10,000 ksi elastic modulus. Determine the maximum pressure that can be applied by the maximum shear stress criterion.

2.8 A 2-in-diameter rod is made from a ductile material that just begins to yield when subjected to the combined axial and torsion loading shown in Figure P2.8. What is the minimum radius of a rod of identical material needed to resist yielding when subjected to the combined action of 1000 in-lb twisting torque T and 2000 in-lb bending couple M as shown?

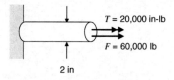

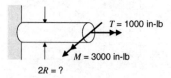

Figure P2.8

2.9 A circular rod is made from a material with 50 ksi yield stress, 10,000 ksi elastic modulus, and Poisson's ratio of $\frac{1}{3}$. The rod is loaded with an axial twisting torque T and a bending moment M (M is perpendicular to T as shown in Figure P2.9). Determine the diameter D necessary to prevent yielding by the maximum principal stress, maximum shear stress, maximum principal strain, and octahedral shear stress criteria. Consider the following cases:

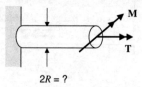

Figure P2.9

(a) $T = 10,000$ in-lb, $M = 0$

(b) $T = M = 10,000$ in-lb

(c) $T = 0$, $M = 10,000$ in-lb

2.10 A cantilever beam is made from a circular rod with radius R, elastic modulus $E = 10,000$ ksi, yield stress $\sigma_{ys} = 50$ ksi, ultimate stress $\sigma_{ult} = 70$ ksi, and Poisson's ratio $v = \frac{1}{4}$. The beam is loaded with the vertical distributed force and axial twisting torque shown in Figure P2.10.

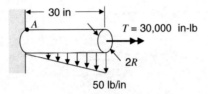

Figure P2.10

(a) Determine the minimum rod radius R by the maximum shear stress criterion.

(b) Determine the minimum rod radius R by the octahedral shear stress criterion.

(c) Determine the minimum rod radius R by the maximum principal stress criterion.

(d) Determine the minimum rod radius R by the maximum principal strain criterion.

2.11 Member AB is pinned at A and supported by a flexible cable at B and a vertical load of 20,000 lb is applied at A, as shown in Figure P2.11. The compression stress–strain properties for material AB are shown. Find the minimum allowable *square* cross-sectional area for member AB if buckling is the failure mode.

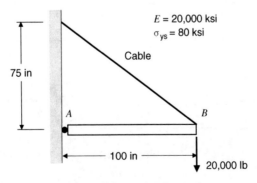

Figure P2.11

2.12 A 72-in-long column is pinned at each end and supported by a lateral brace at its midspan, as shown in Figure P2.12. The column has 4×1.5 in rectangular cross section. The midspan support prevents lateral deflection (allows rotations) but does not offer any resistance to out-of-plane deflection or rotation. The column is made from a material with an elastic modulus of 10,000 ksi and a yield stress of 40 ksi. Find the critical Euler buckling load for this configuration. Justify all equations used.

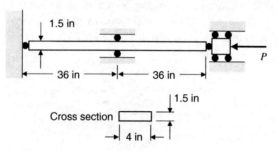

Figure P2.12

2.13   A square rod (b = width of *square* cross section) has a length of 5π in and is loaded in compression as shown. The rod ends are pinned and the material has an elastic module E and density ρ. Develop a formula in terms of modulus and density for the *lightest* rod that can resist the 100 lb applied force. Assume elastic behavior and consider buckling only. Select the rod material from one of those listed in Figure P2.13.

Material	Yield strength (ksi)	Density ρ (lb/in^3)	Elastic modulus E (ksi)
2124-T851 aluminum	65	0.101	10,500
Ti-6-4	130	0.160	15,900
4340 steel	200	0.283	29,000

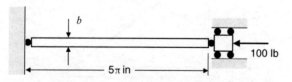

Figure P2.13

2.14   Force P is transferred to the 2-in-diameter aluminum shaft DC through the rigid arm CB and the 0.5-in-diameter steel rod AB shown in Figure P2.14. Linkage AB is fastened to member CB by a frictionless ball and socket joint and is given lateral support at point A by frictionless ball bearings (prevents displacement in x and y directions). The rigid arm CB is welded to shaft DC at point C. The ductile aluminum material has a yield strength of 70 ksi and an elastic modulus of 10,000 ksi. The properties of the steel rod include 30,000 ksi elastic modulus and 100 ksi yield strength. Find the maximum permissible elastic load P. Identify the point and mode of failure. (You may assume that CB does not fail.)

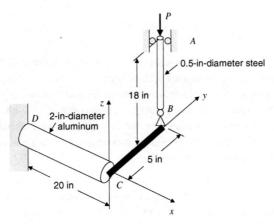

Figure P2.14

2.15   A $3 \times 10 \times 0.04$-in panel is subjected to a compressive stress of 3 ksi applied along its 10-in dimension. The panel is simply supported along all four edges and is made from a material with 65 ksi yield strength, an elastic modulus of 10,500 ksi, and a Poisson's ratio of 0.3. Find the minimum shear flow q applied to the panel edges that will cause buckling when applied in conjunction with the 3 ksi compressive stress.

2.16   A 4×12-in panel is to carry an axial compressive load of 300 lb (applied perpendicular to the 12-in direction) and 200 lb/in shear flow applied along all four outer edges. The panel is to be constructed from a material that has 70 ksi yield stress, 10,700 ksi elastic modulus, and a 0.3 Poisson's ratio. All four panel edges are to be simply supported. Find the minimum panel thickness t to prevent failure. Consider both buckling and yielding (assume von Mises yield criterion) as possible failure modes. Which failure mode controls the design?

2.17   The thin-walled rectangular beam shown in Figure P2.17 is made from a material with 50 ksi yield strength, 10,000 ksi elastic modulus, 3850 ksi shear modulus, and a 0.3 Poisson's ratio. What uniform skin thickness t is required to resist yielding (assume Von mises criterion) and buckling when the beam is subjected to 100 in-kip twisting torque T? Assume "clamped" edge conditions where individual panels are joined to form the hollow rectangular tube. What is the elastic rotation of the free end of the beam for this thickness?

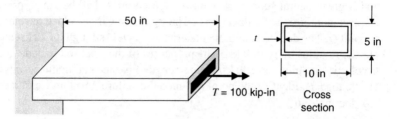

50 in

t

5 in

10 in

$T = 100$ kip-in

Cross section

Figure P2.17

2.18 Different locations of an aircraft structure often employ different materials. Discuss the general issues associated with materials selection in aircraft components. Using the fuselage skin and upper and lower wing surfaces as examples, briefly describe the parameters that influence the choice of materials in these components. Include in your discussion the relationship between applied loading, failure modes, material properties, and other relevant factors in the materials selection decision.

2.19 In structural design it is often possible to obtain performance indices P_i that may be used to optimize various design parameters for a given situation. Consider, for example, a tension member with a *square* cross section and length L. The member is loaded with an axial tension force as shown in Figure P2.19. If one wanted to select a material that would minimize the weight for a given axial deflection caused by the tension force, one would select a material that has the largest ratio of E/ρ, where E is the elastic modulus and ρ is the density of the material. Here $P_i = E/\rho$ may be considered a material performance index for this design.

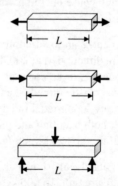

L

L

L

Figure P2.19

(a) Derive the above material selection performance index for the tension member.

(b) Consider the same structural component except that it is loaded in axial *compression* as shown in Fig. P2.19. Again the member is to have length L

and a *square* cross section. Find an analogous material performance index P_i that will select the material for the lightest column that does not buckle. You may assume elastic behavior.

(c) Consider another similar member with length L and a *square* cross section. This component is to be loaded in three-point bending, as shown in Fig. P2.19. Develop a materials performance index P_i for this case that will again identify a material that minimizes the weight for a given elastic deflection.

REFERENCES

1. S. H. Crandall, N. C. Dahl, and T. J. Lardner, *An Introduction to the Mechanics of Solids,* McGraw-Hill, New York, 1978.

2. J. M. Gere, *Mechanics of Materials,* 5th Edition, Books/Cole, Wadsworth, Pacific Grove, California, 2001.

3. F. P. Beer and E. R. Johnston, Jr., *Mechanics of Materials,* Mc-Graw-Hill, New York, 1981.

4. C. T. Sun, *Mechanics of Aircraft Structures,* Wiley, New York, 1998.

5. S. Timoshenko and J. N. Goodier, *Theory of Elasticity,* McGraw-Hill, New York, 1951.

6. A. P. Boresi, *Elasticity in Engineering Mechanics,* Prentice-Hall, Englewood Cliffs, New Jersey, 1965.

7. I. S. Sokolnikoff, *Mathematical Theory of Elasticity,* McGraw-Hill, New York, 1956.

8. E. F. Bruhn, *Analysis and Design of Flight Vehicle Structures,* Jacobs Publishing, Indianapolis, Indiana, 1973.

9. J. R. Newby, Coordinator, *Metals Handbook,* Vol. 8: *Mechanical Testing,* 9th ed., American Society for Metals, Metals Park, Ohio, 1985.

10. G. E. Dieter, Jr., *Mechanical Metallurgy,* Mc-Graw-Hill, New York, 1961.

11. N. E. Dowling, *Mechanical Behavior of Materials,* Prentice-Hall, 2nd ed., Englewood Cliffs, New Jersey, 1999.

CHAPTER 3

INTRODUCTION TO LINEAR ELASTIC FRACTURE MECHANICS

3.1 OVERVIEW

The objective of this chapter is to introduce damage tolerance analysis methods used to determine the influence of preexistent cracks on structural performance. The origin of the initial crack, whether it is a material flaw, induced by manufacturing or service, or assumed by decree, is not of concern here. The scope of this chapter is limited to a simplified overview of basic linear elastic fracture mechanics (LEFM) concepts and is intended primarily as an introduction to more detailed discussions included later in Part II of this volume.

Emphasis here is on the use of the stress intensity factor as a parameter to characterize critical and subcritical crack growth. First, the stress intensity factor is defined in the context of crack tip stress fields. The stress intensity factor is then employed with a simple fracture criterion to define the fracture toughness of a material. The stress intensity factor is then related to the rate of fatigue crack growth and stress corrosion cracking. These subcritical crack growth rate relations may be treated as material properties and integrated to determine component life.

Recall that damage tolerance analysis addresses two points concerning an *initially cracked* structure (see Figure 3.1). First, it is desired to determine the fracture load for a specified crack size. Second, it is necessary to predict the length of time (number of load cycles, days, missions, etc.) required for a "subcritical" crack to grow to the size that causes fracture at the given load. It is assumed here that the crack can extend in a subcritical manner by fatigue and/or stress corrosion cracking.

As described below, the LEFM approach assumes that the stress intensity factor K controls crack growth. Attention is limited here to nominally elastic behavior, although "small" amounts of crack tip plasticity are allowed. Practical implications of

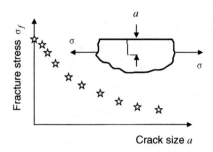

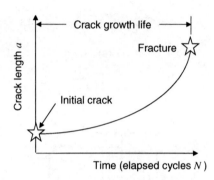

Figure 3.1. Schematic representation of fracture and fatigue behavior of a cracked structural member.

these assumptions as well as procedures to overcome their limitations are discussed in later chapters.

3.2 STRESS INTENSITY FACTOR

The stress intensity factor K is the LEFM parameter that relates remote load, crack size, and structural geometry. The stress intensity factor may be expressed in the form

$$K = \sigma\sqrt{\pi a}\,\beta \qquad (3.1)$$

Here σ is the applied stress, a is the crack length, and β is a dimensionless factor that depends on crack length and component geometry. Stress intensity factor solutions have been obtained for many crack geometries, and several handbook compilations are available [1–4]. Some typical results are given in Figure 3.2 [1–4].

Examining the solutions in Figures 3.2a–g, note that the stress *intensity* factor K is an entirely *different* parameter than the familiar stress *concentration* factor K_t employed in strength of materials. The stress intensity factor is a *crack* parameter and has units of stress × length$^{1/2}$ (conventional units are MPa-m$^{1/2}$ = 0.9102 ksi-in$^{1/2}$).

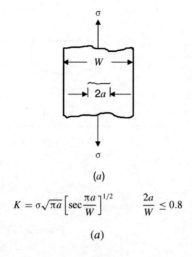

(a)

$$K = \sigma\sqrt{\pi a}\left[\sec\frac{\pi a}{W}\right]^{1/2} \qquad \frac{2a}{W} \le 0.8$$

(a)

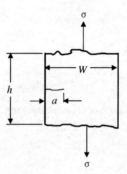

$$K = \sigma\sqrt{\pi a}\,\beta\!\left(\frac{a}{W}\right)$$

$$\beta\!\left(\frac{a}{W}\right) = 1.12 - 0.231\left(\frac{a}{W}\right) + 10.55\left(\frac{a}{W}\right)^2 - 21.73\left(\frac{a}{W}\right)^3 + 30.39\left(\frac{a}{W}\right)^4$$

$$\text{for }\left(\frac{a}{W}\right) \le 0.6 \text{ and } \left(\frac{h}{W}\right) \ge 1.0$$

(b)

Figure 3.2. (*a*) Center-cracked test specimen. (*b*) Edge-cracked strip.

The stress *concentration* factor K_t, on the other hand, is a *dimensionless* term that describes the local increase in stress at a *notch* (K_t = local stress/remote stress).

The formal definition of the stress *intensity* factor lies in the behavior of the linear elastic crack tip stress field. Although a detailed crack tip stress analysis is beyond the scope of this chapter, the nature of crack tip stresses for linear elastic behavior may be indicated by examining the limiting behavior of an elliptical *hole* located in a

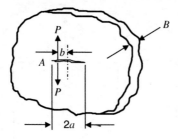

At crack tip A:

$$K = \frac{P}{B\sqrt{\pi a}} \left[\frac{(a+b)}{(a-b)} \right]^{1/2}$$

where P = concentrated force
$\quad\quad\;\; B$ = thickness

(c)

$$K = \sigma\sqrt{\pi a}\,\beta\left(\frac{a}{R}\right) \quad\quad \text{when } \frac{a}{R} \leq 10$$

Single crack:

$$\beta\left(\frac{a}{R}\right) = \frac{0.8733}{0.3245 + (a/R)} + 0.6762$$

Two symmetric cracks:

$$\beta\left(\frac{a}{R}\right) = \frac{0.6865}{0.2772 + (a/R)} + 0.9439$$

(d)

Figure 3.2. (c) Wide center-cracked sheet subjected to concentrated crack face forces. (d) Plate with a radially cracked hole loaded in tension.

large plate loaded in remote tension, as shown in Figure 3.3. The tensile stress at the root of the major axis is given by

$$\sigma_{tip} = \left(1 + 2\sqrt{\frac{a}{\rho}}\right)\sigma \tag{3.2}$$

Here a and c are the major and minor axes of the elliptical notch, σ is the remotely applied tension stress, and ρ is the notch radius of curvature ($\rho = c^2/a$ for an elliptical

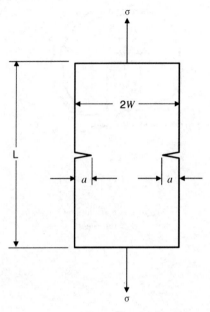

$$K = \sigma\sqrt{\pi a}\beta\left(\frac{a}{W}\right)$$

$$\beta\left(\frac{a}{W}\right) = \frac{1.122 - 0.561(a/W) - 0.205(a/W)^2 + 0.471(a/W)^3 - 0.190(a/W)^4}{\sqrt{1 - a/W}}$$

for any a/W

(e)

Figure 3.2. (e) Tension specimen with two symmetric edge cracks.

hole). Now, defining a crack as the limiting case when the notch radius $\rho \to 0$, the normal stress at the "crack" tip is given by

$$\sigma_{\text{crack tip}} = \lim_{\rho \to 0} \sigma\left(1 + 2\sqrt{\frac{a}{\rho}}\right) = \sqrt{\infty} \qquad (3.3)$$

Note that this simple estimate for the crack tip stress indicates that the *crack tip* stress is "square-root singular," that is, the stress approaches infinity in the special manner $\lim_{\rho \to 0} \rho^{-1/2}$.

The fact that all elastic crack problems have this characteristic square-root singularity (see References 5 and 6 for mathematical proof) leads to the formal definition of the stress intensity factor. If, for example, one determines the stress σ_{tip} at some small distance r ahead of the crack tip and σ_{tip} is found to depend on $r^{-1/2}$, one can define the following *finite* parameter K:

$$K = \lim_{r \to 0} \sqrt{2\pi r}\,\sigma_{\text{tip}} \qquad (3.4)$$

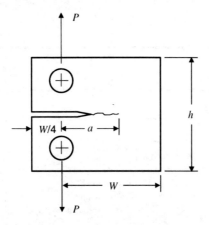

$$K = \frac{P}{B\sqrt{W}} F\left(\frac{a}{W}\right)$$

$$F\left(\frac{a}{W}\right) = \frac{2 + a/W}{1 - (a/W)^{3/2}} \left[0.886 + 4.64\frac{a}{W} - 13.32\left(\frac{a}{W}\right)^2 + 14.72\left(\frac{a}{W}\right)^3 - 5.6\left(\frac{a}{W}\right)^4\right]$$

where $a/W \geq 0.2$

P = concentrated force

B = specimen thickness

$h = 1.2W$

(f)

Figure 3.2. (f) Compact tension specimen.

Considering the three modes of crack opening shown schematically in Figure 3.4, one can expand this stress intensity factor definition. Mode I (opening mode) loading, for example, results when the crack faces move apart in the y direction, as shown in Figure 3.4. The shearing modes II and III result from loading components that cause relative sliding of the crack faces in either the x (sliding mode II) or z (tearing mode III) direction. It can be shown that the elastic stress fields ahead of the crack tips for all three modes of loading depend on the inverse square root of the distance r from the crack tip. This square-root singularity provides the following three stress intensity factor definitions:

$$K_{\mathrm{I}} = \lim_{r \to 0} \sqrt{2\pi r}\,\sigma_y \qquad K_{\mathrm{II}} = \lim_{r \to 0} \sqrt{2\pi r}\,\sigma_{xy} \qquad K_{\mathrm{III}} = \lim_{r \to 0} \sqrt{2\pi r}\,\sigma_{yz} \qquad (3.5)$$

Here K_{I}, K_{II}, and K_{III} are the modes I, II, and III stress intensity factors, r is the distance ahead of the crack tip, and σ_y, σ_{xy}, and σ_{xz} are the tensile, in-plane shear, and out-of-plane shear stresses determined along the line $\theta = 0$ ahead of the crack tip (see Figure 3.4).

Note that the stress intensity factor definitions given by Equation 3.5 only yield useful results if the crack tip stresses have the square root r singularity ($\lim_{r \to 0} r^{-1/2}$).

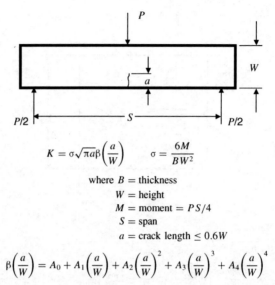

$$K = \sigma\sqrt{\pi a}\,\beta\!\left(\frac{a}{W}\right) \qquad \sigma = \frac{6M}{BW^2}$$

where B = thickness
W = height
M = moment = $PS/4$
S = span
a = crack length $\leq 0.6W$

$$\beta\!\left(\frac{a}{W}\right) = A_0 + A_1\!\left(\frac{a}{W}\right) + A_2\!\left(\frac{a}{W}\right)^2 + A_3\!\left(\frac{a}{W}\right)^3 + A_4\!\left(\frac{a}{W}\right)^4$$

	A_0	A_1	A_2	A_3	A_4
Three-point bend, $S/W = 8$	1.11	−1.55	7.71	−13.5	14.2
Three-point bend, $S/W = 4$	1.09	−1.73	8.20	−14.2	14.6
Four-point (pure) bend	1.12	−1.39	7.32	−13.1	14.0

(g)

Crack (x–y) plane

$$K = \frac{\sigma\sqrt{\pi a}}{E(K)}\left[\sin^2\phi + \left(\frac{a}{c}\right)^2\cos^2\phi\right]^{1/4} \qquad E(k) = \int_0^{\pi/2}\sqrt{1 - \left(\frac{a}{c}\right)^2\sin^2\theta}\; d\theta$$

If $a = c$ = "penny crack," $E(k) = \pi/2$ $\qquad\qquad K = \dfrac{2\,\sigma\sqrt{\pi a}}{\pi}$

(h)

Figure 3.2. (g) Edge-cracked bending specimens (three-point and four-point bending). (h) Flat elliptical crack (major and minor axis lengths c and a) embedded in an infinite body and subjected to remote tension stress σ.

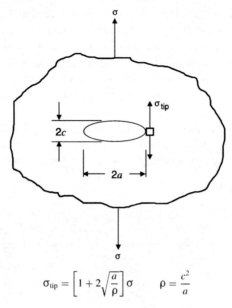

$$\sigma_{tip} = \left[1 + 2\sqrt{\frac{a}{\rho}}\right]\sigma \qquad \rho = \frac{c^2}{a}$$

Figure 3.3. Schematic view of elliptical hole in a large plate loaded with remote tensile stress σ, showing location and magnitude of maximum stress.

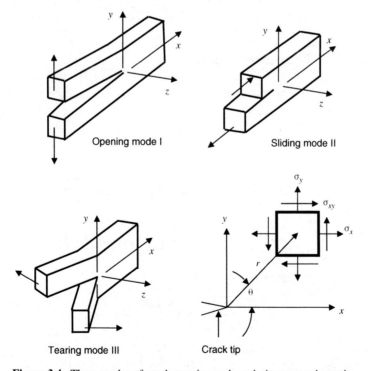

Figure 3.4. Three modes of crack opening and crack tip stress schematic.

If the stresses are proportional to r^{-1}, for example, Equation 3.5 would give infinite values for K_I, K_{II}, and K_{III}. If, on the other hand, the stresses were proportional to $r^{1/2}$ instead of $r^{-1/2}$, the three stress intensity factors $K_I = K_{II} = K_{III} = 0$ by Equation 3.5. Thus, the definitions for the stress intensity factors are based on the fact that *all elastic* crack problems yield crack tip stresses that vary inversely with the square root of the distance r from the crack tip (i.e., the stresses are dominated by a term that behaves as $\lim_{r \to 0} r^{-1/2}$).

Based on the stress intensity factor definitions given by Equations 3.5, the elastic stresses ($\sigma_x, \sigma_y, \sigma_z, \sigma_{xy}, \sigma_{xz}, \sigma_{yz}$) and the x-, y-, and z-direction displacements (u, v, w) in the vicinity of a crack tip are given below for the three modes of loading:

Mode I (Opening Mode)
- Crack tip stresses:

$$\sigma_x = \frac{K_I}{\sqrt{2\pi r}} \cos \frac{\theta}{2} \left(1 - \sin \frac{\theta}{2} \sin \frac{3\theta}{2} \right)$$

$$\sigma_y = \frac{K_I}{\sqrt{2\pi r}} \cos \frac{\theta}{2} \left(1 + \sin \frac{\theta}{2} \sin \frac{3\theta}{2} \right)$$

$$\sigma_{xy} = \frac{K_I}{\sqrt{2\pi r}} \sin \frac{\theta}{2} \cos \frac{\theta}{2} \cos \frac{3\theta}{2} \tag{3.6}$$

$$\sigma_{xz} = \sigma_{yz} = 0$$

$$\text{Plane stress} \rightarrow \sigma_z = 0$$

$$\text{Plane strain} \rightarrow \sigma_z = \nu(\sigma_x + \sigma_y)$$

- Displacements:

$$\text{Plane strain } (w = \varepsilon_z = 0)$$

$$u = \frac{K_I}{G} \left(\frac{r}{2\pi} \right)^{1/2} \cos \frac{\theta}{2} \left(1 - 2\nu + \sin^2 \frac{\theta}{2} \right)$$

$$v = \frac{K_I}{G} \left(\frac{r}{2\pi} \right)^{1/2} \sin \frac{\theta}{2} \left(2 - 2\nu - \cos^2 \frac{\theta}{2} \right)$$

$$\text{Plane stress } (\sigma_z = 0) \tag{3.7}$$

$$u = \frac{K_I}{G} \left(\frac{r}{2\pi} \right)^{1/2} \cos \frac{\theta}{2} \left(\frac{1 - \nu}{1 + \nu} + \sin^2 \frac{\theta}{2} \right)$$

$$v = \frac{K_I}{G} \left(\frac{r}{2\pi} \right)^{1/2} \sin \frac{\theta}{2} \left(\frac{2}{1 + \nu} - \cos^2 \frac{\theta}{2} \right)$$

Mode II (Sliding Mode)
- Stresses:

$$\sigma_x = \frac{-K_{II}}{\sqrt{2\pi r}} \sin \frac{\theta}{2} \left(2 + \cos \frac{\theta}{2} \cos \frac{3\theta}{2} \right)$$

$$\sigma_y = \frac{K_{II}}{\sqrt{2\pi r}} \sin \frac{\theta}{2} \cos \frac{\theta}{2} \cos \frac{3\theta}{2}$$

$$\sigma_{xy} = \frac{K_{II}}{\sqrt{2\pi r}} \cos \frac{\theta}{2} \left(1 - \sin \frac{\theta}{2} \sin \frac{3\theta}{2} \right) \tag{3.8}$$

$$\sigma_{xz} = \sigma_{yz} = 0$$

Plane stress $\rightarrow \sigma_z = 0$

Plane strain $\rightarrow \sigma_z = \nu(\sigma_x + \sigma_y)$

- Displacements (plane strain, $w = 0 = \varepsilon_z$)

$$u = \frac{K_{II}}{G} \left(\frac{r}{2\pi} \right)^{1/2} \sin \frac{\theta}{2} \left(2 - 2\nu + \cos^2 \frac{\theta}{2} \right)$$

$$v = \frac{K_{II}}{G} \left(\frac{r}{2\pi} \right)^{1/2} \cos \frac{\theta}{2} \left(-1 + 2\nu + \sin^2 \frac{\theta}{2} \right) \tag{3.9}$$

Mode III (Tearing Mode)
- Stresses

$$\sigma_{xz} = \frac{-K_{III}}{\sqrt{2\pi r}} \sin \frac{\theta}{2}$$

$$\sigma_{yz} = \frac{K_{III}}{\sqrt{2\pi r}} \cos \frac{\theta}{2} \tag{3.10}$$

$$\sigma_x = \sigma_y = \sigma_z = \sigma_{xy} = 0$$

- Displacements

$$u = v = 0$$

$$w = \frac{K_{III}}{G} \left(\frac{2r}{\pi} \right)^{1/2} \sin \frac{\theta}{2} \tag{3.11}$$

In Equations 3.6–3.11, G is the elastic shear modulus, ν is Poisson's ratio, and (r, θ) are the polar coordinates for the particular point where the stresses or displacements are evaluated. Note that Equations 3.5–3.11 are limited to points "near" the crack tip (i.e., for $r < 10\%$ of the crack length) and assume linear elastic behavior. Discussion of crack tip plasticity and its ramifications on crack growth is beyond the scope of this section (see Chapters 5–7). It will be assumed here, however, that yielding is confined to a small zone near the crack tip and that Equations 3.6–3.11 give an accurate description of the stresses and displacements near the crack tip. The manner in which stress intensity factors may be used to characterize crack growth is described in the following sections.

3.3 FRACTURE

This section describes residual strength calculations used to determine the fracture stress as a function of crack size for a given component. Simply stated, the LEFM criterion employs the experimental observation that many "brittle" materials fracture when the stress intensity factor reaches a "critical" value:

$$K = \sigma\sqrt{\pi a}\beta = K_c = \text{constant at fracture} \tag{3.12}$$

Here K_c is a material property called the *fracture toughness* of the material and is the limiting value of the stress intensity factor that causes catastrophic fracture in all components made from the same material. Note that since K relates load, crack length, and structural geometry, this simple fracture criterion allows one to correlate fracture measurements from laboratory specimens with failure of different structural components. (It is assumed in Equation 3.12 that all components are subjected to the same mode of crack opening. In general, modes I, II, and III loadings are not expected to give the same fracture toughness value). Fracture toughness values for many structural materials are reported in material handbooks [7, 8], and some typical values are given in Table 3.1. This fracture criterion is shown schematically in Figure 3.5. Note that the K_c fracture criterion overestimates the test data for small crack lengths and would, in fact, predict the erroneous result of an infinite fracture stress for zero crack length. This small crack length limitation is associated with crack tip plasticity issues and is discussed more thoroughly in Chapters 5 and 6.

Example 3.1 Assume that the wide panel shown in Figure 3.6 contains a 1.0-in- (2.5-cm-) diameter hole with a radial crack located perpendicular to the applied

TABLE 3.1 Typical Fracture Toughness Values Measured in Thick Plates

Material (thick plate)	2024-T351 Aluminum	7075-T651 Aluminum	Ti-6Al-4V Titanium	300M Steel (235 ksi = 1620 MPa yield)	18 Nickel (200 ksi = 1380 MPa yield)
K_c					
ksi-in$^{1/2}$	31	26	112	47	100
MPa-m$^{1/2}$	34.1	28.6	123	57.6	110

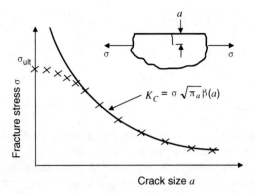

Figure 3.5. Schematic relation between fracture stress and crack size in precracked specimens loaded to failure.

tensile load. It is known that 0.6-in- (1.5-cm-) long cracks can occur at the hole. A 4.0-in- (10.2-cm-) wide *edge-cracked* laboratory specimen made from an identical sheet of material fractures at a stress of 5 ksi (34.7 MPa) when the crack length is 2.0 in (5.1 cm). Both sheets are 1.0 in (2.5 cm) thick and the material has 65 ksi (450 MPa) yield strength. Determine the residual strength of the panel with the cracked hole.

In *order to apply the fracture criterion given by Equation 3.12, the fracture tough-ness K_c* must be computed for the structural material. Employing the stress intensity factor solution from Figure 3.2b for the edge-cracked strip gives

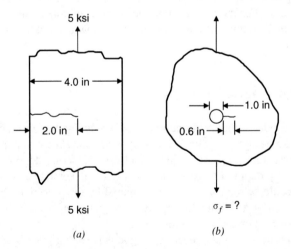

Figure 3.6. Example problem with (a) an edge-cracked strip and (b) a radially cracked hole loaded to fracture.

$$K = \sigma\sqrt{\pi a} \left[1.12 - 0.231 \left(\frac{a}{w}\right) + 10.55 \left(\frac{a}{w}\right)^2 - 21.72 \left(\frac{a}{w}\right)^3 \right.$$
$$\left. + 30.39 \left(\frac{a}{w}\right)^4 \right] \tag{3.13}$$

Letting $\sigma = 5$ ksi (34.7 MPa), $a = 2.0$ in (5.1 cm), and $w = 4.0$ in (10.2 cm), the fracture toughness is found to be $K_c = 35.7$ ksi-in$^{1/2}$ (39.2 MPa-m$^{1/2}$). Now, since all components made from this sheet of material fracture when the stress intensity factor achieves this limiting fracture toughness value, the residual strength for the member with the cracked hole can be computed. Combining Equation 3.12 and the stress intensity factor solution [9] for the cracked hole (see Figure 3.2d) gives

$$K = K_c = \sigma\sqrt{\pi a} \left(\frac{0.8733}{0.3245 + a/R} + 0.6762 \right) \tag{3.14}$$

Now, $K_c = 35.7$ ksi-in$^{1/2}$ (39.3 MPa-m$^{1/2}$) from before, $a = 0.6$ in (1.5 cm), and $R = 0.5$ in (1.25 cm). Solving Equation 3.14 gives $\sigma = 21.0$ ksi (145 MPa) as the fracture stress (i.e., residual strength) of the plate with the cracked hole.

Although the previous example has been greatly simplified, it does describe the general procedure for computing the residual strength of a given component. Note the critical crack size could also have been calculated had the stress been fixed or other geometries considered provided the appropriate stress intensity factors are known. It should be noted here, however, that a more thorough consideration of crack tip plasticity leads to the fact that the fracture toughness K_c depends on material thickness (see Section 6.5.1). Plasticity considerations also limit the specimen sizes and crack lengths that can be analyzed by the fracture criterion of Equation 3.12 (see Chapters 5 and 6). For the present time, however, it will be assumed that crack tip plasticity is "small" relative to the crack length and the material behaves in a "brittle" elastic manner.

3.4   FATIGUE CRACK GROWTH

This section introduces the LEFM approach for predicting fatigue crack growth lives for members subjected to cyclic loading. It is assumed the component of interest contains a preexistent crack of length a_0 and it is desired to determine the number of load cycles N_f required to grow the initial crack to some final size a_f. The final crack length could be the fracture size computed by the procedure discussed in the preceding section or a smaller size specified by some other criterion (e.g., ease of repair, incorporation of a safety factor).

The fracture mechanics approach to fatigue is based on work by Paris et al. [10, 11], who demonstrated that the cyclic range in stress intensity factor ΔK controls the fatigue crack growth rate da/dN. Here ΔK is the difference between the maximum and minimum stress intensity factors during a particular cycle of loading:

$$\Delta K = K_{max} - K_{min}$$

$$= (\sigma_{max} - \sigma_{min})\sqrt{\pi a}\beta$$

$$= \Delta\sigma\sqrt{\pi a}\beta \tag{3.15}$$

In Equation 3.15, $\Delta\sigma$ is the cyclic stress range, a is the current crack length, and β is a dimensionless function of crack size, as before. It is assumed that the crack extends only a small amount during the load cycle, so that both K_{max} and K_{min} are computed for the same crack length. It should also be noted that $K_{min} \geq 0$, since negative stress intensity factors have no physical meaning. Compressive stresses would close the crack tip and eliminate the stress singularity, so that the stress intensity factor defined by Equation 3.5 would yield $K = 0$. Thus, $\Delta\sigma$ in Equation 3.5 is also limited to the *tensile* portion of the applied stress range.

The fact that ΔK controls the rate of fatigue crack growth, and thus cyclic life, can be demonstrated in several ways. Anderson and James [12], for example, describe a series of fatigue crack growth tests with large center-cracked panels. As shown schematically in Figure 3.7a, one group of specimens were loaded *remotely* with a constant cyclic stress $\Delta\sigma$, whereas the remaining panels were symmetrically loaded across the crack faces with a cyclic *force* ΔP (crack face loading).

The specimens were placed in a fatigue machine and tested so that the cyclically applied load amplitude was fixed at either $\Delta\sigma$ or ΔP. Crack lengths were measured at periodic cyclic intervals and plotted as a function of elapsed cycles N. Crack-length-versus-cycles curves are illustrated in Figure 3.7a.

Note the different crack growth behavior for the two types of loadings. The remotely stressed cracks grew at an increasing rate as the crack length increased. (The slope da/dN of the crack-length-versus-cycles curve increased as the test progressed.) The crack face loading specimens gave entirely different results, however, as the growth rate da/dN *decreased* with longer crack lengths. Although this difference in crack growth behavior may seem surprising at first—one group of cracks grew at a faster rate whereas the other cracks slowed down as the test progressed—the results can be explained in terms of the cyclic stress intensity factor ΔK.

Stress intensity factors for the two specimens are given in Figures 3.2a, c. Assuming that the plate width is large in comparison to the crack size ($a/w \rightarrow 0$), those results simplify to

$$\Delta K = \Delta\sigma\sqrt{\pi a} \tag{3.16}$$

for the remotely stressed plate and to

$$\Delta K = \frac{\Delta P}{B\sqrt{\pi a}} \tag{3.17}$$

for the crack face loading (the distance $b = 0$ in Figure 3.2c). While Equation 3.17 is not expressed in the conventional form given by Equation 3.1, it is a valid stress intensity factor solution and does give the correct units for K (i.e., stress $\times$ length$^{1/2}$). Comparing Equations 3.16 and 3.17, note the significant difference in

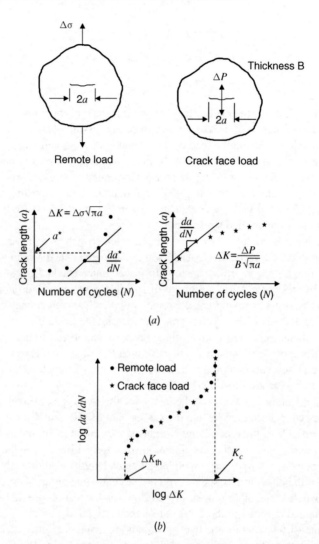

Figure 3.7. (*a*) Schematic view of constant-amplitude fatigue crack growth tests conducted with remotely stressed and crack-face-loaded specimens. (*b*) Schematic representation of log da/dN versus log ΔK data obtained from remote and crack-face-loaded specimens considered in Figure 3.7*a*.

how ΔK depends on crack length for these two specimens. The stress intensity factor range *increases* as the crack grows for the remotely stressed plate (Equation 3.16), while ΔK *decreases* with increasing crack size for the crack face loading specimen (Equation 3.17).

The results from the two sets of experiments agree with each other when the fatigue crack growth rate is plotted versus the cyclic stress intensity factor (see the curves of log da/dN versus log ΔK in Figure 3.7*b*). Here da/dN is calculated from the curve of crack length a versus elapsed cycles N for a particular crack size and ΔK is computed for that crack length. Note that the data from the two different crack geometries lie on the same da/dN-versus-ΔK curve, indicating that the cyclic stress intensity factor ΔK *is the parameter that controls fatigue crack growth rate.*

Actual fatigue crack growth rate test data for an annealed 304 stainless steel [13] are given in Figure 3.8. The different symbols in Figure 3.8 indicate results for *different-shaped* specimens (see Figure 3.9) machined from the same material. The various specimen types were subjected to constant-amplitude loading and the crack length measured as a function of elapsed cycles N. The fatigue crack growth rate da/dN was computed at various crack lengths, as before, and plotted versus the corresponding range in cyclic stress intensity factor (using the appropriate

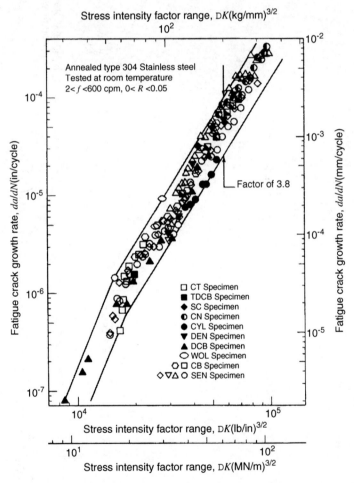

Figure 3.8. Fatigue crack growth data for type 304 stainless steel at room temperature obtained from 10 different specimen designs [13]. Reproduced by permission of the International Atomic Energy Agency.

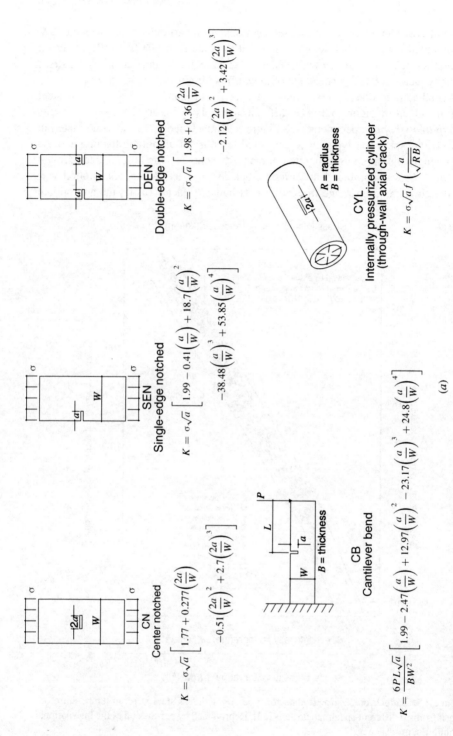

Figure 3.9. Test specimens used to generate the fatigue crack growth data shown in Figure 3.8 [13]. Reproduced by permission of the International Atomic Energy Agency. (Note that the stress intensity factor K solutions given here are presented in a slightly different format than that of Equation 3.1.)

CN
Center notched

$$K = \sigma\sqrt{a}\left[1.77 + 0.277\left(\frac{2a}{W}\right) - 0.51\left(\frac{2a}{W}\right)^2 + 2.7\left(\frac{2a}{W}\right)^3\right]$$

SEN
Single-edge notched

$$K = \sigma\sqrt{a}\left[1.99 - 0.41\left(\frac{a}{W}\right) + 18.7\left(\frac{a}{W}\right)^2 - 38.48\left(\frac{a}{W}\right)^3 + 53.85\left(\frac{a}{W}\right)^4\right]$$

DEN
Double-edge notched

$$K = \sigma\sqrt{a}\left[1.98 + 0.36\left(\frac{2a}{W}\right) - 2.12\left(\frac{2a}{W}\right)^2 + 3.42\left(\frac{2a}{W}\right)^3\right]$$

CB
Cantilever bend

B = thickness

$$K = \frac{6PL\sqrt{a}}{BW^2}\left[1.99 - 2.47\left(\frac{a}{W}\right) + 12.97\left(\frac{a}{W}\right)^2 - 23.17\left(\frac{a}{W}\right)^3 + 24.8\left(\frac{a}{W}\right)^4\right]$$

CYL
Internally pressurized cylinder
(through-wall axial crack)

R = radius
B = thickness

$$K = \sigma\sqrt{a}\,f\left(\frac{a}{\sqrt{RB}}\right)$$

(a)

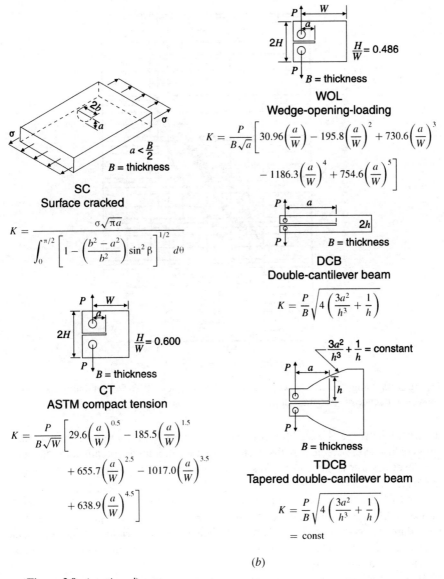

$$\frac{H}{W} = 0.486$$

B = thickness

WOL
Wedge-opening-loading

$$K = \frac{P}{B\sqrt{a}}\left[30.96\left(\frac{a}{W}\right) - 195.8\left(\frac{a}{W}\right)^2 + 730.6\left(\frac{a}{W}\right)^3\right.$$
$$\left. - 1186.3\left(\frac{a}{W}\right)^4 + 754.6\left(\frac{a}{W}\right)^5\right]$$

B = thickness

DCB
Double-cantilever beam

$$K = \frac{P}{B}\sqrt{4\left(\frac{3a^2}{h^3} + \frac{1}{h}\right)}$$

$$\frac{3a^2}{h^3} + \frac{1}{h} = \text{constant}$$

B = thickness

TDCB
Tapered double-cantilever beam

$$K = \frac{P}{B}\sqrt{4\left(\frac{3a^2}{h^3} + \frac{1}{h}\right)}$$

$$= \text{const}$$

$a < \frac{B}{2}$

B = thickness

SC
Surface cracked

$$K = \frac{\sigma\sqrt{\pi a}}{\displaystyle\int_0^{\pi/2}\left[1 - \left(\frac{b^2 - a^2}{b^2}\right)\sin^2\beta\right]^{1/2} d\theta}$$

$$\frac{H}{W} = 0.600$$

B = thickness

CT
ASTM compact tension

$$K = \frac{P}{B\sqrt{W}}\left[29.6\left(\frac{a}{W}\right)^{0.5} - 185.5\left(\frac{a}{W}\right)^{1.5}\right.$$
$$\left. + 655.7\left(\frac{a}{W}\right)^{2.5} - 1017.0\left(\frac{a}{W}\right)^{3.5}\right.$$
$$\left. + 638.9\left(\frac{a}{W}\right)^{4.5}\right]$$

(b)

Figure 3.9. (*continued*)

K equation given in Figure 3.9 for that particular specimen geometry). Again, note that fatigue crack growth rates for *ten* different crack configurations lie on a single $da/dN-\Delta K$ curve. It should be pointed out, however, that there is often considerable scatter in the test data, indicative of the well-known variability in fatigue behavior (note the scatter bands in Figure 3.8). The $da/dN-\Delta K$ curves can also be a function of mean stress, as shown in Figure 3.10 for 7075-T6 aluminum data [8], where

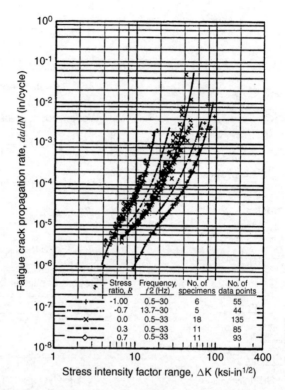

The table within the figure legend:

Stress ratio, R	Frequency, f2 (Hz)	No. of specimens	No. of data points
-1.00	0.5–30	6	55
-0.7	13.7–30	5	44
0.0	0.5–33	18	135
0.3	0.5–33	11	85
0.7	0.5–33	11	93

Figure 3.10. Fatigue crack growth data for 0.090-in-thick 7075-T6 aluminum alloy sheet showing influence of stress ratio R [8].

different results are obtained for different stress ratios R (R = minimum/maximum stress in load cycle). Note here that faster crack growth rates are obtained for the same ΔK level when R is increased. (This stress ratio effect is discussed more thoroughly in Chapter 7.)

The LEFM approach to fatigue is based on the fact that the experimentally determined da/dN–ΔK curve can be effectively treated as a material property. Standard procedures for obtaining the da/dN–ΔK data are recommended by the American Society for Testing and Materials [14] and handbook data are available for common structural materials [7, 8].

When collected over a wide range of crack growth rates, da/dN–ΔK curves for many materials have the characteristic sigmoidal shape shown schematically in Figure 3.7b. A vertical asymptote is observed for this $R = 0$ case when $K_{max} = \Delta K = K_c$ since fracture occurs at that point. There may also be an asymptote at low ΔK levels, designated as the fatigue threshold stress intensity factor ΔK_{th}. Below ΔK_{th} cracks do not extend by cyclic loading ($da/dN = 0$), and the specimen would have "infinite" life. Measuring ΔK_{th} can be difficult, however, involving long test times and many other practical problems (see Chapter 7). The ΔK_{th} value may be considered the fatigue crack growth analog to the endurance limit measured in *uncracked* fatigue specimens (see Section 9.2.2).

A linear relation between log da/dN and log ΔK is often observed between the upper and lower asymptotes. Paris et al. [10, 11] expressed the crack growth behavior in that region by the simple power law

$$\frac{da}{dN} = C \, \Delta K^m \qquad (3.18)$$

Here C and m are empirical constants obtained for a particular set of data. The exponent m is a dimensionless quantity which typically lies in the range $2 < m < 9$.

Many other more general crack growth equations have been employed to relate da/dN with ΔK. One expression suggested by Forman et al. [15], for example, also includes the stress ratio term R and another empirical constant K_c to reflect the upper asymptote in da/dN as the peak K per cycle approaches the fracture toughness of the material:

$$\frac{da}{dN} = \frac{C \, \Delta K^m}{(1 - R)K_c - \Delta K} \qquad (3.19)$$

This expression has been successfully used to represent $da/dN–\Delta K$ curves for different stress ratios by a single mathematical expression. Numerous other fatigue crack growth models are available (see Section 7.3). Thus, fatigue crack growth can be related with the cyclic stress intensity factor by equations of the form

$$\frac{da}{dN} = F(K) \qquad (3.20)$$

Here $F(K)$ is a mathematical expression that fits da/dN over an appropriate range of ΔK values, including the upper and lower asymptotes. The empirical model may also account for other loading variables, such as mean stress and temperature.

Returning now to the original objective of predicting the fatigue crack growth *life*, it is a simple task to integrate Equation 3.20 for the total cycles N_f required to grow an initial crack of length a_0 to some final size a_f. Solving Equation 3.20 for the cyclic life gives

$$N_f = \int_{a_0}^{a_f} \frac{da}{F(K)} \qquad (3.21)$$

Example 3.2. Fatigue Crack Growth Life Calculation   As an example, compute the fatigue crack growth life for an edge crack located in a semi-infinite strip (Figure 3.2b configuration with $a/W \rightarrow 0$). Assume the initial crack size a_0, the constant-amplitude stress $\Delta\sigma$, and the final crack size a_f are known. In addition, assume the Paris law of Equation 3.18 adequately describes fatigue crack growth, where C and m are known material constants. Now, the stress intensity factor equation obtained from Figure 3.2b for the edge crack simplifies to

$$K = \sigma\sqrt{\pi a} \times 1.12 \qquad (3.22)$$

Combining Equations 3.18, 3.21, and 3.22 and integrating give

$$N_F = \int_{a_0}^{a_f} \frac{da}{F(K)}$$

$$= \int_{a_0}^{a_f} \frac{da}{C\,\Delta K^m} = \int_{a_0}^{a_f} \frac{da}{C[1.12\,\Delta\sigma\sqrt{\pi a}]^m}$$

$$N_f = \frac{1}{C(1.12\,\Delta\sigma\sqrt{\pi})^m(1 - 0.5m)}\left[a_f^{1-0.5m} - a_0^{1-0.5m}\right] \tag{3.23}$$

Note that a closed form solution for the fatigue crack growth life has been obtained for this particular example. The loading is represented in Equation 3.23 by the constant-amplitude stress $\Delta\sigma$, the material is specified by the constants C and m (and the choice of Equation 3.18 for the crack growth model), and the component geometry is reflected by the crack sizes a_0, a_f and by the edge-cracked stress intensity factor (Equation 3.22). Since most practical problems involve more complicated stress intensity factor equations and/or fatigue crack growth models, however, it is usually not possible to integrate Equation 3.21 in closed form as in this example. In those cases, a numerical integration scheme may be readily employed. Moreover, variable-amplitude load histories (where $\Delta\sigma$ is not constant) can be considered by cycle-by-cycle integration methods (see Chapter 7).

As a final note, it is important to recognize limitations to the stress intensity factor based approach described here. It is, of course, assumed that K is a valid crack parameter and that crack tip plasticity effects are negligible. Large peak loads applied during the fatigue cycling can introduce large plastic zones, however, which significantly influence subsequent fatigue crack growth (cause fatigue crack retardation as described in Section 7.4). Although beyond the scope of this introduction, more sophisticated life analysis procedures have been developed to analyze peak overloads, mean stress, temperature, and environmental influences that may occur in service. Additional details of those procedures are presented in Chapter 7.

3.5   STRESS CORROSION CRACKING

The chemical and thermal environment subjected to a component can significantly influence crack growth under both static and cyclic loading. Environmentally assisted crack growth resulting from a sustained *static* load is known as stress corrosion cracking, while the combined action of a cyclic load and an "aggressive" environment is commonly called corrosion fatigue. This section briefly outlines the fracture mechanics approach to stress corrosion cracking. (Some aspects of corrosion fatigue crack growth are discussed in Section 7.8.)

The stress corrosion cracking phenomenon can be described with the aid of Figure 3.11. Imagine that a series of specimens are machined from a single sheet of metal and contain cracks of various lengths. The members are immersed in a tank of salt water (or some other environment of interest) and subjected to a fixed load. The cracks in some specimens grow and eventually cause fracture, with the total failure time being dependent on the initial crack size. Plotting the *initially* applied stress intensity

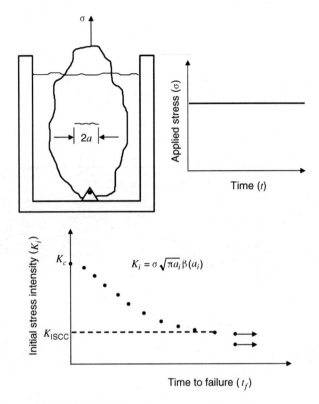

Figure 3.11. Schematic view of stress corrosion cracking experiment.

factor K_i (computed with the applied load and initial crack size) versus the time t_f to specimen failure gives the curve of K_i versus failure time t_f shown schematically in Figure 3.11. Note that specimens initially loaded to the fracture toughness K_c value fracture immediately but that as the initial applied K is reduced for other specimens, time to failure increases, until a "threshold" value of stress intensity factor, labeled K_{ISCC}, is reached. (The subscripts ISCC denote mode I stress corrosion cracking). Specimens loaded below K_{ISCC} do not fracture but have "infinite" stress corrosion lives. Thus, the K_{ISCC} value is an important measure of a material's ability to resist stress corrosion cracking. Although K_{ISCC} will vary for different alloys and chemical environments, stress corrosion cracking threshold values are available for many common structural material–environment combinations [7, 8].

If crack lengths are measured as a function of elapsed time, instead of recording only total time to failure, the stress corrosion data can be expressed in a crack growth rate format similar to that employed for fatigue. This approach is shown schematically in Figure 3.12, where the crack growth rate da/dt at a particular crack length a^* is computed from the crack-length-versus-*time* data and plotted versus the stress intensity factor for the corresponding crack. (Here K is computed for the current crack length and the sustained load using the appropriate stress intensity factor equation for the particular specimen geometry.) Stress corrosion cracking data [7] for 300M steel

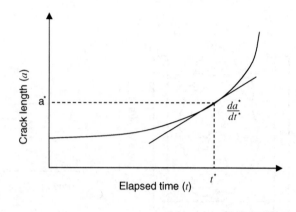

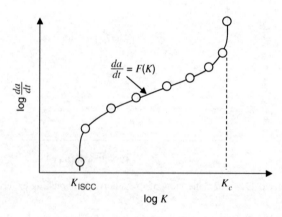

Figure 3.12. Schematic representation of correlation of stress corrosion cracking growth rate with applied stress intensity factor.

tested in distilled water (the aggressive environment) are shown in Figure 3.13. Note that in this case the crack growth rate da/dt is expressed in units of length per time instead of length per cycle as for fatigue.

Again the curve of $\log da/dt$ versus $\log K$ assumes a sigmoidal shape between a lower (K_{ISCC}) and upper (K_c) asymptote. As before, these data could be represented by the empirical equation

$$\frac{da}{dt} = f(K) \tag{3.24}$$

Here $f(K)$ is some convenient mathematical function chosen to represent the test data. Now, the total time t_f required to grow a crack from length a_0 to a_f is given by

$$t_f = \int_{a_0}^{a_f} \frac{da}{f(K)} \tag{3.25}$$

Condition/Ht: 1600F OQ 575F 2+2 hr
Form: 0.1-in Sheet
Specimen Type: DCB
Specimen Thickness: 0.1 in
Orientation: *T–L*
Yield Strength: 245 ksi
Ultimate Strength:
RMS % Error: 52.15

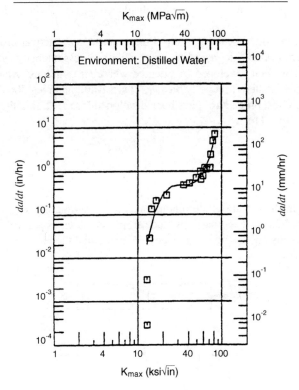

Figure 3.13. Stress corrosion cracking data for 300M steel in distilled water [7].

Note that different crack geometries and material property curves are treated in a manner analogous to computing fatigue crack growth lives.

It is important here to also note the significant effect environment has when combined with cyclic loading. In general, corrosion fatigue crack growth rates can be considerably faster than those observed for cyclic loading in an inert environment. The influence the environment plays on fatigue life depends on the cyclic frequency, the shape of the curve of applied load versus time (i.e., the waveform), the temperature, the environment, the crack orientation (with respect to material axes), and, of course, the particular material of interest (see Section 7.8). Since so many variables can influence corrosion fatigue, it is best to collect data as closely to anticipated service conditions as possible.

3.6 CONCLUDING REMARKS

This chapter has provided an introduction to the LEFM approach for analyzing cracked structures. The main goal here has been to overview concepts that are treated in more depth in Part II of this volume. Although small amounts of crack tip plasticity are allowed, elastic behavior is nominally assumed. Fatigue crack growth or fracture problems involving "large"-scale plasticity must be analyzed by other crack parameters (*R*-curve, *J*-Integral, crack opening displacement, etc., which are discussed in Section 6.7).

In spite of the small-scale plasticity limitation, many practical problems can be analyzed to a reasonable degree of accuracy with the stress intensity factor approach. The method has been developed to a degree where stress intensity factor solutions [1–4] and LEFM material property data [7, 8] are available in handbook form. In addition, standard test procedures have been developed for measuring the crack growth material properties [14, 16].

PROBLEMS

3.1 A 0.5-in-thick plate of aluminum has a fracture toughness of 25 ksi-in$^{1/2}$. A series of 2.0-in-wide bars are made from this plate, as shown in Figure P3.1 (all dimensions are given in inches). Assume that the bars are long relative to their width.

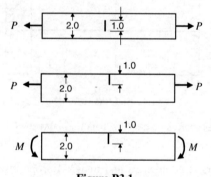

Figure P3.1

(a) Find the remote force *P* required to fracture one of the bars if it contains a center crack whose total length $2a = 1$ in.

(b) Find the remote force *P* required to fracture another bar that contains a 1.0-in-long edge crack.

(c) Find the bending moment *M* required to fracture a bar that contains a 1.0-in-long edge crack.

3.2 A residual strength diagram plots the load-carrying capacity of a structure as a function of crack size. Plot the residual strength (i.e., fracture force *P*) for the center-cracked specimen considered in problem 3.1a. Assume the center-cracked length $2a$ lies in the range $0.05 \leq 2a \leq 1.5$ in.

3.3 Prepare a residual strength diagram for the edge-cracked specimen considered in problem 3.1b. Assume the edge-cracked length a lies in the range $0.05 \leq a \leq 1.1$ in.

3.4 Prepare a residual strength diagram for the edge-cracked specimen considered in problem 3.1c. In this case, plot the maximum moment M that can be applied as a function of crack size. Assume the edge-cracked length a lies in the range $0.05 \leq a \leq 1.2$ in.

3.5 Assume that the long edge-cracked beam considered in problem 3.1c is 1 in wide and 0.25 in thick and is made from a material with a fracture toughness $K_c = 50$ ksi-in$^{1/2}$. Determine the maximum bending moment M that can be applied for crack lengths in the range $0.1 \leq a \leq 0.6$ in. Plot the permissible bending moment versus crack size.

3.6 A 3-in-wide bar is 1 in thick and is to be made from a steel that has a fracture toughness $K_c = 93$ ksi-in$^{1/2}$ and a fatigue threshold stress intensity factor $\Delta K_{th} = 3.0$ ksi-in$^{1/2}$. The bar is loaded in remote tension, and it is assumed that this bar will develop *symmetric* edge cracks in service, as shown in Figure P3.6.

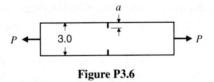

Figure P3.6

(a) Determine a plot of the maximum force P that can be applied without causing fracture as a function of final crack size a, where a is restricted to the range $0.05 < a < 1.4$ in.

(b) Plot on the same set of axes the maximum cyclic force ΔP that can be applied as a function of initial crack size if there is to be no crack growth in the component. Assume the stress ratio $R = $ min/max load $= 0$ and the initial crack size a is again restricted to the range $0.05 < a < 1.4$ in.

Note that the upper curve sets the static maximum load that can be applied without fracture while the lower curve sets the maximum cyclic force for infinite life (i.e., the endurance limit). Applied loads between these two limits result in components with finite fatigue lives.

3.7 Several long tension members are 4 in wide and 1 in thick. They are made from a material with a fracture toughness $K_c = 40$ ksi-in$^{1/2}$ and a threshold stress intensity factor $\Delta K_{th} = 2.5$ ksi-in$^{1/2}$. These members contain cracks as shown in Figure P3.7. Determine the tensile force P required to *fracture* each member for the following cases:

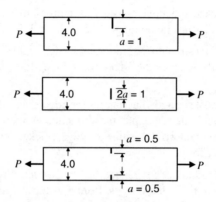

Figure P3.7

(a) One of the members contains a through-thickness *center crack* whose total crack length $2a$ is 1.0 in.

(b) Another member contains a single edge crack whose total length a is 1.0 in.

(c) Another member contains two symmetric edge cracks that are each 0.5 in long.

(d) Which member is capable of sustaining the largest applied tensile force P? Noting that each of these cases had the same total crack area (and net area cross section), what do these results indicate about the influence of crack area on fracture?

3.8   Another group of tension members is made from the same material considered in problem 3.7. These members are cracked to the configurations indicated in parts a, b, and c of problem 3.7, *except that the crack sizes are doubled in each case*. Determine the maximum value of the cyclic force ΔP that can be applied *without* the cracks extending by fatigue for each case (i.e., determine the endurance limit). Which member can sustain the largest cyclic force ΔP without fatigue failure? Assume the stress ratio R ($R = $ minimum/maximum load) is zero.

3.9   Repeat problem 3.6 for the center-cracked member and material considered in problem 3.7a. Assume the half-crack length $0.05 \leq a \leq 1.5$ in.

3.10 Repeat problem 3.6 for the single edge-cracked member and material considered in problem 3.7b. Assume the crack length $0.05 \leq a \leq 2.5$ in.

3.11 Repeat problem 3.6 for the double edge-cracked member and material considered in problem 3.7c. Assume that each crack length $0.05 \leq a \leq 1.5$ in.

3.12 A series of 10-in-wide by 1-in-thick plates are machined from the same piece of material. These plates have the center-cracked, single edge-cracked, and double edge-cracked configurations shown in Figure P3.7, except they are 10 in wide rather than 4 in wide. The center-cracked specimen fractures at a remotely applied tensile stress $\sigma = 20$ ksi when the initial center-cracked length $2a = 4$ in.

   (a) What is the maximum remotely applied stress that can be applied to a single edge-cracked specimen if the crack length $a = 4$ in?

   (b) What is the maximum stress in a symmetric double edge cracked specimen if each edge-cracked length $a = 2$ in?

   (c) What is the maximum length a for a pair of symmetric edge cracks if the remote applied stress $\sigma = 25$ ksi?

3.13 A three-point bend specimen with a span–width ratio $S/W = 4$ (see Figure P3.13) is used to measure the fracture toughness in a steel plate that is expected to have a fracture toughness $K_c = 75$ ksi-in$^{1/2}$. If the specimen width $W = 2.0$ in, the thickness $B = 0.5$ in, and the crack length $a = 1.0$ in, what force P must be applied to fracture the specimen?

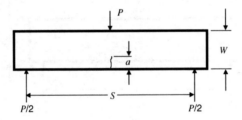

Figure P3.13

3.14 Assume that the three-point bend specimen considered in problem 3.13 is made from a material with a fracture toughness of 35 ksi-in$^{1/2}$ and a threshold stress intensity $\Delta K_{th} = 2$ ksi-in$^{1/2}$. Repeat problem 3.6 for crack lengths in the range $0.1 \le a \le 1.1$ in.

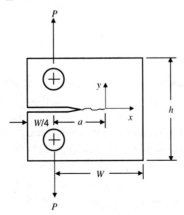

Figure P3.15

3.15 The *compact tension specimen* shown in Figure P3.15 has a thickness $B = 0.2$ in, a width $W = 3.0$ in, a height $h = 3.6$ in, and an initial crack length $a = 2.0$ in. Determine the fracture toughness K_c for this material if the specimen fractures when the applied force $P = 600$ lb.

3.16 Assume that the compact tension specimen considered in problem 3.15 is made from a material with an elastic modulus $E = 10 \times 10^6$ psi and a Poisson's ratio $v = 0.3$. If an applied force $P = 100$ lb is applied to the specimen, determine the y-direction strain ε_y at a point located 0.1 in directly ahead of the crack tip. Assume elastic loading and plane stress conditions.

3.17 An edge-cracked bar is loaded in three-point bending as shown in Figure P3.17. The bar is 2 in wide and 1 in thick and contains a 0.8-in-long edge crack. Determine the normal and shear stresses $\sigma_x, \sigma_y,$ and τ_{xy} along the perimeter of a 0.05-in-radius centered on the crack tip. Determine these stresses in terms of the applied force P for $15°$ increments of the polar angle θ and plot the stresses versus θ. Assume plane stress conditions.

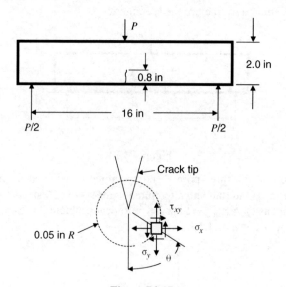

Figure P3.17

3.18 A wide panel contains a center crack whose length $2a = 6$ in. The panel is subjected to a remotely applied stress $\sigma = 10$ ksi as shown in Figure P3.18.
  (a) Assuming elastic behavior, determine the values of the stresses $\sigma_x, \sigma_y,$ and τ_{xy} at points A, B, and C located a distance $r = 0.2$ in from the crack tip.
  (b) Determine the value of the maximum shear stress at point B. (*Hint:* Draw Mohr's circle.)
  (c) Consider a point located directly ahead of the crack tip (i.e., $\theta = 0$). Plot the value of the normal stress σ_y when the distance r from the crack tip is in the range $0 < r \le 0.3$ in.

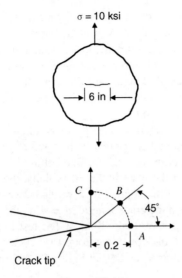

Figure P3.18

3.19 Prepare a *residual strength diagram* for a *wide* plate that contains a 0.5-in-diameter hole loaded with a remote tensile stress σ. Assume the plate is made from a material with a fracture toughness $K_c = 50$ ksi-in$^{1/2}$ and an ultimate tensile stress of 75 ksi and that a *single* radial crack emanates from one side of the hole, as shown in Figure P3.19. Recall that a residual strength diagram is a plot of the maximum load-carrying capacity (in this case, the remote tensile stress σ) of the component versus crack size. Consider crack sizes $a \leq 2.0$ in.

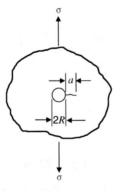

Figure P3.19

3.20 A fatigue crack growth experiment is conducted with a *wide* center-cracked panel made from material X. The panel is subjected to a remotely applied constant-amplitude stress which varies between zero and 20 ksi. Some measured fatigue crack growth data from this test are shown in Figure P3.20.

(a) Determine the fracture toughness for material X and the constants C and m in the Paris fatigue crack growth model.

(b) Another center-cracked panel made from material X has a width $W = 2$ in and is subjected to a remote stress σ as shown in Figure P3.20. Prepare a residual strength diagram for this plate (i.e., plot the stress required to fracture this plate as a function of crack size). Assume the initial crack length $0 < 2a < 1.6$ in.

(c) Another wide center-cracked panel is subjected to the wedge loading shown in Figure P3.20. Here a cyclic force P (units are pounds) is applied along the centerline near the top and bottom faces of the center crack. The thickness B of the plate is 0.5 in. Initially, the total center crack length $2a = 1$ in, and the applied load varies between 0 and 15 kips (1 kip = 1000 lb). When the crack length $2a = 4$ in, the force is increased so that it varies between zero and 30 kips. The test is continued until the crack length $2a = 6$ in. How many total load cycles are applied to the plate? (*Hint:* Determine N_1 and N_2 in Figure P3.20.)

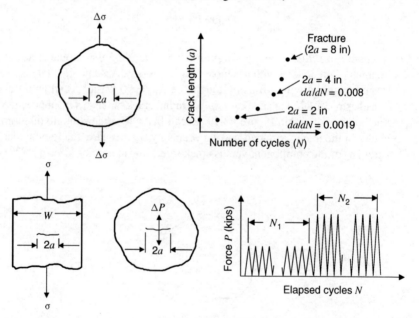

Figure P3.20

3.21 The purpose of this problem is to simulate the use of a repair operation to extend the life of a cracked structure. Consider a *wide* plate that contains a single edge crack that is initially 0.2 in long. The plate is subjected to a *remote* cyclic tensile stress that varies between zero and 10 ksi. When the crack has grown to a length $a = 1.0$ in, the cyclic loading is stopped and an attempt is made to repair the crack by making it shorter, as shown in Figure P3.21. To this end, 0.5 in of material is machined from the cracked edge of the plate, effectively *shortening* the current 1.0 in crack length to 0.5 in. Cycling is then continued to failure at the original zero to 10 ksi remotely applied stress.

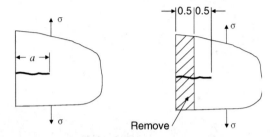

Figure P3.21

What is the total life of the cracked specimen? How much *additional life* has the repair (i.e., machining operation) added to the original life of the test specimen? Assume the following:

(a) Material fracture toughness $K_c = 40$ ksi-in$^{1/2}$.

(b) Machining operation does not affect the remotely applied cyclic stress of 10 ksi (i.e., the plate is *very wide*).

(c) Fatigue crack growth in the plate material is given by the expression

$$\frac{da}{dN} = 5 \times 10^{-9}(\Delta K)^4$$

where the units of da/dN are inches per cycle and ΔK is measured in ksi-in$^{1/2}$.

3.22 The da/dN–ΔK data shown in Figure 3.8 were obtained from a single lot of 304 stainless steel employing 10 different specimen geometries. Note that although there is no specimen dependence in the data, indicating that the da/dN–ΔK data are a true *material* property, there is considerable "scatter" associated with this material property. The purpose of this problem is to determine the influence of this material variability on fatigue life.

(a) Note that a scatter band consisting of four line segments has been drawn through these data. Fit four power laws of the form $da/dN = C(\Delta K)^m$ through the upper and lower bound segments shown on the sketch (i.e., use the "eyeball" method to estimate the constants C and m in the Paris laws that fit the given line segments). Record in a table the values for C and m and the range in ΔK over which they apply. What are the units of C and m?

(b) Use your fitted power laws to compute the upper and lower bound fatigue crack growth rates for $\Delta K = 12$ and 40 ksi-in$^{1/2}$.

(c) Consider a *wide* plate that contains a center crack with initial length $2a = 0.5$ in. If the plate is subjected to a remote tensile stress σ that varies between 0.5 and 12.5 ksi, determine the number of cycles N required to grow the crack from a size $2a = 0.5$ in to $2a = 10.0$ in. Use the upper and lower bound da/dN–ΔK curves and compare the lives you obtain.

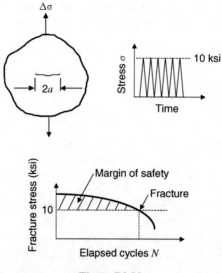

Figure P3.23

3.23 A residual strength diagram describes the maximum load-carrying capacity of a structure. This maximum load capacity can be plotted against either crack size or structural life. The goal of this problem is to prepare residual strength diagrams for a *wide* plate that contains a center crack of length $2a$ as shown in Figure P3.23. The plate is loaded in remote tension and is made from a material which has a fracture toughness of 75 ksi-in$^{1/2}$ and whose fatigue crack growth properties are given by the expression

$$\frac{da}{dN} = 5 \times 10^{-10} \, \Delta K^4$$

Here the units of da/dN are measured in inches per cycle and ΔK is expressed in ksi-in$^{1/2}$.

(a) Determine the fracture stress for the range of crack lengths $0.1 < a < 5.0$ and plot the fracture stress versus crack size.

(b) Consider another center-cracked plate made from the same material which has an initial crack size $a = 0.1$ in ($2a = 0.2$ in). This plate is subjected to a remote stress that varies between zero and 10 ksi. Determine the fatigue life for this plate. Then, prepare a diagram which shows how the residual strength (maximum load capacity) decreases during the life of the structure. Explicitly plot fracture stress versus cycles for the plate that is subjected to the 0–10 ksi remote stress.

3.24 Assume that load restrictions have been placed on a fleet of transport aircraft to extend their fatigue lives. The maximum allowable takeoff weight was reduced from 62,000 to 55,000 tons, representing a 11.3% load reduction. Make a "back of the envelope" estimate of the increase in fatigue life that could be anticipated from such a load reduction.

For purposes of this problem, assume the following:
- Crack geometry = edge crack in a semi-infinite plate
- Initial crack size $a_i = 0.25$ in
- Original cyclic stress = 15 ksi (assume constant amplitude, R = min/max stress = 0)
- Reduced cyclic stress = 13.3 ksi (again, assume constant amplitude, $R = 0$)
- Fracture toughness = 50 ksi-in$^{1/2}$
- Fatigue crack growth rates given by $da/dN = 3.9 \times 10^{-10}(\Delta K)^{4.175}$, where the units of da/dN are inches per cycle and ΔK are ksi-in$^{1/2}$

(a) Determine the percent change in fatigue life gained by the proposed 11.3% load reduction.

(b) What percent increase in life would have been obtained at the original cyclic stress if the initial crack size was reduced to 0.05 in (e.g., improve the nondestructive inspection capability)?

REFERENCES

1. G. C. Sih, *Handbook of Stress Intensity Factors for Researchers and Engineers,* Institute of Fracture and Solid Mechanics, Lehigh University, Bethlehem, Pennsylvania, 1973.

2. H. Tada, P. Paris, and G. Irwin, *The Stress Analysis of Cracks Handbook,* Paris Productions Incorporated, St. Louis, Missouri, 1985.

3. D. P. Rooke, and D. J. Cartwright, *Compendium of Stress Intensity Factors,* Her Majesty's Stationary Office, London, 1976.

4. Y. Murakami, (Editor-in-Chief), *Stress Intensity Factors Handbook*, Pergamon, New York, New York, 1987.

5. M. L. Williams, "On the Stress Distribution at the Base of a Stationary Crack," *Journal of Applied Mechanics,* Vol. 24, pp. 109–114, 1957.

6. J. Eftis, N. Subramonian, and H. Liebowitz, "Crack Border Stress and Displacement Equations Revisited," *Engineering Fracture Mechanics*, Vol. 9, No. 1, pp. 189–210, 1977.

7. D. A. Skinn, J. P. Gallagher, A. P. Berens, P. D. Huber, and J. Smith, *Damage Tolerant Design Handbook, a Compilation of Fracture and Crack Growth Data for High Strength Alloys,* CINDAS/JSAF CRDA Handbooks Operation, Purdue University, West Lafayette, Indiana, May 1994.

8. *Metallic Materials and Elements for Aerospace Vehicle Structures,* MIL-HDBK-5g, Department of Defense, Washington, D.C., 1994.

9. A. F. Grandt, Jr., "Stress Intensity Factors for Some Thru-Cracked Fastener Holes," *International Journal of Fractures,* Vol. 11, No. 2, pp. 283–294, April 1975.

10. P. C. Paris, M. P. Gomez, and W. E. Anderson, "A Rational Analytic Theory of Fatigue," *Trend in Engineering,* University of Washington, Vol. 13, No. 1, p. 9, January 1961.

11. P. C. Paris, "Fatigue—An Interdisciplinary Approach," in *Proceedings of the Tenth Sagamore Conference,* Syracuse University Press, Syracuse, New York, 1964, p. 107.

12. W. E. Anderson, and L. A. James, "Estimating Cracking Behavior of Metallic Structures," Proc. Paper 7213, *Journal of the Structural Division,* ASCE, Vol. 96, No. ST4, pp. 773–790, April 1970.

13. L. A. James, "Fatigue-Crack Propagation in Austenitic Stainless Steels," *Atomic Energy Review,* Vol. 14, No. 1, 1976, pp. 40–41.

14. "Standard Test Method for Constant-Load-Amplitude Fatigue Crack Growth Rates," ASTM Standard E647, in *Annual Book of ASTM Standards, Volume 03.01,* American Society for Testing and Materials, West Conshohocken, Pennsylvania, 2000.

15. R. G. Forman, V. E. Kearney, and R. M. Engle, "Numerical Analysis of Crack Propagation in a Cyclic-Loaded Structure," *Journal of Basic Engineering,* Vol. 89D, pp. 459–464, 1967.

16. "Standard Test Method for Plane-Strain Fracture Toughness of Metallic Materials," ASTM Standard E399, in *Annual Book of ASTM Standards, Volume 03.01,* American Society for Testing and Materials, West Conshohocken, Pennsylvania, 2000.

CHAPTER 4

NONDESTRUCTIVE EVALUATION

4.1 OVERVIEW

Chapter 3 introduced LEFM concepts for determining the influence of preexistent cracks on structural performance. This chapter, in a similar vein, overviews nondestructive techniques for finding structural damage before it can lead to catastrophic fracture. Clearly, component inspection is a key task that must be performed together with damage tolerant analysis to ensure continued structural integrity.

The objective here is to summarize the requirements for successful NDE and to preview various methods for locating structural and material anomalies. Discussion is limited to a general overview of the inspection process as it relates to damage tolerant design and is intended to set the context for more detailed discussion of particular inspection methods provided in Part III.

Early detection of damage is critical to maintain safe operation of structures whose failure could lead to loss of life or property. Inspection is required both during initial manufacturing to ensure that the new structure is free of unacceptable anomalies and later in life to detect service-induced damage. As discussed previously, damage tolerance design concepts are based on the assumption that a preexistent crack exists in the structure of interest, and the safe operating life is the period required for this initial crack to grow to a size that causes failure. In general, the initial crack size assumptions are based on inspection capabilities (i.e., the largest crack that could go undetected) and frequently control allowable stress levels in a structural design. Thus, inspection limits have a direct influence on component size, weight, material requirements, and service life.

It should be noted that a number of terms have been used historically to describe the inspection process:

- nondestructive testing (NDT),
- nondestructive inspection (NDI), and
- nondestructive evaluation (NDE).

The latter title, nondestructive evaluation, implies that the results of the test or inspection are used to evaluate the suitability of the object of interest. This interpretative aspect is a more comprehensive description of the process and will be used here. The goals of NDE are presented below, followed by a brief summary of various inspection methods. Proof testing is discussed first, followed by a concise description of the primary and secondary inspection methods employed in industry. The chapter concludes with a discussion of factors that influence inspection accuracy and methods for quantifying the reliability of NDE.

4.2 GOALS FOR NONDESTRUCTIVE EVALUATION

Before discussing individual inspection methods, it is important to first consider the desired outcomes of NDE and the general factors that influence the required time between inspections.

4.2.1 Detection Requirements

The ideal inspection method should be economical, simple to apply, sensitive to small anomalies, and reliable. For service applications, the technique must be suitable for field use by maintenance personnel. As shown schematically in Figure 4.1 [1], the NDE process involves measuring and interpreting the response of a stimulus that is applied to the test object. The stimulus may, for example, be some form of electromagnetic or ultrasonic energy that interacts with anomalies in the test piece. This interaction leads to an "indication" that is then interpreted by the inspector. As shown in Figure 4.1, this interpretation involves distinguishing between "relevant," "nonrelevant," and "false" indications.

Ideally, interpretation of the indication leads to a simple go/no-go decision regarding acceptance of the test piece. This goal is shown schematically in Figure 4.2, where the probability of rejecting a cracked component is plotted against crack size. As indicated, the objective is to reject all members that contain crack sizes $a \geq a^*$ (the inspection limit) and to accept all remaining components (zero probability of rejection). The inspection limit a^* is the maximum *allowable* crack size specified for the particular component in question. Recalling from Chapter 3 that cracks can grow in service by fatigue, creep, and/or environmental attack, a^* is the smallest crack that could grow to a size that causes failure during a specified time t^* (see Figure 4.3). Note that when evaluating NDE capability, the key question is not to determine the smallest crack size that can be found by a given inspection but rather to specify the *largest crack* that can be *missed* and grow to *failure* during the desired life of the component.

Assuming that an appropriate inspection limit a^* can be defined, the inspector's goal is to find and reject all components with cracks sizes $a \geq a^*$ and to accept all

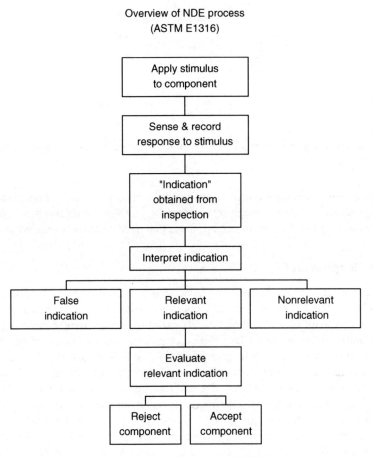

Figure 4.1. Overview of the NDE [1].

components whose maximum crack size $a < a^*$. Due to a variety of factors, how-ever, the "ideal" and "actual" probability of rejection curves will differ in practice. Some "good" components, for example, may be incorrectly rejected, leading to the Type I errors shown in Figure 4.2. This situation represents an economic burden as acceptable members are discarded. The Type II errors shown in Figure 4.2 occur when

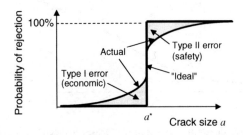

Figure 4.2. Schematic comparison of ideal and actual probability of rejection curves.

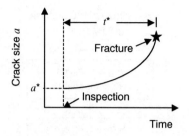

Figure 4.3. Crack growth curve showing definition of inspection limit a^* and time period t^*.

members with crack sizes larger than the inspection limit a^* *are not rejected* by the NDE process. Type II errors represent potential *safety problems*, since they could lead to premature failure from faulty components that have mistakenly entered service.

4.2.2 Inspection Period

It is important to recognize that the inspection size limit a^* must be defined with respect to a particular time period t^*, as shown in Figure 4.3. In practice, t^* is usually modified by an appropriate safety factor to provide two or more inspection opportunities to locate anomalies before they grow to become critical within the next inspection interval (e.g., t^* is frequently specified as half of the time it would take for a missed crack to actually grow to fracture). Note that this time period depends on the applied loads, component material, and final desired crack size and could have a number of interpretations. It may, for example, be the total desired service life, the time it takes for a crack to grow to the maximum size that can be readily repaired, the time interval before another inspection is performed, or the warranty period. As indicated in Figure 4.4, repeated inspection and repair actions could, in principle, keep a structure in service indefinitely (i.e., retirement for cause).

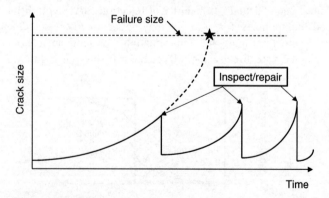

Figure 4.4. Schematic diagram showing how repeated inspections could be incorporated in a retirement-for-cause life management process.

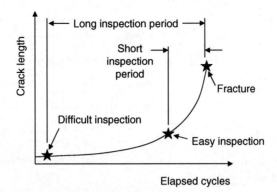

Figure 4.5. Trade-off between inspection period and initial crack size.

Since fatigue life is often quite sensitive to initial crack size, one may need to compare the relative merits of a few detailed, time-consuming inspections versus additional but less complex examinations. Suppose, for example, that a "difficult" inspection can reliably find "small" crack sizes but requires lengthy inspection processes and extensive down time for the structure. On the other hand, an "easy" NDE method (e.g., a simple walk-around visual inspection) may only locate "large" cracks to the required degree of confidence but can be repeated quite often. As shown schematically in Figure 4.5, the "difficult" NDE procedure results in a longer period of safe operation between inspections than the easy method, leading to a trade-off between the costs associated with inspection complexity and frequency.

Example 4.1. Effect of Initial Crack Size on Life   The edge-cracked member discussed previously in Example 3.2 may be used to demonstrate the relation between fatigue life and initial crack size. Recall that Equation 3.21 gives the fatigue life of a cracked member. Solving this expression for an edge crack in a *wide* plate made from a material whose fatigue crack growth properties can be described by a Paris law (Equation 3.18) yields the closed form solution for constant-amplitude fatigue life N_f given by Equation 3.23:

$$\frac{da}{dN} = C \, \Delta K^m \tag{3.18}$$

$$N_f = \int_{a_0}^{a_f} \frac{da}{F(K)} \tag{3.21}$$

$$N_f = \frac{1}{C(1.12 \, \Delta\sigma\sqrt{\pi})^m (1 - 0.5m)} \left(a_f^{1-0.5m} - a_0^{1-0.5m} \right) \tag{3.23}$$

Here C and m are the empirical material constants in the Paris law used to characterize fatigue crack growth and a_f and a_0 are the final and initial crack lengths.

In order to obtain a quantitative demonstration of the initial crack size on fatigue life, assume that the panel is subjected to a constant-amplitude remote stress of 25 ksi

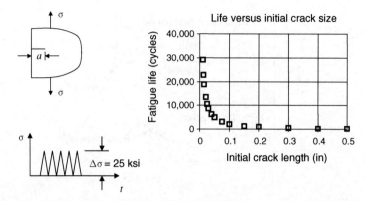

Figure 4.6. Example showing effect of initial crack size on fatigue life for an edge-cracked specimen.

($R = 0$) and is made from a material with a fracture toughness of 50 ksi-in$^{1/2}$ and with fatigue crack growth properties given by the following Paris law:

$$\frac{da}{dN} = 3.9 \times 10^{-10}(\Delta K)^{4.175}$$

Here the units of the cyclic stress intensity factor ΔK are ksi-in$^{1/2}$ and the fatigue crack growth rate is given in inches per cycle. Recall that the stress intensity factor for an edge crack in remote tension is given by $K = 1.12\sigma(\pi a)^{1/2}$. Assuming that fracture occurs at the peak 25 ksi stress when the fracture toughness $K = 50$ ksi-in$^{1/2}$ gives the final crack size a_f to be 1.015 in. Now, Equation 3.23 may be solved for the fatigue life N_f as a function of initial crack size a_0. As shown by the results given in Figure 4.6, fatigue life (and the minimum required inspection period) increases dramatically for small crack sizes. The LEFM concepts introduced in Chapter 3 and developed further in Part II can be used to formulate similar scenarios for other types of crack geometries and loading and can establish the minimum inspection limits for the particular problem of interest.

4.3 PROOF TESTING

Having outlined the general requirements for an effective NDE procedure, we now consider an overview of various inspection methods. Before describing the traditional nondestructive inspection methods, however, consider what is perhaps the simplest method to evaluate a component's suitability for service—determining whether it will resist an overload (proof load). If the member does not fail during the overload, it could reasonably be expected to be safe when the smaller service load is applied. Although often not considered a form of *nondestructive* testing (since damage is found only after a component fails), proof testing provides the engineer with a powerful tool to assess the integrity of a given structure. In addition, discussing proof testing in this introduction to NDE also provides a means to reinforce some of the fracture mechanics concepts developed in the previous chapter.

Proof testing is the practice of overloading a structure to demonstrate that it can safely withstand the expected service load. The fact that a structure successfully passes a proof test can be combined with knowledge of fracture mechanics to determine the maximum crack size that *could* be present. First we define the following quantities:

proof stress σ_p, the stress applied to a component during proof test;

operating stress σ_0, the nominal value of stress during service;

the proof test factor α, equal to σ_p/σ_0, usually greater than 1;

fracture toughness K_c, of the component material; and

a_0, the largest initial flaw size *possible* in the structure.

Consider a cracked component that is subjected to a proof test. If the member does not fracture when the proof stress σ_p is applied, we know that the stress intensity factor K_p at the proof load has not exceeded the fracture toughness K_c as given by the equation

$$K_p = \sigma_p \sqrt{\pi a_0} \beta(a_0) < K_c \qquad (4.1)$$

Here $\beta(a_0)$ is the dimensionless stress intensity factor coefficient for the potential component crack (recall Equation 3.1 and Figure 3.2). Now conservatively assume that the structural crack was almost large enough to cause fracture at the proof load. Then

$$K_c \approx K_p = \sigma_p \sqrt{\pi a_0} \beta(a_0)$$

and

$$a_0 [\beta(a_0)]^2 = \frac{1}{\pi} \left(\frac{K_c}{\sigma_p} \right)^2 = \frac{1}{\pi} \left(\frac{K_c}{\alpha \sigma_0} \right)^2 \qquad (4.2)$$

Specifying a particular crack geometry of interest defines $\beta(a_0)$ and gives the maximum crack length a_0 that could be in the structure, since larger cracks would have fractured during the proof load. Although the value determined for a_0 is an upper bound on the largest possible initial crack, there is still the possibility for smaller cracks that did not propagate during the proof test. The structure may, in fact, be crack free or it was not possible to proof load the component of interest without failing another member.

The proof test parameters may be adjusted to decrease a_0 (i.e., to eliminate the possibility for larger cracks). One could increase the proof stress σ_p, although there is a limit in how large the proof stress can be before other structural damage takes place (i.e., yield occurs if $\sigma_p > \sigma_{ys}$). Since many materials become more brittle at colder temperatures, conducting the proof test at a lower test temperature could reduce the fracture toughness K_c and decrease a_0.

Although proof testing is usually limited to relatively simple components (e.g., pressure vessels, tension bars), large-scale structures have been proof tested. The F-111 aircraft fleet, for example, was proof tested at $-40°F$ ($-40°C$) in a large facility constructed for that purpose [2]. Periodic proof tests of the F-111's were, in fact, a key step in keeping that fleet operational for three decades following early structural integrity problems associated with poor damage tolerance discovered during the late 1960s [3]. The advantages and disadvantages of proof testing are summarized below:

Advantages

- The structure is readily screened for gross defects.
- Previous inspections (ultrasonic, X-ray, etc.) are reliably checked.
- An upper limit for the initial crack size is readily determined.
- The proof test overload may "retard" subsequent fatigue crack growth (this point is discussed further in Chapter 7).

Disadvantages

- The structure may fracture during proof test (i.e., the method is not "nondestructive").
- Proof testing can be expensive, particularly for large components.
- It may be difficult to proof load all components in a structure to the same level or to accurately determine the loads applied to certain areas.
- It may be difficult to proof load the structure in a manner similar to service.
- Preexistent cracks may extend subcritically during the proof load cycle.

Example 4.2. Proof Test Example Calculation Calculation of the proof stress needed to assure a given life may be demonstrated by considering the wide *edge-cracked* panel discussed earlier in Example 4.1. In this case, assume that the panel is subjected to a constant-amplitude $R = 0$ stress of 15 ksi and is made from the same material as before. The goal here is to determine the proof stress that must be applied to guarantee that the panel is safe for an additional 1000 cycles of 0–15 ksi loading.

The first step is to determine the crack size that will cause fracture at the 15 ksi peak stress. In this case fracture occurs when the stress intensity factor equals the fracture toughness K_c:

$$K = 1.12\sigma(\pi a_f)^{1/2} = K_c = 50 \text{ ksi-in}^{1/2}$$

Solving for the fracture crack size at the 15 ksi peak stress gives $a_f = 2.82$ in. Now, the initial crack size that will give the desired 1000-cycle life can be obtained from Equation 3.23:

$$N_f = \frac{1}{C(1.12\,\Delta\sigma\sqrt{\pi})^m(1 - 0.5m)}\left(a_f^{1-0.5m} - a_0^{1-0.5m}\right)$$

Solving this expression when $N_f = 1000$ cycles, $\Delta\sigma = 15$ ksi, $a_f = 2.82$ in, and the Paris law constants $C = 3.9 \times 10^{-10}$ and $m = 4.175$ gives an initial crack size $a_0 = 1.071$ in. Now assuming that fracture "almost" occurs during the proof test gives

$$K_{\text{proof}} = 1.12\sigma_{\text{proof}}(\pi a_0)^{1/2} = K_c = 50 \text{ ksi-in}^{1/2}$$

Solving for the required proof stress gives $\sigma_{\text{proof}} = 24.3$ ksi, yielding a proof test factor $\alpha = \sigma_{\text{proof}}/\sigma_{\text{applied}} = 24.3/15 = 1.62$. Thus, all components that withstand a single application of the 24.3 ksi proof stress are guaranteed to last *at least* an additional 1000 cycles when subjected to the subsequent 15 ksi constant-amplitude operating stress. Note that this calculation assumes the life-limiting feature is the through-thickness edge crack employed for the fatigue and fracture analysis. Other crack geometries could be considered by selecting a different β factor (see Figure 3.2), while alternate safe operating lives could be obtained by selecting a different magnitude of proof stress.

4.4   OVERVIEW OF PRIMARY NONDESTRUCTIVE EVALUATION METHODS

The next two sections describe the major methods to locate life-limiting damage in engineering structures. Discussion here is brief, as these inspection techniques are developed more thoroughly in subsequent chapters. Attention is first directed to the "big six" NDE methods: visual, dye penetrant, radiography, ultrasonic, eddy current, and magnetic particle. A brief summary of other specialized and emerging NDE techniques is then given.

4.4.1   Visual Inspection

Visual inspections are the first line of defense for locating component anomalies and are the most commonly employed NDE procedure. Here visible light is the inspection stimulus, and the human eye is the inspection sensor. Visual inspections may be conducted with the naked eye but often employ low-level magnification devices and hand-held light sources. Lighting is especially critical, as shadows and reflections are significant aids for locating surface anomalies. While the human eye and brain combine as a most effective inspection system that is applicable to many types of materials and components, detection is limited to surface damage and to relatively large crack sizes. Thus, visual inspections are usually combined with some other NDE method when small cracks or internal damage is to be detected.

4.4.2   Dye Penetrant Inspection

Liquid penetrant inspections are a relatively simple, yet powerful technique for aiding visual detection of surface-breaking defects in many types of materials. This NDE method involves flooding the surface of the test object with a low-viscosity fluid. The part is allowed to soak for a period of time (dwell) as the penetrant enters into cracks and other tight crevices by capillary action. The penetrant is then carefully cleaned from the test piece, followed by application of a developer to highlight any penetrant that remains between crack faces. The developer provides both a "blooming" effect to enhance the penetrant remnants and a visually contrasting surface to view the crack penetrant indication. A variety of pentetrant types are available, including fluorescent versions that glow under an ultraviolet or black light. Penetrants can be applied with small spray cans, allowing field-level applications of this popular inspection method.

4.4.3 Radiography

Radiography is one of the oldest NDE methods for locating *internal* and external defects. The technique employs the characteristic of many materials to differentially absorb and pass high-energy radiation. Variations in density lead to different absorption rates as the radiation passes through the test object and creates a shadow image of the internal structure on the recording plane. The source of radiation may be X-rays that are generated in a vacuum tube, gamma rays emitted from radiographic isotopes, or neutrons obtained from a nuclear reactor or particle accelerator. The recording medium is often photographic film but could also be a fluorescent screen or some other type of electronic sensor. Radiography is especially suited for locating non-planar defects such as voids or porosity but can also identify planar flaws that are located parallel to the test beam. Health hazards posed by human exposure to X-rays and gamma rays require compliance to strict safety procedures and significantly add to inspection costs and limit application of this NDE technique.

4.4.4 Ultrasonic Inspection

Ultrasonic inspection is a powerful NDE method that is capable of measuring various material properties as well as locating and quantifying the size of both surface and internal defects. High-frequency (200-kHz to 25-MHz) sound waves are introduced by a transducer that is coupled with the test piece by a fluid or by various other contact methods. The sound waves propagate in the member as compression, shear, and/or guided waves (e.g., surface, Lamb, or Love waves). Ultrasonic sensors placed at various structural locations then sense interactions of the sound waves with component surfaces and boundaries, and provide information regarding both the size and location of potential anomalies. A number of ultrasonic interrogation schemes are possible (e.g., pulse-echo, through-transmission) allowing this method to be adapted to many types of materials and particular applications. Ultrasonic inspections are relatively simple and safe and may be used with many types of materials.

4.4.5 Eddy Current Inspection

Eddy current testing generates small electrical currents in the test piece by means of a varying electromagnetic field from a test coil placed near the object of interest. These eddy currents develop, in turn, a reverse magnetic field that is sensed by the test coil. The eddy currents and, thus, the induced electromagnetic field are sensitive to many test object parameters, including electrical conductivity, magnetic permeability, specimen thickness, residual stress, microstructure, geometric factors, and discontinuities or cracks. This sensitivity to a wide range of parameters makes eddy current testing a versatile NDE method that can be tailored to many different applications. Although suitable for field-level testing, the technique is limited to materials that are electrical conductors and to examining surface or near-surface anomalies.

4.4.6 Magnetic Particle Inspection

Magnetic particle inspection is a subset of the more general class of magnetic inspection techniques. Employing magnets or an applied electrical current first magnetizes the ferromagnetic test piece. Anomalies in the test piece cause magnetic flux "leakage" fields that are then detected by a variety of visual or electronic sensors. The magnetic particle technique relies on visual detection of small particles that migrate to the flux leakages associated with flaws in the test piece. These magnetic particles may be applied dry or wet and come in a variety of colors (including fluorescent). Magnetic particle inspection is sensitive to surface or near-surface cracks and, although limited to ferromagnetic materials, has many practical applications in industry.

4.5 OVERVIEW OF OTHER NONDESTRUCTIVE EVALUATION METHODS

In addition to the major inspection methods discussed in the previous section, there are a variety of other techniques that may be used to interrogate the structure for evidence of damage. While some of these methods are limited to specialized applications, others have the potential for broad use with a variety of structures and materials. Indeed, research in the NDE field has increased dramatically in recent years as scientists have sought new inspection methods to help engineers ensure structural integrity for both new and old structures. This section briefly introduces a few additional inspection methods that are then developed further in Chapter 15.

4.5.1 Acoustic Emission

Many types of material damage (e.g., fatigue cracks) emit detectable bursts of sound when subjected to an applied load. Acoustic emission inspection involves detecting these sound bursts and determining their source of origin. While acoustic emission inspections cannot determine flaw size, they are useful for quickly locating potentially damaged areas of a structure that should be examined by other NDE techniques. The acoustic emission method has been especially useful with composite materials.

4.5.2 Thermal Imaging

Surface temperatures measured on a component that is rapidly heated (or cooled) may also be used for inspection purposes. In this case a heat sensor such as an infrared camera records the induced thermal gradients. Local irregularities in the temperature caused by structural or material anomalies are then detected by comparison to reference standards. Thermal imaging has been used to locate hidden damage such as corrosion or disbonds between two sheets that form a lap joint. Increasing developments with infrared technology suggest further applications for this promising inspection method.

4.5.3   Advanced Optical Methods

There are a number of optical techniques for measuring in-plane or out-of-plane deformations on the surface of a test piece that is subjected to a small change in load. Local perturbations in these displacements indicate the presence of voids, cracks, or other forms of surface or near-surface damage. Superimposing holograms taken of the object in its original and deformed states may, for example, generate optical interference fringes that represent a greatly magnified view of the surface displacements caused by a small thermal pulse or change in pressure. Examination of these interference fringes identifies local strain concentrations associated with a flaw in the test piece. Another advanced optical method employs Moiré techniques to superimpose a finely graduated reference grid over a similar grid mounted to the test piece. When the test piece is subjected to a small load, its grid deforms, resulting in an interference pattern between the reference and deformed grids, and provides an extremely sensitive measure of in-plane deformations. Again, local concentrations of displacement are readily identified by the Moiré fringe patterns and indicate the presence of flaws at or near the surface of the test piece.

4.5.4   Other Inspection Methods

Many other forms of stimuli can be used to locate component anomalies. Leak testing, for example, relates escaped fluids or gases from a pressure vessel to damage in that component. The leaks may be sensed by pressure gages, trace sensitive devices, mass spectrometers, or the sound of escaping fluids or gases. Alterations in a component's dynamic behavior are another indication of inhomogeneities or changes in material properties. In this case a known force vibration is applied to the test piece, and its dynamic response is compared with a reference standard. Other inspections employ microwaves (or radar waves) for locating moisture in dielectric materials and to measure component thickness. Many other specialized inspection methods may be employed for particular NDE applications. Indeed, only the inspector's imagination limits what forms of stimuli can be used to interrogate structures for evidence of anomalies.

4.6   INSPECTION RELIABILITY

Inspections must, of course, yield accurate and reliable results for them to play their key role in ensuring structural integrity. This section considers issues that influence inspection accuracy and briefly discusses procedures for quantifying NDE reliability.

Many factors determine whether or not an inspection results in the correct accept/reject decision:

- selection of the proper inspection method for the given situation;
- correct application of the technique (i.e., proper procedures, training, calibration of equipment, etc.);
- the test environment (ease of access to test piece, working conditions, temperature, etc.);
- variability in equipment performance, test material, component dimensions, crack shape, and so on; and
- human factors.

4.6.1 Human Factors

Human factors are often the largest variables in the effectiveness of NDE systems, resulting in considerable effort to automate as much of the inspection procedure as possible. Human factors include the inspector's proficiency, response to the work environment, and mental attitude. Rigorous training and certification are required to ensure that an individual is well versed in the details of the inspection method and is capable of obtaining appropriate results given the proper equipment and procedures. Continuous education is often needed to keep abreast of technological advances.

The work environment plays a significant role in the inspector's performance. Field-level NDE may, for example, be conducted outdoors in inclement weather or inside cold and drafty shops (or, alternatively, in hot and humid buildings). These facilities are often a noisy hubbub of activity that makes concentration difficult. Moreover, access to test pieces may involve climbing tall scaffolds or working in tight, cramped quarters.

The inspector's mental attitude is an especially important human factor. Is the operator alert and focused on the task at hand? Are there adequate work breaks? Does the inspector have confidence in the available equipment and his or her ability to perform the test? Is the individual in good health, well rested, and free of drugs or stimulants that could cloud judgment? Does management inspire quality performance from the inspector and instill confidence that discovered anomalies will be corrected before the structure is returned to service? Boredom is another attitude factor that often leads to inspection errors when many repeated details must be examined. A large transport aircraft may, for example, have many thousands of fastener holes that must be examined. Since only a very small portion will be damaged, the inspector must avoid being led to a false sense of security by the vast majority of measurements that fail to result in crack indications.

Finally, "prior knowledge" is a key issue when locating structural damage. If, for example, it is known that a particular location has a history of cracking, it is more likely that the inspector's complete attention will be focused on examination of that area. Prior knowledge is a two-edged sword, however, when overattention to one detail causes the inspector to miss large anomalies in adjacent areas that usually do not exhibit damage.

4.6.2 Probability of Detection

Since NDE systems are usually driven to their limits when detecting small anomalies, repeated inspections by the same or different operators can lead to inconsistent results. Thus, the goal for a simple go/no-go decision may not be achievable in practice, and NDE often needs to be considered in terms of detection probability. Consider the potential results of an inspection of N specimens shown in Figure 4.7 [4]. Here N_1 specimens contain $A_1 + A_2$ cracks, while N_2 specimens are uncracked ($N = N_1 + N_2$). Note that the four possible results of the N_1 cracked and N_2 uncracked specimens shown in Figure 4.7 can be used to define the following quantities:

- Probability of detection (POD) = sensitivity of detection = A_1/N_1
- Probability of recognition (POR) = A_4/N_2
- False-call probability (FCP) = A_3/N_2
- Accuracy of the observer = $(A_1 + A_4)/N$

Repeated inspections with groups of cracked and uncracked specimens can be used to establish these various probabilities. The results will depend on operator performance, equipment capabilities, and system operating parameters. Note, for example, that adjusting equipment settings to focus on smaller and smaller flaws will likely increase the number of Type I false calls. In the limit, one could increase sensitivity so much that false calls are obtained every time, and although failures would be avoided by rejecting all components, the inspection would have little value. Ideally, one seeks a system with a high probability of detection with a low FCP.

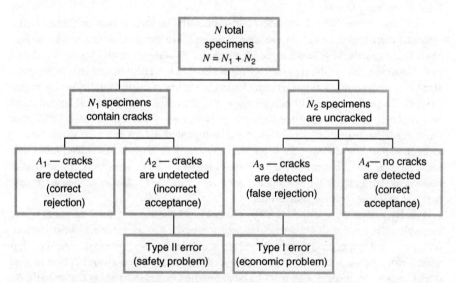

Figure 4.7. Possible results of an inspection of N specimens containing $A_1 + A_2$ cracks [4].

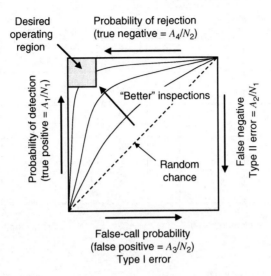

Figure 4.8. Comparison of ROC curves showing desired area of operation.

One measure of system efficiency is the relative operating characteristic (ROC) curve shown in Figure 4.8 [4, 5]. Here POD is plotted versus FCP. Individual curves would be obtained from experiments involving operators with various proficiency levels who examine specimens with a particular crack size and given inspection procedure. The dotted line in Figure 4.8 (i.e., POD = FCP) represents random chance, whereas successively better inspections are given by the curves on the left. Note that the desired operating point is in the upper left-hand corner where the inspector demonstrates a high probability of detection with a low FCP. The ROC curve was originally developed in World War II to quantify the proficiency of radar operators and, although used in the medical field, is not employed as often for other NDE applications.

A more common measure of NDE performance is to specify the POD as a function of crack size. Typical results for 2219-T87 aluminum specimens inspected by various NDE methods are shown in Figure 4.9 [5]. Note that the measured POD depends on crack size and inspection method and often exhibits considerable scatter due to the variations in equipment and inspector limitations described previously.

Statistical procedures can, in principle, be used to establish confidence limits on the POD–crack size curve [6–8]. Figure 4.10, for example, schematically shows the mean curve (50% confidence) along with the 95% confidence bound. An NDE crack size a_{nde} may be formally defined from these data as the crack size that can be determined to a given probability of detection and confidence level. The a_{nde} value shown in Figure 4.10 is based on 90% probability of detection with 95% confidence. The 90/95 initial crack size was specified as the initial crack size criterion for the B-1 damage tolerance design requirements in the 1970s [5] and is typical of that employed for other damage tolerance designs [9]. Alternative probability/confidence levels may be appropriate for other applications.

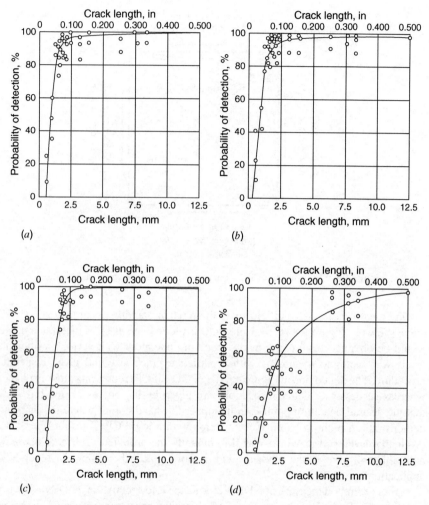

Figure 4.9. Probability of detection versus crack size for four different NDE methods using the same set of specimens: (*a*) penetrant inspection; (*b*) ultrasonic inspection; (*c*) eddy current inspection; (*d*) X-ray inspection. (From p. 682, Ref. 5.)

Although the entire POD–crack length curve can be used to establish the 90/95 crack size limit, it is possible to demonstrate this capability by inspecting groups of specimens with a single crack size. If an inspector is given at least two uncracked specimens for every cracked specimen, the 90/95 inspection capability can be met if the inspector satisfies the cracks detected/missed criteria shown in Table 4.1 [5]. Thus, if 29 specimens are examined, all cracks of the given size must be found. For example, if 46 specimens are tested, there can be one error while still meeting the 90/95 criteria.

The complete POD–crack length *a* curve may be established by "hit-or-miss" or "signal response" data [6]. In the *hit-or-miss* approach, for example, specimens that

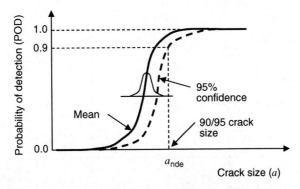

Figure 4.10. Schematic probability of detection curve showing definition of inspection crack size a_{nde} based on 90% probability of detection with 95% confidence.

contain cracks of different sizes (including at least three uncracked specimens for every cracked one) are examined. The inspector only records whether or not a crack was found (i.e., a hit or a miss) and does not attempt to provide length measurements for detected cracks. These hit-or-miss (or go/no-go) type results are typical of many historical NDE testing. Although beyond the scope of the present chapter, it is possible to generate a POD–crack length curve from hit-or-miss data (see References 6 and 8 for a discussion of the required mathematics). A sample POD–crack length a curve obtained by Berens [6] from the results of fluorescent penetrant inspections by three inspectors on 35 cracks in flat plates is given in Figure 4.11. The results from the hit-or-miss measurements are shown as open symbols, and the mathematically determined mean and the lower 95% confidence bound for the complete POD–crack length response are shown as solid and dotted lines, respectively. Note that in this case the mean 90% POD (90/50) crack size is approximately 0.28 in (7 mm), whereas the 90/95 crack size cannot be established with the given data (i.e., the 90/95 crack length is larger than the crack sizes examined in the tests). Many other results obtained from hit-or-miss inspection data are given in Reference 10, a handbook that summarizes inspection results for various inspection methods.

TABLE 4.1   Number of Cracks That Can Be Missed versus Number of Specimens Tested in Order to Demonstrate 90% Probability of Detection with 95% Confidence

Number of Observations	Number of Misses
29	0
46	1
61	2
75	3
89	4
103	5

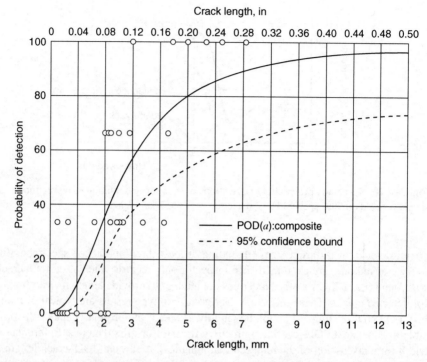

Figure 4.11. Example of mean and 95% confidence POD–crack length curves obtained from hit-or-miss data for fluorescent penetrant inspection. (From p. 692, Ref. 6.)

The *signal response* approach for obtaining POD versus crack length employs additional data from the inspection demonstration program. In this case, the inspector records crack size measurement information, rather than simply indicating whether the crack is present or not. The crack size measurement is in the form of the strength of the measure response signal (e.g., the peak voltage measured in an eddy current or ultrasonic test or the measured size of a penetrant indication). The apparent crack length may, of course, be related to the actual physical crack size by appropriate calibration. Again, rigorous statistical methods are available [6] to determine the complete POD–crack length curve from the signal strength data.

The signal response approach has several advantages over the hit-or-miss method. Fewer tests are needed to establish the POD–*a* curve by the signal response method. The data can also be processed to reflect various signal thresholds for a given measurement, different "decision threshold" crack sizes, as well as saturation limits in the detected signal. By relating the signal response data to physical crack lengths, it may also be possible to estimate results for other crack types that are difficult to introduce into test specimens. Examples of signal strength data and POD curves are given in Reference 6. It should be noted, however, that many of those data are obtained from tests performed in a clean, 70°F, relaxed environment, whereas results from more realistic work environments may not be as good.

Planning and conducting valid NDE demonstration experiments to quantify the reliability of a given inspection setup involve a number of important considerations [6, 10]. Since service cracks can show great variability in, for example, size and material characteristics, the test specimens should contain damage that is representative of that expected in the given application. Although artificially introducing naturally occurring damage (fatigue cracks, corrosion, etc.) in test specimens can be difficult, natural damage should be used rather than machined slots whenever possible.

There are also important sample size issues for correct statistical interpretation of the results (here sample size is determined by the number of cracks in the inspection and not by the number of inspections). The same specimen should not be "resampled" because an important element is crack-to-crack variation, and repeated inspections of the same crack are not independent measurements. The hit-or-miss procedure for determining the POD curve requires 60–80 cracked specimens (with at least three times as many uncracked members), whereas the signal response approach requires fewer (40–60) cracked specimens (again, at least three times as many uncracked specimens). The majority of cracks in the test samples should be near the expected threshold detection size for the particular method. Although not often utilized for POD results, the use of double-blind testing, as employed for evaluating medical devices, could also be an important consideration [11].

Since important human factors issues arise during demonstration testing, the inspection data should be collected under conditions that match those expected in service. Inspectors may be more alert, for example, when they know they are being evaluated or the boredom factor may be reduced since the inspector knows that some specimens are likely cracked. Moreover, specimen accessibility is often more convenient than in actual service applications.

4.7 CONCLUDING REMARKS

The objective of this chapter has been to introduce the field of NDE and to set the context for subsequent development of particular inspection methods in Part III. As indicated, there are many techniques available to locate life-limiting anomalies. Inspection techniques can, for example, be characterized in several ways [12].

- *Active Techniques*   Energy is introduced to the test object, and anomalies result in an observable change in input energy (e.g., eddy current, ultrasonics, radiograpy, magnetic techniques).
- *Passive Techniques*   The test object is observed in the "as is" condition, and anomalies are detected by some response from the tested specimen (e.g., visual inspection, penetrants, leak testing, residual magnetic techniques, noise analysis).
- *Surface Techniques*   Only surface-breaking anomalies can be found (e.g., penetrants).
- *Near-Surface Techniques*   Techniques are limited to finding defects that occur near the surface of the test piece. Although special enhancements can increase

depth of penetration, there are still surface limitations (e.g., eddy current or magnetic methods).

- *Volumetric Techniques*   Techniques are capable of finding internal anomalies that are well below the surface of the test piece (ultrasonic, acoustic emission, radiography).

The assurance provided to structural integrity by inspections often represents a significant financial investment. It is reported, for example, that NDE costs for aircraft can be as much as 15% of the value of primary structure and up to 6% of that of secondary structure [11]. Although NDE is a well-established technology, it remains a subject that is often misunderstood or underappreciated. Thus, it is fitting to conclude this introduction with a summary of common myths and misunderstandings regarding NDE (reproduced in its entirety from page 3–4 of Reference 10):

- "No Flaws Criteria"—NDE is not absolute and "flaws" that are below the detection capabilities threshold may be present after application and discrimination by an NDE procedure.

- "Assumption that the NDE capability is at the smallest flaw detected"—The significant characteristic output of an NDE process is not the "Smallest flaw detected" but the "Largest flaw missed."

- "Assumption that the detection and discrimination capability of an NDE procedure is at the "calibration level"—The discrimination capability of an NDE procedure is rarely at the "calibration" / reference artifact used in set-up of the NDE procedure. Increase of the amplifier gain does not change the discrimination level, but may increase the "noise" response and thereby increase the "false call" level.

- "Cracks and Slots are Equal"—NDE response from artifacts such as slots, saw cuts and electrodischarge machine (EDM) notches are rarely the same as the responses from cracks of an equivalent size.

- "All Cracks are Created Equal"—Cracks of the same size that are initiated and grown under various conditions may produce wide variation in their respective NDE responses.

- "Cracks are equally detectable under all conditions (laboratory/factory/field)"— Crack response may vary with equipment and application conditions. Attention to "calibration, scanning and personnel qualifications" are required to support quantitative field operations.

- "Critical crack size applicable everywhere"—Specification of a critical crack size in general requires information of zoning and expected location of an "assumed crack." The NDE process qualification and application cost will be adjusted to meet specific requirements. Specification of a critical flaw size in one location will not generally be applicable to all locations and flaw orientations.

- "All NDE personnel perform at the same level"—Personnel training, qualification, and certification ensure performance at the highest possible capability for a specific

NDE procedure. Variations in personnel skill and dexterity will produce variations in a specific NDE procedure performance level.

PROBLEMS

4.1 Briefly critique the following statement: An important aspect of nondestructive evaluation is to determine the minimum crack size that can be found by a given inspection method.

4.2 What is the "stimulus" used for the following types of NDE? How would you expect the interaction of these stimuli with the component anomalies to be sensed and recorded?
  (a) Visual inspection
  (b) X-ray inspection
  (c) Ultrasonic inspection
  (d) Magnetic particle inspection
  (e) Dye penetrant inspection
  (f) Eddy current inspection
  (g) Thermal image inspection

4.3 Give examples of "nonrelevant" indications in the context of Figure 4.1.

4.4 Give examples of "false" indications in the context of Figure 4.1.

4.5 Give examples of human factors that could influence the inspection process.

4.6 Give examples of human factors that could influence an individual's performance in the following activities:
  (a) Playing a basketball game
  (b) Performing a musical solo
  (c) Hunting or fishing
  (d) Studying for an exam
  (e) Taking an exam
  Discuss any similarities between the human factors identified above with those that could influence an inspector's NDE performance.

4.7 Compare and contrast the human factors that could affect an athlete's performance in a team sport versus an individual sport (i.e., baseball versus golf). Discuss any similarities to the role of human factors in nondestructive inspection.

4.8 Define Type I and Type II inspection errors.

4.9 Discuss the considerations that determine the maximum acceptable crack size for a given situation.

4.10 Discuss the considerations that determine the maximum period between inspections for a given situation.

4.11 Give a schematic diagram of a "probability-of-detection" curve and use this curve to discuss the definition of a "90/95" NDE crack size.

4.12 Discuss the considerations that determine whether an inspection plan should require many simple inspections or one complicated inspection.

4.13 Redo Example 4.1 assuming that the component in question is a *wide* plate that contains a *centrally located* through-thickness crack of length 2a (see Figure 3.2a).

4.14 Redo Example 4.2 assuming that the component in question is a *wide* plate that contains a *centrally located* through-thickness crack of length 2a (see Figure 3.2a).

4.15 A material has a fracture toughness K_c of 50 ksi-in$^{1/2}$. When the stress ratio $R \geq 0$, fatigue crack growth may be expressed in terms of the following "Walker equation":

$$\frac{da}{dN} = 1.51 \times 10^{-9} \big[(1 - R)^{0.641} K_{max} \big]^{3.70}$$

where da/dN = fatigue crack growth rate (in/cycle)
K_{max} = maximum stress intensity factor during cycle of loading (ksi-in$^{1/2}$)
R = stress ratio (minimum/maximum stress per cycle)
A *wide* panel made from this material is cycled between 5 and 20 ksi remote stress. Experience indicates that these panels will develop through-thickness *center cracks* of length 2a after a period of use. Determine the minimum proof stress necessary to guarantee that the panel is safe for an additional 5000 cycles of 5–20 ksi remote cyclic stress.

4.16 A *wide* tension member fails a proof test when subjected to a remote tensile stress of 20 ksi. (This is well below the desired operating stress.) Posttest examination of the component indicated that it contained an *edge crack* of length $a = 1.0$ in. It has been determined that for the structure to have the desired service life, it cannot contain any edge cracks larger that 0.2 in in length. What proof stress must be withstood to guarantee that the component has no cracks larger than 0.2 in? The component is made from a material with 70 ksi yield stress and 80 ksi ultimate stress.

4.17 Briefly discuss whether or not proof testing should be considered a form of NDE.

4.18 Considerable research has been conducted in recent years to develop new and improved inspection methods. When evaluating new technology, it is important to have specific criteria for judging when the research has matured to a state that is acceptable for practical applications. Briefly list and discuss the key criteria you feel should be used to evaluate the technical maturity of a new nondestructive inspection method.

4.19 Which of the six NDE methods described in Section 4.6 might be suitable for the following applications?

  (a) Locating weld cracks in a steel pipeline

  (b) Locating cracks in the adhesive bond used to form a lap joint of two wide sheets of fiberglass

  (c) Locating dry rot in the wooden timbers of an old house

  (d) Locating cracks along the bore of an open hole in an aluminum plate

  (e) Locating cracks in titanium fan blades

  (f) Monitoring the development of a human fetus

4.20 Recall the failure modes discussed in Section 1.3 of Chapter 1: elastic deformation, inelastic deformation, buckling, fatigue, creep, corrosion, and brittle fracture. In order to employ NDE to help prevent structural failure by these mechanisms, the inspector needs to know what type of critical damage to detect.

  (a) Briefly list and discuss the types of preexistent material or structural damage that would *cause or initiate* each of these failure modes.

  (b) Briefly list and discuss the types of damage that would *result from* the occurrence of each of these failure modes and which should be located by NDE.

REFERENCES

1. "Standard Terminology for Nondestructive Examinations," ASTM Designation E1316, in *1995 Annual Book of ASTM Standards, Volume 3.03, Nondestructive Testing*, American Society for Testing and Materials, West Conshohocken, Pennsylvania, 1995.

2. W. D. Buntin, "Concept and Conduct of Proof Test of F-111 Production Aircraft," paper presented to the Royal Aeronautical Society, London, October 27, 1971.

3. B. J. Sutherland, "F-111 Service Experience—Use of High Strength Steel," AIAA Paper 95-1515, 36th AIAA/ASME/ASCE/AHS/ASC Structures, Structural Dynamics, and Materials Conference, New Orleans, Louisiana, April 1995.

4. R. Halmshaw, *Nondestructive Testing,* Edward Arnold, Baltimore, Maryland, 1991.

5. W. D. Rummel, G. L. Hardy, and T. D. Cooper, "Applications of NDE Reliability to Systems," in *ASM Handbook,* Vol. 17: *Nondestructive Evaluation and Quality Control*, ASM International, Materials Park, Ohio, 1989, pp. 674–688.

6. A. P. Berens, "NDE Reliability Data Analysis," in *ASM Handbook*, Vol. 17: *Nondestructive Evaluation and Quality Control*, ASM International, Materials Park, Ohio, 1989, pp. 689–701.

7. S. I. Vukelich, C. L. Petrin, Jr., and C. A. Annis, Jr., *A Recommended Methodology for Quantifying NDE/NDI Based on Aircraft Engine Experience,* AGARD-LS-190, Advisory Group for Aerospace Research & Development, April 1993.

8. B. D. Olin and W. G. Meeker, "Application of Statistical Methods to Nondestructive Evaluation," *Technometrics,* Vol. 38, No. 2, pp. 95–130, May 1996.

9. *Standard NDE Guidelines and Requirements for Fracture Control Programs,* MSFC-STD 1249, National Aeronautics and Space Administration, Washington, D.C., September 11, 1985.

10. W. D. Rummel and G. A. Matzkanin, *Nondestructive Evaluation (NDE) Capabilities Data Book,* Nondestructive Testing Information Analysis Center (NTIAC), Texas Research Institute Austin, Austin, Texas, May 1996.

11. R. L. Crane, personal communication, 2002.

12. D. E. Bray and R. K. Stanley, *Nondestructive Evaluation, a Tool in Design, Manufacturing, and Service,* rev. ed., CRC Press, Boca Raton, Florida, 1997.

PART II

DAMAGE TOLERANCE ANALYSIS

The objective of the five chapters in Part II is to provide further details of the capabilities and limitations of the crack growth methodology introduced in Chapter 3. Chapter 5 discusses the influence of crack tip plasticity on crack growth and establishes limits for application of the LEFM approach. Chapter 6 expands crack-induced fracture concepts for determining residual strength to include the consequences of crack tip plasticity, and Chapter 7 develops the fracture mechanics approach to consider practical issues associated with fatigue crack growth. Chapter 8 overviews methods to obtain stress intensity factor solutions for fracture mechanics analyses, and Chapter 9 concludes Part II with a discussion of the various damage mechanisms that can lead to structural cracking during service, including fatigue crack nucleation, corrosion, and discrete source damage.

CHAPTER 5

CRACK TIP PLASTICITY

5.1 OVERVIEW

This chapter discusses the plasticity that results from the singular stress field ahead of a crack tip. Although a rigorous plasticity solution is not attempted, several engineering estimates are provided for the crack tip plastic zone size and shape. These plastic zone models have two important uses. First, they allow one to establish limits for the application of LEFM concepts. If, for example, crack tip plasticity is confined to a "small" region relative to the crack length, the stress intensity factor may be confidently used to analyze fracture and subcritical crack growth, as described in Chapter 3. The second major application of these plastic zone models is to provide a physical explanation for several peculiar crack growth behaviors. As discussed in later chapters, the effect of specimen thickness on fracture and fatigue crack growth is readily explained in terms of the plane stress and plane strain plastic zone sizes obtained here with the von Mises yield criterion. These plastic zone models are used later to predict load interaction effects associated with variable-amplitude loading and to explain specimen size requirements and precracking procedures associated with crack growth testing.

As discussed in Section 3.2, the definition of the stress intensity factor K is based on the theoretical result that all elastic crack problems yield square-root singular stresses near the crack tip. It is obvious that no real material can withstand infinite stresses, however, and plastic deformation will occur at actual crack tips. Since subsequent chapters employ stress intensity factor relationships to analyze fracture, fatigue crack growth, and stress corrosion cracking, it is important to *estimate* the extent of crack tip plasticity in order to assess the validity of K as a crack characterization parameter. Three such plastic zone models are described below. All are limited to

123

small-scale yielding (a rigorous plasticity analysis is not attempted here). These plastic zone estimates will be used to establish limitations to the LEFM approach and to explain several peculiar aspects of crack growth behavior.

5.2 IRWIN CIRCULAR PLASTIC ZONE

As discussed in Chapter 3, the stresses *near* the tip of a mode I crack are given by the equation

$$\sigma_y = \frac{K_I}{\sqrt{2\pi r}} \cos\frac{\theta}{2}\left(1 + \sin\frac{\theta}{2}\sin\frac{3\theta}{2}\right)$$

$$\sigma_x = \frac{K_I}{\sqrt{2\pi r}} \cos\frac{\theta}{2}\left(1 - \sin\frac{\theta}{2}\sin\frac{3\theta}{2}\right)$$

$$\sigma_{xy} = \frac{K_I}{\sqrt{2\pi r}} \sin\frac{\theta}{2}\cos\frac{\theta}{2}\cos\frac{3\theta}{2} \tag{5.1}$$

$$\sigma_{xz} = \sigma_{yz} = 0$$

$$\sigma_z = \begin{cases} 0 & \text{for plane stress} \\ \nu(\sigma_x + \sigma_y) & \text{for plane strain} \end{cases}$$

Here K_I is the mode I stress intensity factor for the particular crack configuration of interest and (r, θ) are the polar coordinates for a point near the crack tip. Equation 5.1 gives the following value for the y component of normal stress along the line $\theta = 0$ ahead of a crack tip (see schematic representation of Figure 5.1):

$$\sigma_y = \frac{K_I}{\sqrt{2\pi r}} \tag{5.2}$$

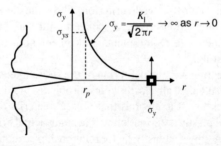

Figure 5.1. Schematic representation of y component of normal stress σ_y along line directly ahead of crack tip ($\theta = 0$).

Note that this result indicates that at the crack tip (i.e., $r = 0$) the stress σ_y is infinite. Solving Equation 5.2 for the distance r_p when the normal stress σ_y equals the tensile yield stress σ_{YS} gives

$$r_p = \frac{1}{2\pi}\left(\frac{K_I}{\sigma_{YS}}\right)^2 \tag{5.3}$$

Irwin [1] suggested that for "small-scale" yielding crack tip plasticity is confined to a circular zone of radius r_p ahead of the crack front and proposed an "effective" crack length a^* whose tip acts at the center of the plastic zone, as shown schematically in Figure 5.2:

$$a^* = a + r_p \tag{5.4}$$

One argument for making the total plastic zone size be $2r_p$ is based on the assumed redistribution of stresses ahead of the crack tip after yielding, as shown in Figure 5.3 [2]. The stress distribution for the "effective" crack is now given in terms of the distance x ahead of the crack tip:

$$\sigma_{y,\,\text{effective}} = \begin{cases} \sigma_{YS} & \text{for } 0 \le x \le 2r_p \\[2mm] \dfrac{K^*}{\sqrt{2\pi(x - r_p)}} & \text{for } 2r_p \le x \end{cases} \tag{5.5}$$

Here K^* is computed for the effective crack length a^*:

$$K^* = \sigma\sqrt{\pi a^*}\beta(a^*) \tag{5.6}$$

Note that for this redistribution of stresses to have occurred, the areas A_1 and A_2 in Figure 5.3 must be equal. Computing the area A_1 gives

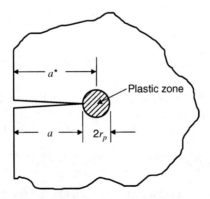

Figure 5.2. Schematic representation of effective crack length a^* with crack tip located at center of plastic zone.

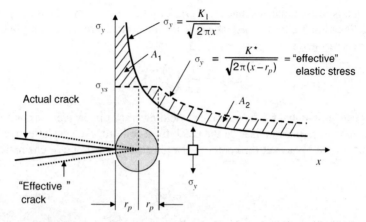

Figure 5.3. Redistribution of normal stress ahead of crack tip associated with crack tip yielding.

$$A_1 = \int_0^{r_p} \sigma_y \, dx - \sigma_{YS} \, r_p$$

$$= \int_0^{r_p} \frac{K_1}{\sqrt{2\pi x}} \, dx - \sigma_{YS} \, r_p$$

$$= 2K_1 \sqrt{\frac{r_p}{2\pi}} - \sigma_{YS} \, r_p \qquad (5.7a)$$

Similarly, area A_2 is given by

$$A_2 = \sigma_{YS} \, r_p - \int_{r_p}^{2r_p} \frac{K_1}{\sqrt{2\pi x}} dx + \int_{2r_p}^{\infty} \left(\frac{K^*}{\sqrt{2\pi(x - r_p)}} - \frac{K_1}{\sqrt{2\pi x}} \right) dx$$

$$= \sigma_{YS} \, r_p - \frac{2K_1}{\sqrt{2\pi}} \left(\sqrt{2r_p} - \sqrt{r_p} \right) - \frac{2}{\sqrt{2\pi}} \left(K^* \sqrt{r_p} - K_1 \sqrt{2r_p} \right) \qquad (5.7b)$$

Now assuming small-scale plasticity, $r_p \ll a$, and $K^* \approx K_1$ gives $A_2 = \sigma_{YS} \, r_p$. Now requiring $A_1 = A_2$ and solving for r_p give

$$A_1 = 2K_1 \sqrt{\frac{r_p}{2\pi}} - \sigma_{YS} \, r_p = A_2 = \sigma_{YS} \, r_p \Rightarrow r_p = \frac{1}{2\pi} \left(\frac{K_1}{\sigma_{YS}} \right)^2 \qquad (5.8)$$

which is the same result obtained earlier in Equation 5.3 for the plastic zone radius.

Thus, to account for plasticity by the Irwin model, one simply considers the "effective" crack length $a^* = a + r_p$. Note that an iterative procedure is required since r_p depends on K_I and K_I depends on $a + r_p$. This plastic zone correction is only appropriate, however, for *small-scale* yielding.

We may now define small-scale yielding (i.e., "brittle" behavior) as the case when the crack length is much larger than the plastic zone radius ($a \gg r_p$, i.e., $a > 10r_p$). We may, in turn, use this plastic zone size estimate as a criterion for when to use LEFM. Note that by this criterion, LEFM is valid when

- K_I is small (e.g., small loads, as in fatigue applications),
- the tensile yield stress σ_{YS} is large (high-strength materials are often more "brittle" than low-strength materials), and
- the crack length a is physically large.

By the same token, we might expect LEFM to break down when the opposite conditions are true (i.e., large applied loads, low-strength materials, and small cracks). This important point is discussed more in later sections.

Example 5.1 A 0.1-in- (2.5-mm-) thick sheet of aluminum alloy 7475-T61 has a yield stress of approximately 64 ksi (411 MPa) and a fracture toughness $K_c = 75$ ksi-in$^{1/2}$ (82.4 MPa-m$^{1/2}$). Thus, the plastic zone size for this material when fracture occurs is

$$r_p = \frac{1}{2\pi} \left(\frac{K_c}{\sigma_{YS}} \right)^2 = \frac{1}{2\pi} \left(\frac{75}{64} \right)^2 = 0.22 \text{ in (5.6 mm)}$$

Now, the small-scale yielding criterion at fracture leads to the following results:

- If $a > 2.2$ in (56 mm), LEFM is valid and fracture may be determined by the K_c fracture criterion.
- If $a < 2.2$ in (56 mm), applications of LEFM are questionable.

Note that K may still be applicable for *fatigue* problems in this material with small crack lengths, since fatigue crack growth must occur at smaller values of K to avoid fracture. If, for example, the applied $\Delta K = 10$ ksi-in$^{1/2}$ (11 MPa-m$^{1/2}$), the crack tip plastic zone radius is

$$r_p = \frac{1}{2\pi} \left(\frac{10}{64} \right)^2 = 0.004 \text{ in (0.1 mm)}$$

Thus, we could safely analyze fatigue crack lengths as small as 0.04 in (1 mm) at this ΔK level. The smallest crack size that can be considered by LEFM for *fracture*, however, is $a_{min} = 10r_p = 2.2$ in (56 mm). Note that this does not mean a smaller crack will not cause fracture—just that it will fail in a "ductile" fracture mode that cannot be predicted by the K_c fracture criterion.

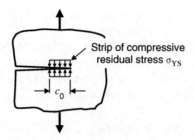

Figure 5.4. Schematic view of Dugdale plastic zone showing narrow strip of compressive stresses located at crack tip.

5.3 DUGDALE (STRIP) MODEL

Consider another simple plasticity model proposed by Dugdale [2, 3] to estimate the crack tip plastic zone size. This model assumes that yielding leads to compressive residual stresses that act along a narrow strip behind the crack tip and effectively eliminate the crack tip singularity, as shown schematically in Figure 5.4. Here the strip of compressive residual stress is c_0 long and has a magnitude equal to the tensile yield stress. Linear superposition may now be employed to estimate the strip length c_0 required to eliminate the singular stresses at the elastic crack tip. Note that we will be using "elastic" arguments to estimate the amount of "plasticity" in this case.

First, break the strip model into two elastic components as shown in Figure 5.5. Note that the stress intensity factor K is a linear elastic parameter, and since we are assuming elastic behavior in constructing this model, the K's are also additive, leading to the equation

$$K_A = K_B + K_C \tag{5.9}$$

The goal here is to eliminate the crack tip singularity in member A by adding the appropriate length c_0 of compressive stress, so that

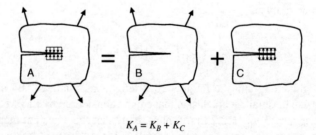

$$K_A = K_B + K_C$$

Figure 5.5. Decomposition of cracked member with strip yield plastic zone into separate remote loading and crack face loading problems.

$$K_A = 0 \qquad K_B = -K_C \qquad (5.10)$$

Now determine the stress intensity factor for members B and C. Let K_B be the stress intensity factor for the original cracked body (without a plasticity correction) and assume that it is known for the problem of interest. Now compute K_C for the strip problem by using the point load solution discussed previously for the problem shown in Figure 5.6. (Here K_C designates the stress intensity factor for member C and not the fracture toughness.) The stress intensity factor for the point load problem is given by

$$K = \frac{P}{B\sqrt{\pi a}} \sqrt{\frac{a+b}{a-b}} \qquad (5.11)$$

Expressing Equation 5.11 in terms of the distance c from the crack tip gives the following results:

$$a = b + c \qquad a + b = 2a - c \qquad a - b = c$$

$$(5.12)$$

$$K = \frac{P}{B\sqrt{\pi a}} \sqrt{\frac{2a-c}{c}}$$

Now assuming that the distance $c \ll a$ (i.e., assume small plastic zone size) gives

$$K = \frac{P}{B\sqrt{\pi a}} \sqrt{\frac{2a}{c}} = \frac{P}{B} \sqrt{\frac{2}{\pi c}} \qquad (5.13)$$

This result may now be used to find the stress intensity factor K_C for member C by treating the "point load" P as the force resulting from the stress acting over the infinitesimal length dc of the crack face, as shown in Figure 5.7. Letting the force $P = -\sigma_{YS} B \, dc$ (i.e., force equals stress times area) and integrating Equation 5.13 over the length c_0 of the Dugdale strip give

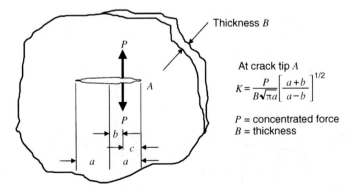

Figure 5.6. Center crack loaded with force P near crack tip. The stress intensity factor solution for this geometry is used to generate K for the strip load problem.

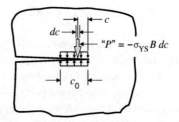

Figure 5.7. Schematic view of strip yield model showing use of crack force P loading to represent distributed force over length c_0.

$$K_C = -\int_0^{c_0} \frac{\sigma_{YS}B}{B} \sqrt{\frac{2}{\pi c}} \, dc \tag{5.14}$$

Performing this integration gives

$$K_C = -2\sqrt{\frac{2}{\pi}} \sigma_{YS} \sqrt{c_0} \tag{5.15}$$

Finally, requiring $K_B = -K_C$ gives

$$K_B = K = -K_C = -\left(-2\sqrt{\frac{2}{\pi}} \sigma_{YS} \sqrt{c_0}\right) \tag{5.16}$$

Solving for the Dugdale plastic zone size c_0 gives

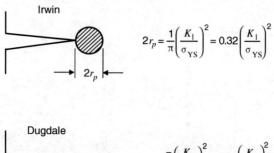

Figure 5.8. Comparison of Irwin circular plastic zone and Dugdale strip yield model.

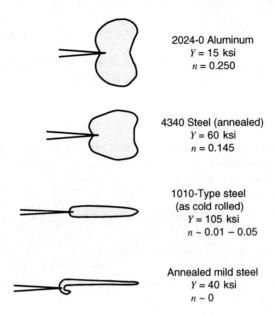

2024-0 Aluminum
$Y = 15$ ksi
$n = 0.250$

4340 Steel (annealed)
$Y = 60$ ksi
$n = 0.145$

1010-Type steel
(as cold rolled)
$Y = 105$ ksi
$n \sim 0.01 - 0.05$

Annealed mild steel
$Y = 40$ ksi
$n \sim 0$

Figure 5.9. Plane stress plastic zone shapes observed on surface of polished specimens by metallographic etching technique [4] (not drawn to same scale).

$$c_0 = \frac{\pi}{8}\left(\frac{K}{\sigma_{YS}}\right)^2 \tag{5.17}$$

Now compare the Dugdale and Irwin plastic zones shown schematically in Figure 5.8 and given by Equations 5.3 and 5.17. Note that both of these mode I models indicate that the crack tip plastic zone is proportional to $(K/\sigma_{YS})^2$ and give approximately the same size (compare the coefficients $1/\pi = 0.32$ with $\pi/8 = 0.39$). The models do differ in shape, however, as Irwin assumes a circular plastic zone, whereas Dugdale assumes plasticity is confined to a narrow strip. Figure 5.9 shows experimentally measured plastic zones obtained by Hahn et. al. [4]. Note that materials that exhibit work hardening give circular plastic zones as suggested by Irwin, whereas elastic–perfectly plastic materials (i.e., the strain-hardening exponent $n = 0$) confine yielding to a narrow strip like the Dugdale model.

5.4 VON MISES PLASTIC ZONE

Another estimate for the extent of crack tip plasticity may be obtained from the von Mises yield criterion described in Section 2.6.2 (see Equation 2.34). By this criterion, yielding occurs for a particular state of stress $(\sigma_x, \sigma_y, \sigma_z, \sigma_{xy}, \sigma_{yz}, \sigma_{xz})$ when

$$2\sigma_{YS}^2 = (\sigma_x - \sigma_y)^2 + (\sigma_x - \sigma_z)^2 + (\sigma_y - \sigma_z)^2 + 6(\sigma_{xy}^2 + \sigma_{xz}^2 + \sigma_{yz}^2) \quad (5.18)$$

Since the elastic stresses are known in the vicinity of the crack tip, Equation 5.18 can be used to determine the region where crack tip yielding begins.

5.4.1 Mode III Case

Consider, for example, application of the von Mises yield criterion to the case of mode III cracking (recall Figure 3.4). The elastic stresses for this out-of-plane shear loading are given by the equations

$$\sigma_{xz} = -\frac{K_{III}}{\sqrt{2\pi r}} \sin\frac{\theta}{2} \qquad \sigma_{yz} = \frac{K_{III}}{\sqrt{2\pi r}} \cos\frac{\theta}{2}$$

$$\sigma_x = \sigma_y = \sigma_z = \sigma_{xy} = 0 \tag{5.19}$$

Substituting these crack tip stresses into the von Mises yield criterion of Equation 5.18 gives the following:

$$2\sigma_{YS}^2 = (\sigma_x - \sigma_y)^2 + (\sigma_x - \sigma_z)^2 + (\sigma_y - \sigma_z)^2 + 6(\sigma_{xy}^2 + \sigma_{xz}^2 + \sigma_{yz}^2)$$

$$= (0-0)^2 + (0-0)^2 + (0-0)^2 + 6(0^2 + \sigma_{xz}^2 + \sigma_{yz}^2)$$

$$= 6\left[\left(\frac{-K_{III}}{2\pi r}\sin\frac{\theta}{2}\right)^2 + \left(\frac{K_{III}}{2\pi r}\cos\frac{\theta}{2}\right)^2\right]$$

$$= 6\frac{K_{III}^2}{2\pi r}\left(\sin^2\frac{\theta}{2} + \cos^2\frac{\theta}{2}\right) = 6\frac{K_{III}^2}{2\pi r}$$

Solving for the distance r_p ahead of the crack tip where yielding begins leads to the equation

$$r_p = \frac{3}{2\pi}\left(\frac{K_{III}}{\sigma_{YS}}\right)^2 \tag{5.20}$$

Thus, yielding occurs for all points defined by $r \le r_p$, whereas elastic behavior exists for $r > r_p$. Note that in this instance the plastic zone radius is independent of the polar coordinate θ, so that the plastic zone has a circular shape reminiscent of Irwin's circular plastic zone for mode I loading.

5.4.2 Mode I Case

In mode I loading, the elastic crack tip stresses are given by Equation 5.1. Inserting these stresses in the von Mises yield criterion gives the polar coordinates (r, θ) for the boundary where yielding begins. For *plane stress* conditions (defined as the state of stress for which $\sigma_z = 0$), the resulting elastic–plastic boundary is given by

$$r_p^* = \frac{1}{2\pi} \left(\frac{K_I}{\sigma_{YS}} \right)^2 \cos^2 \frac{\theta}{2} \left(1 + 3 \sin^2 \frac{\theta}{2} \right) \qquad (5.21)$$

For *plane strain*, specified by $\varepsilon_z = 0$, Hooke's law gives $\sigma_z = \nu(\sigma_x + \sigma_y)$, where ν is Poisson's ratio and the normal stresses σ_x and σ_y near the crack tip are again given by Equations 5.1. Now, inserting the mode I crack tip stresses for the plane strain case into the von Mises yield criterion (Equation 5.18) gives the following result for the plastic zone radius:

$$r_p^* = \frac{1}{2\pi} \left(\frac{K_I}{\sigma_{YS}} \right)^2 \cos^2 \frac{\theta}{2} \left[(1 - 2\nu)^2 + 3 \sin^2 \frac{\theta}{2} \right] \qquad (5.22)$$

Comparing schematic plots of Equations 5.21 and 5.22 in Figure 5.10, one notes that the plane stress plastic zone is considerably larger than that for plane strain. If, for example, $\nu = \frac{1}{3}$ and one evaluates r_p^* directly ahead of the crack tip ($\theta = 0$), one obtains the following results:

$$r_p^* = \frac{1}{2\pi} \left(\frac{K_I}{\sigma_{YS}} \right)^2 \qquad \text{for plane stress}$$

$$r_p^* = \frac{1}{2\pi} \left(\frac{K_I}{\sigma_{YS}} \right)^2 \frac{1}{9} \qquad \text{for plane strain}$$

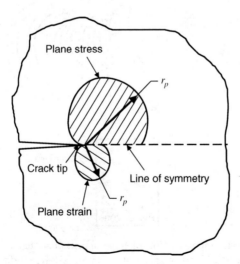

Figure 5.10. Comparison of plane stress and plane strain plastic zone shapes determined by von Mises yield criterion.

Note that in this case the plane stress plastic zone is nine times larger than the plane strain case. In addition, note that at $\theta = 0$ the plane stress plastic zone (Equation 5.21) reduces to the circular plastic zone radius given by the Irwin model (Equation 5.3). The von Mises plastic zone is not completely circular in shape, however, but has the "butterfly" shape commonly observed for materials that exhibit strain hardening. (Compare Figure 5.10 with the experimentally measured plastic zones given in Figure 5.9 for 2024-0 aluminum and for 4340 steel.)

The fact that the von Mises yield criterion indicates that crack tip yielding depends on the state of stress is a significant result that will be useful later for explaining thickness effects in fracture. Consider, for example, a through-thickness crack that is located in a thick plate. On the free surfaces of the plate, plane stress conditions exist ($\sigma_z = 0$), and the mode I plastic zone is given by Equation 5.21. If the plate is thick enough, however, plane strain conditions ($\varepsilon_z = 0$) can develop in the center of the plate, so that the smaller plane strain plastic zone forms (Equation 5.22). Thus, we have a variation in the plastic zone size as we progress through the thickness of the plate along the crack tip. The "dumbbell" plastic zone shown schematically in Figure 5.11 represents this transition from a large crack tip plastic zone on the free surface to lesser amounts of yielding in the specimen interior. Experimental evidence for this through-thickness variation in plastic zone size is given in Reference 4. As discussed later, this through-thickness variation in the crack tip plastic zone explains several specimen thickness dependent fracture and fatigue crack growth behaviors.

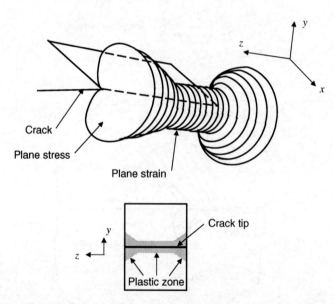

Figure 5.11. Schematic representation of three-dimensional "dumbbell"-shaped crack tip plastic zone showing transition from plane stress yielding at free surfaces of plate to smaller plane strain yield zone at center of plate (crack lies in x–z plane and is growing in the x direction).

5.5 CYCLIC PLASTIC ZONE

The plastic zone models discussed in the preceding sections have dealt with static conditions, where the member is not allowed to unload. When the member is unloaded, however, as in a fatigue test, reversed yielding can occur, leading to a different plastic zone size. It has been shown [5] that the crack tip plastic zone size for the cyclic case is approximately one-fourth that caused by monotonic loading. Thus, LEFM conditions (i.e., small plastic zone size relative to crack length) will often apply for fatigue crack growth problems. There will remain, however, several instances where the small-scale yield assumption breaks down in fatigue (e.g., small fatigue cracks, overloads). These situations are discussed in later chapters.

5.6 CONCLUDING REMARKS

This chapter deals with characterizing the plastic deformation around crack tips and is intended to set the framework for subsequent discussions of the influence of crack tip plasticity on fatigue and fracture. Although the elastic crack tip stress field is singular, these large stresses are confined to a fairly small region, and as a result, the extent of yielding may also be limited in size. When the plastic zone size is small relative to the crack length, the stress intensity factor K (which is a linear elastic parameter) may be confidently used to characterize crack growth. If yielding extends over a larger area, however, LEFM assumptions are no longer valid, and a more sophisticated plasticity analysis may be required to predict fatigue and fracture.

Small-scale yielding estimates for mode I cracking under plane stress conditions are given by Equations 5.3 (Irwin), 5.17 (Dugdale), and 5.21 (von Mises). Although these results differ in the manner in which they model plasticity and lead to different plastic zone shapes, they give approximately the same plastic zone size. The von Mises yield criterion was also used to determine the mode I plastic zone for the plane strain case (Equation 5.22) and indicates much less plasticity for plane strain than for plane stress conditions. Thus, there will be much more yielding near the crack tip in a thin sheet (plane stress) than at the center of a thick plate (plane strain). Moreover, since plane stress conditions exist on the surface of a thick plate, the crack tip plastic zone size varies through the thickness of the plate (Figure 5.11).

As discussed later, crack tip plasticity considerations may also be used to explain several crack growth phenomena. In Chapter 6, for example, fracture toughness (K_c) is shown to vary with specimen thickness in a manner consistent with the three-dimensional plastic zone size shown in Figure 5.11. Load interaction effects associated with variable-amplitude fatigue crack growth (i.e., crack retardation) are explained in Chapter 7 and modeled in terms of the crack tip plastic zone. Moreover, many experimental test procedures are associated with controlling the amount of plasticity that can develop in crack growth specimens. Thus, an appreciation for the crack tip plastic zone size is essential to establish limits for application of LEFM and to understand many specialized aspects of crack growth.

PROBLEMS

5.1 Define a "brittle" material in the context of LEFM.

5.2 The ASTM standard procedure for measuring fracture toughness requires that the crack length used for the test specimen exceeds 2.5 $(K_{Ic}/\sigma_{YS})^2$ where K_{Ic} is the fracture toughness and σ_{YS} is the 0.2% offset yield strength of the material (see ASTM standard E399). What is the basis for this crack length requirement?

5.3 A cracked specimen is made from an aluminum alloy with a tensile yield strength of 47 ksi and a fracture toughness of 31 ksi-in$^{1/2}$. Determine the smallest crack length whose fracture conditions can be analyzed by LEFM. What is the smallest fatigue crack that can be analyzed by LEFM if $\Delta K = 20\%$ of K_c (for $R = 0$ cyclic loading)?

5.4 Repeat problem 5.3 for another aluminum alloy that has a tensile yield stress of 67 ksi and a fracture toughness of 26 ksi-in$^{1/2}$.

5.5 Repeat problem 5.3 for a steel that has a tensile yield stress of 47 ksi and a fracture toughness of 235 ksi-in$^{1/2}$.

5.6 A wide panel contains an edge crack of length a and is subjected to a remotely applied tensile stress. The panel is made from a material with a yield stress $\sigma_{YS} = 50$ ksi and a fracture toughness $K_c = 50$ ksi-in$^{1/2}$.

   (a) Assuming LEFM conditions, determine the crack size that will cause fracture if the remotely applied stress is 35 ksi. Comment on the validity of LEFM for this calculation.

   (b) Redo part a employing the effective crack length approach proposed by Irwin. Discuss the result.

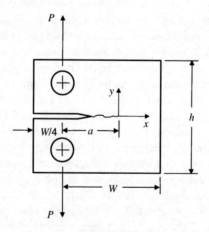

Figure P5.7

5.7 The compact tension specimen shown in Figure P5.7 has a thickness $B = 0.2$ in, a width $W = 3.0$ in, a height $h = 3.6$ in, and an initial crack length

$a = 2.0$ in, and is made from a material with a tensile yield stress of 50 ksi. The applied force $P = 500$ lb. Determine the following assuming plane stress conditions and that yield occurs by the maximum-shear-stress criterion (Tresca yield criterion, see Equation 2.31):

(a) Distance x directly ahead of the crack tip where yielding begins

(b) Distance y directly above the crack tip where yielding begins

5.8 Repeat problem 5.7 by using the von Mises yield criterion.

5.9 Consider a wide center-cracked sheet that is made from a material with a tensile yield strength $\sigma_{YS} = 50$ ksi and Poisson's ratio $v = \frac{1}{3}$. The sheet is subjected to a remote tensile stress of and the center-cracked length $2a = 2$ in.

(a) Determine the Irwin and Dugdale plastic zone sizes for the cracked sheet. Do LEFM conditions exist for this problem? Why?

(b) Determine the *von Mises* plastic zone size r_p for values of the crack tip polar coordinate $0 < \theta < 180°$. Find the plastic zone size for both plane stress and plane strain conditions and prepare a plot that clearly compares the shape of these two plastic zone (i.e., plot r_p versus θ). Compare the values for $\theta = 0$ with the Irwin and Dugdale results found in part a.

(c) Determine the *Tresca (maximum-shear-stress criterion)* plastic zone for θ values of 0, 90°, and 135° for the same problem considered in part b (plane stress and plane strain cases). Although you do not need to develop a general solution for all θ, use the Tresca yield condition to find r_p for the three given values of θ. Assume both plane stress and plane strain conditions and compare these values with the von Mises results in part b.

Hint: Recall the Tresca yield criterion discussed in Section 2.6.2. Determine the maximum shear stress at the three points near the crack tip given for this problem (e.g., use Mohr's circle) and use the Tresca yield condition to find r_p.

5.10 (a) Use the von Mises yield criterion to determine the plastic zone distance r_p ahead of a crack tip ($\theta = 0$) for mode II loading (sliding mode). Compare the results for plane stress and plane strain conditions.

(b) Determine the plastic zone distance r_p for mode II loading when $\theta = \pi/2$. Again compare results for both plane stress and plane strain conditions. *Hint:* Consider the mode II crack tip stress equations given in Section 3.2.

5.11 Repeat problem 5.10 using the Tresca yield condition.

5.12 A cracked specimen is made from a material that has 50 ksi yield stress and whose fatigue crack growth properties are given by the following power law: $da/dN = 10^{-9} \, \Delta K^4$. The specimen is *cycled under K control* such that the $R = 0$ stress intensity factor range ΔK is kept constant at a value of 10 ksi-in$^{1/2}$. At time B, a single overload cycle of $\Delta K = 20$ ksi-in$^{1/2}$ is applied. The load is then returned to the original constant-amplitude $\Delta K = 10$ ksi-in$^{1/2}$ value, and the crack is grown an additional length, as shown in Figure P5.12. Note that this new crack growth extends through the plastic zones that occurred at times A and B.

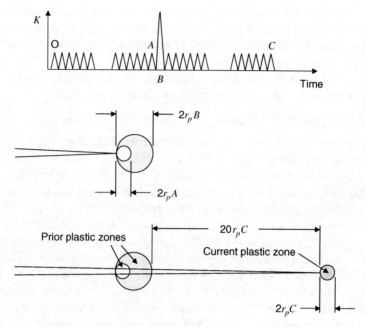

Figure P5.12

(a) Using the Irwin plastic zone model, determine the plastic zone diameter $2r_p A$ at time A (before the overload is applied).

(b) Determine the plastic zone diameter $2r_p B$ immediately after application of the overload at time B.

(c) How many cycles are required for the new crack growth to exceed 10 overload plastic zone diameters $20r_p C$ at time C?

5.13   Reconsider problem 5.12 except that this time a *wide center-cracked* specimen is subjected to a remote constant-amplitude cyclic *stress* that varies between 0 and 10 ksi. At time B, the peak *stress* is increased to 20 ksi for one cycle and then returned to the 0–10-ksi constant amplitude.

(a) If the crack length $2a = 4$ in at time A, determine the plastic zone diameter.

(b) Determine the plastic zone diameter immediately after the 20 ksi overload *stress* is applied at time B. (You may neglect any crack growth that occurs during the overload cycle in your calculation.)

(c) Determine the number of cycles required for the crack to extend an additional 10 overload plastic zone diameters at time C.

REFERENCES

1. G. R. Irwin, "Plastic Zone Near a Crack and Fracture Toughness," in *Mechanical and Metallurgical Behavior of Sheet Materials*, Proceedings of the Seventh Sagamore Ordnance Materials Research Conference, pp. IV-63–IV-78, 1960.

2. D. R. Broek, *Elementary Engineering Fracture Mechanics,* 3rd ed., Martinus Nijhoff, The Hague, 1982.

3. D. S. Dugdale, "Yielding of Steel Sheets Containing Slits," *Journal of the Mechanics and Physics of Solids,* Vol. 8, pp. 100–104, 1960.

4. G. T. Hahn, A. R. Rosenfield, L. E. Hulbert, and M. F. Kanninen, *Elastic-Plastic Fracture Mechanics,* Technical Report AFML-TR-67-143, Air Force Materials Laboratory, Wright-Patterson AFB, Ohio, January 1968.

5. S. Suresh, *Fatigue of Materials,* Cambridge University Press, Cambridge, UK, 1991.

CHAPTER 6

FRACTURE AND RESIDUAL STRENGTH

6.1 OVERVIEW

This chapter extends the discussion of crack-induced fracture begun in Chapter 3 to include the consequences of crack tip plasticity discussed in Chapter 5. The main objective is to determine the residual strength of cracked components, one of the key components of the three-legged "damage tolerance stool" described in Chapter 1. First, however, a brief historical perspective to fracture is provided, followed by a description of the original Griffith–Irwin energy balance that leads to the strain energy release rate parameter G. The relation between G and the stress intensity factor K is then obtained for linear elastic conditions. Finally, consequences of crack tip plasticity in limiting the critical stress intensity factor approach are discussed along with a description of several procedures to treat fracture under conditions of large-scale yielding.

6.2 HISTORICAL PERSPECTIVE

Although fracture mechanics developed during the mid-1950s to solve fracture problems associated with preexistent structural damage, the importance of initial flaws on fracture has long been recognized. DaVinci, for example, noted in the 1500s that long wires were weaker than short wires, an observation in contrast to Galileo's subsequent belief that strength depended only on cross-sectional area. Since the longer wires had a greater probability for containing defects, DaVinci's experiments provided early evidence for the "weakest link" concept and for the importance of initial damage on component performance.

A. A. Griffith published two classic papers [1, 2] in the early 1920s that showed that the fracture strength of glass was controlled by preexistent cracks. Following energy arguments, he demonstrated that the fracture stress σ_f in a wide remotely loaded sheet with a center crack of length $2a$ is given by

$$\sigma_f \sqrt{\pi a} = \sqrt{2E\gamma_s} = \text{constant} \qquad (6.1)$$

Here E is the elastic modulus and γ_s is the free surface energy per unit area. Current fracture mechanics concepts are based on subsequent modifications of Griffith's original work with brittle materials by Irwin [3] and Orowan [4][†] that include plastic deformation encountered in structural metals (i.e., they added γ_p, the plastic work per unit surface area, to γ_s in Equation 6.1). An important result of Griffith's pioneering effort, however, was the conclusion that the fracture stress in a brittle material is inversely proportional to the square root of crack size.

6.3   CRACK EXTENSION FORCE G

As defined in Chapter 3, the stress intensity factor K is based on the crack tip stress field and may be employed to characterize fracture and subcritical crack growth. Historically, however, the energy release rate (or crack extension force) method of analysis preceded the crack tip stress field (K) method. (It is shown later that both methods lead to equivalent results for elastic problems.) Consider a linear elastic spring loaded as shown in Figure 6.1. Here the force in the spring is given as $ky = P$, where k is the spring constant and y is the displacement of the load point. The work W done in stretching the spring a distance y_0 is given by

$$W = \int_0^{y_0} P \, dy = \int_0^{y_0} ky \, dy$$

$$= \tfrac{1}{2}ky_0^2 = \tfrac{1}{2}P_0 y_0 \qquad (6.2)$$

where $P_0 = ky_0$. Note that the work done is equal to the potential energy U added to the spring and is given by the area under the load displacement diagram. Also in this case

$$\frac{dU}{dy_0} = ky_0 = P_0 = \text{force in spring} \qquad (6.3)$$

[†]References 1–4, along with approximately 100 other early fracture mechanics papers have been collected and republished in the following two-volume SPIE Milestone Series edited by R. J. Sanford and B. J. Thompson: *Selected Papers on Foundations of Linear Elastic Fracture Mechanics*, SEM Classic Papers Volume CP1, SPIE Milestone Series Volume MS 137, SPIE Optical Engineering Press, Bellingham, Washington, 1997; and *Selected Papers on Crack Tip Stress Fields*, SEM Classic Papers Volume CP2, SPIE Milestone Series Volume MS 138, SPIE Optical Engineering Press, Bellingham, Washington, 1997.

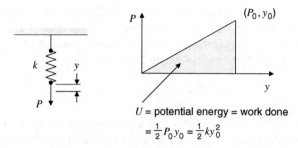

U = potential energy = work done

$$= \frac{1}{2}P_0 y_0 = \frac{1}{2}k y_0^2$$

Figure 6.1. Schematic representation of the work done by loading a linear elastic spring.

For a body in equilibrium under a set of external forces P_i, the principal of virtual work defines a generalized force P_i by the equation

$$P_i = \frac{\partial U}{\partial y_i} \qquad (6.4)$$

Here y_i is the generalized displacement corresponding to the generalized force P_i.

By analogy, Irwin defined a generalized *crack extension force* G as the derivative of the potential energy stored in a cracked body with respect to a *crack area displacement* [5, 6]. Thus, G is defined as

$$G = \frac{\partial U}{\partial A} \qquad (6.5)$$

Here G is the crack extension force or *energy release rate,* with units of force per length, U is the elastic strain energy stored in a cracked member, and A is the crack area. Note that G is a "generalized" force. It is not a "push" or a "pull" but is the rate of change of stored elastic strain energy with respect to an increment of crack extension that corresponds to an increase in crack area ∂A. Irwin proposed the strain energy release rate G as a parameter to characterize fracture (i.e., $G = G_c = $ constant at fracture).

If one can determine the elastic strain energy as a function of crack length, G is readily computed. The strain energy release rate G is also related to the change in "compliance" of a cracked body as the crack extends [5]:

$$G = \frac{1}{2}P_0^2 \frac{\partial C}{\partial A} = \frac{1}{2}\frac{P_0^2}{B}\frac{\partial C}{\partial a} \qquad (6.6)$$

Here P_0 is the remotely applied force, B is the specimen thickness, a is the crack length, and C is the compliance of the cracked member given as

$$C = \frac{\text{load point deflection}}{\text{load}} = \frac{e}{P} \qquad (6.7)$$

Note that the compliance C is the inverse of the spring constant k. As demonstrated by the following two examples, Equation 6.6 provides a means for measuring or computing G for a given crack configuration.

Example 6.1. G Calculation for Edge-Cracked Strip   The goal here is to compute G for the cracked member shown in Figure 6.2. For analysis purposes, assume the cracked legs act like two cantilever beams of height h and length a that are clamped together at the crack tip. Now the displacement e of each end is given by simple cantilever beam theory:

$$e = \frac{Pa^3}{3EI}$$

Here E is the elastic modulus, B is the thickness, and I is the cross-sectional moment of inertia of each leg:

$$I = \frac{Bh^3}{12}$$

$$2e = \frac{2Pa^3}{3EI} = \frac{2Pa^3}{3E(Bh^3/12)} = \frac{8Pa^3}{EBh^3}$$

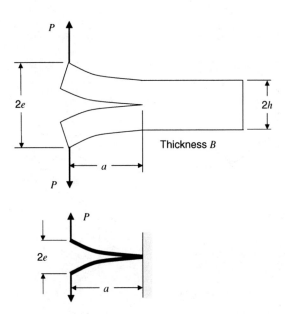

Figure 6.2. Cracked member modeled as two cantilever beams.

Now, the compliance C relating the load and the load point displacement e is given by

$$C = \frac{2e}{P} = \frac{8a^3}{EBh^3}$$

and

$$G = \frac{1}{2}\frac{P^2}{B}\frac{\partial C}{\partial a} = \frac{1}{2}\frac{P^2}{B}\left(\frac{24a^2}{EBh^3}\right)$$

$$G = \frac{12P^2a^2}{EB^2h^3}$$

Example 6.2. Experimental Procedure   Since compliance can be determined experimentally, Equation 6.6 also provides a convenient technique to measure G for various crack geometries. As shown schematically in Figure 6.3, measure P versus e for various crack lengths a_i and compute the compliance C_i as a function of crack length. Here C_1 is the compliance for the a_1 crack length curve, and so on. Now compute the rate of change of compliance with respect to crack length $\partial C/\partial a$ for various crack lengths and insert into Equation 6.6 to obtain G. Note that this experimental approach requires considerable care in obtaining the load–displacement record, since evaluation of G requires two separate derivative calculations: $C = e/P$ and $\partial C/\partial a$.

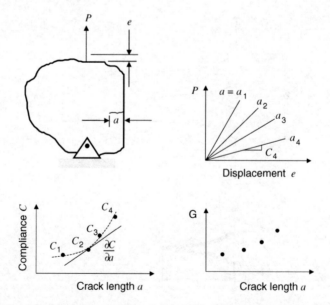

Figure 6.3. Compliance measurements used to establish the strain energy release rate G–crack length relationship.

Similar calculations for G can also be accomplished numerically if the compliance versus crack length can be computed (e.g., as per subsequent Example 6.3 or by finite element techniques).

6.4 RELATION BETWEEN G AND K

Both the stress intensity factor K and strain energy release rate G have been proposed to characterize crack growth (i.e., correlate with fracture or subcritical crack growth behavior). The stress intensity factor is obtained from crack tip stress analysis considerations as described in Chapter 3, whereas the strain energy release rate discussed in the previous section describes the change in strain energy caused by a growing crack. Irwin [6] showed that the two parameters are related when linear elasticity prevails.

To develop this elastic relationship between G and K, imagine that we "close" a portion α of the tip of crack in a loaded body as shown in Figure 6.4. Assume linear elastic behavior, "fixed" load grips (i.e., no external work is done in closing the crack tip), and plane strain conditions exist. These assumptions lead to the fact that the change δU in stored potential energy is equal to the work δW expended in closing the crack tip ($\delta W = \delta U$). Now

$$G = \lim_{\alpha \to 0} \frac{\delta U}{\alpha} = \lim_{\alpha \to 0} \frac{\delta W}{\alpha} \tag{6.8}$$

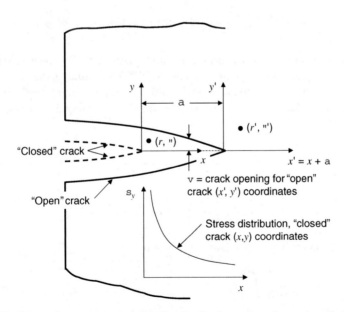

Figure 6.4. Schematic representation of stress distribution and crack opening displacements associated with closing a portion of the crack tip.

where α is the increment of crack "advance" (or closure in this case) per unit thickness. The work δW necessary to close the crack tip is

$$\delta W = \int_0^\alpha \frac{1}{2}\sigma_y(2v)\,dx \tag{6.9}$$

Here σ_y is the normal stress ahead of the "closed" crack tip and v is the y displacement at the corresponding point for the "open" crack (total opening is $2v$). Equations 6.8 and 6.9 lead to

$$G = \lim_{\alpha \to 0} \frac{1}{\alpha} \int_0^\alpha \sigma_y v\,dx \tag{6.10}$$

where the crack tip stress σ_y is given by

$$\sigma_y = \frac{K_I}{\sqrt{2\pi r}}\cos\frac{\theta}{2}\left(1 + \sin\frac{\theta}{2}\sin\frac{3\theta}{2}\right) \tag{6.11}$$

Now evaluating σ_y along the line ahead of the "closed" crack tip (i.e., $\theta = 0$) gives

$$\sigma_y = \frac{K_I}{\sqrt{2\pi x}} \tag{6.12}$$

and

$$v = 2\left(\frac{1+v}{E}\right)K_I\sqrt{\frac{r'}{2\pi}}\sin\frac{\theta'}{2}\left[2(1-v) - \cos^2\frac{\theta'}{2}\right] \tag{6.13}$$

Here v is the plane strain y displacement from Equation 3.7, E is the elastic modulus, v is Poisson's ratio, and the polar coordinates (r', θ') are based on the (x', y') coordinate system located at the tip of the open crack. The crack opening profile is obtained when $\theta' = \pi$, and $x' = x - \alpha$, which leads to the crack opening displacement v given by the equation

$$v = 2\left(\frac{1+v}{E}\right)K_I\sqrt{\frac{\alpha - x}{2\pi}}2(1-v)$$

$$= 2\sqrt{\frac{2}{\pi}\frac{(1-v^2)}{E}}K_I\sqrt{\alpha - x} \tag{6.14}$$

Now, the strain energy release rate G is given by

$$G = \lim_{\alpha \to 0} \frac{1}{\alpha}\int_0^\alpha \sigma_y v\,dx$$

$$= \lim_{\alpha \to 0} \frac{1}{\alpha}\int_0^\alpha \left(\frac{K_I}{\sqrt{2\pi x}}\right)\left(2\sqrt{\frac{2}{\pi}\frac{(1-v^2)}{E}}K_I\sqrt{\alpha - x}\right)dx$$

$$= \lim_{\alpha \to 0} \left(\frac{2}{\alpha} \frac{(1 - v^2)}{E} \frac{K_I^2}{\pi} \int_0^\alpha \sqrt{\frac{\alpha - x}{x}} \, dx \right) \qquad (6.15)$$

Evaluating the integral in Equation 6.15 gives

$$G = \lim_{\alpha \to 0} \left(\frac{2}{\alpha} \frac{(1 - v^2)}{E} \frac{K_I^2}{\pi} \alpha \frac{\pi}{2} \right) = \frac{1 - v^2}{E} K_I^2$$

which leads to the final relation between G and K for *plane strain* conditions:

$$G = \frac{1 - v^2}{E} K_I^2 \qquad (6.16)$$

Following a similar development for *plane stress* conditions (i.e., thin sheets), it can be shown that

$$G = \frac{K_I^2}{E} \qquad (6.17)$$

Similar considerations for modes II and III loading lead to the following results [7]:

$$G_{II} = (1 - v^2) \frac{K_{II}^2}{E} \qquad (6.18)$$

$$G_{III} = (1 + v) \frac{K_{III}^2}{E} \qquad (6.19)$$

The total energy release rate for combined mode cracking is given by

$$G = G_I + G_{II} + G_{III} = \frac{1 - v^2}{E} \left(K_I^2 + K_{II}^2 + \frac{K_{III}^2}{1 - v} \right) \qquad (6.20)$$

These relations suggest that either G or K can be used to characterize crack growth for nominally elastic conditions (i.e., small-scale plasticity). While both parameters are suitable for this situation, the stress intensity factor K is used more often in practice and is the term emphasized in this volume. There are situations, however, particularly when dealing with mixed-mode cracking, where the strain energy release rate is the preferred quantity.

Example 6.3. Computing K from G   Compute the mode I stress intensity factor K_I for the split-beam problem considered in Example 6.1 and shown in Figure 6.2. Recall that

$$G = \frac{12 P^3 a^2}{E B^2 h^3}$$

Now for *plane stress* conditions

$$K_I = \sqrt{GE} = \left(\frac{12 P^2 a^2}{E B^2 h^3} E \right)^{1/2}$$

$$= \frac{2\sqrt{3} P a}{B h^{3/2}}$$

6.5   FRACTURE TOUGHNESS VERSUS PLASTIC ZONE SIZE

This section employs crack tip plastic zone considerations to explain the dependence of fracture toughness on specimen thickness and material yield stress. Recall that the crack tip plastic zone size r_p was found in Chapter 5 to be a function of the stress intensity factor, yield strength σ_{YS}, and state of stress. From Equations 5.21 and 5.22, the von Mises plastic zone radius r_p was given by

$$r_p = \frac{R}{2\pi} \left(\frac{K_I}{\sigma_{YS}} \right)^2$$

where R is defined as

$$R = \begin{cases} \cos^2 \frac{\theta}{2} \left(1 + 3 \sin^2 \frac{\theta}{2} \right) & \text{for plane stress} \\ \cos^2 \frac{\theta}{2} \left((1 - 2v)^2 + 3 \sin^2 \frac{\theta}{2} \right) & \text{for plane strain} \end{cases}$$

Here θ is the polar coordinate defined at the crack tip and v is Poisson's ratio. Note that the distance r_p to the plastic zone boundary increases as the load K increases or as the yield strength σ_{YS} decreases. In addition, the crack tip plastic zone is larger for conditions of plane stress than for plane strain. The consequences of this plastic zone on fracture may be anticipated by the intuitive argument that a "tough" material should exhibit more crack tip plasticity than a "brittle" material. Thus, as discussed below, factors that increase the crack tip plastic zone size at fracture also increase the value of fracture toughness K_c (or G_c).

6.5.1   Toughness versus Specimen Thickness

Experimental data indicate that the fracture toughness K_c (or G_c) for a given material depends on specimen thickness as shown in Figure 6.5. The schematic data shown here are obtained from fracture specimens made from identical material but with different thicknesses. Results for several different alloys are given in Figure 6.6 [8]. (A cautionary note from Reference 8 states that these data should be considered as "average" values.)

Note here that the observed K_c data decrease with increased thickness until a minimum value is reached. This minimum fracture toughness is given the special designation K_{Ic} and is called the *plane strain fracture toughness*. An empirical "rule of thumb" indicates that the minimum value of fracture toughness K_{Ic} is obtained if

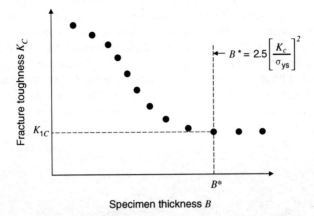

Figure 6.5 Dependence of fracture toughness K_c on specimen thickness B.

$$B \geq 2.5 \left(\frac{K_c}{\sigma_{YS}} \right)^2 \qquad (6.21)$$

Here B is the specimen thickness, σ_{YS} is the material yield strength, and K_c is the value of the fracture toughness at given specimen thickness B. Another empirical relation [9] for K_c as a function of K_{Ic}, specimen thickness B, and yield strength is given by

$$K_c = K_{Ic} \left[1 + \frac{1.4}{B^2} \left(\frac{K_{Ic}}{\sigma_{YS}} \right)^4 \right]^{1/2} \qquad (6.22)$$

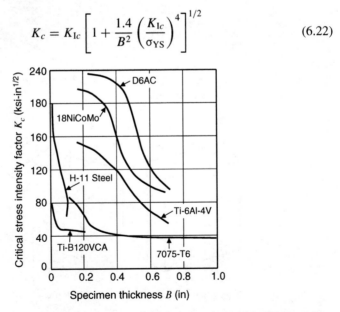

Figure 6.6. Examples of toughness–thickness behavior for several materials [8]. Note that a cautionary statement in Reference 8 asserts that these values should be considered as *average*.

Note that the dependence of fracture toughness on specimen thickness is consistent with changes in crack tip plastic zone sizes. Plane stress conditions, for example, are appropriate for thin sheets and lead to large plastic zones, and, consequently, larger values of K_c. Plane strain conditions, however, are met at the center of a thick plate, giving a smaller plastic zone and the smaller toughness K_{Ic}. As the crack tip plastic zone size "transitions" from plane stress to plane strain conditions for intermediate thicknesses, K_c should also vary as shown in Figures 6.5 and 6.6.

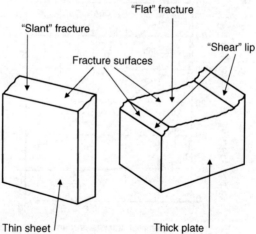

Figure 6.7. Schematic representation of fracture surface appearance showing change in crack plane with specimen thickness (thickness ranges from $\frac{1}{16}$ to $\frac{1}{2}$ in).

The appearance of the specimen fracture surface also shows a *fracture mode transition* as specimen thickness is changed. Figure 6.7, for example, shows the fracture surfaces for several 2024-T851 aluminum specimens. These compact tension members contained the same fatigue precrack length and were identical in all aspects except for specimen thickness, which varied between $\frac{1}{16}$ and $\frac{1}{2}$ in (1.6 and 12.7 mm). Note that the fracture surface in the "thin" sheet is inclined, or "slanted," at 45° to the load axis, suggesting ductile shear in the large plane stress crack tip plastic zone. Since the crack opening is no longer pure mode I (some mode II sliding is also present), this thin sheet fracture is sometimes called *plane stress fracture* or *mixed-mode fracture*. The state of stress for much of the "thick" specimen approximates plane strain, however, and the crack tip plastic zone size is smaller, resulting in a "flat" fracture surface or "cleavage" failure. This flat fracture is indicative of mode I crack growth. Hence the thick-plate fracture toughness designations of K_{Ic} and "plane strain" fracture. The small "shear lips" shown for the thick plate in Figure 6.7 are caused by the larger plane stress deformation at the specimen surface. Intermediate specimen thicknesses yield neither true plane stress or plane strain conditions and exhibit varying degrees of slant and flat crack growth on the fracture surface.

Since unconservative fracture predictions could result if the K_c value measured in a "thin sheet" was used to predict fracture in a "thick plate," considerable emphasis is placed on experimental procedures required to ensure measurement of the plane strain fracture toughness K_{Ic} [10]. If one knows the component thickness, however, it is appropriate to use the K_c value determined for that thickness for fracture analyses. It should also be noted that designs that employ thin sections have a decided advantage in their ability to resist catastrophic fracture due to the potentially larger K_c value.

6.5.2 Toughness versus Yield Stress

Selection of the "best" material for structural applications requires careful consideration of the material properties needed for a given situation. Desirable material characteristics include high strength (yield and/or ultimate), fracture resistance (K_{Ic} or K_c), stiffness (modulus of elasticity), resistance to fatigue (crack growth and/or crack nucleation), corrosion resistance (K_{ISCC}), low density, and low cost. Unfortunately, however, it is rare to find a single material with all of the desired characteristics, so trade-offs are usually required.

This compromise is especially true when considering the interaction between fracture toughness and yield strength. Since the crack tip plastic zone decreases as the tensile yield stress increases, one expects fracture toughness K_c to also decrease as the yield strength increases for a given class of materials. This general trend is shown by the aircraft alloy data given in Figure 6.8 [11]. First, several steels and aluminums are compared with a titanium alloy on the basis of their strength-to-weight ratio (i.e., tensile ultimate stress in psi divided by density in lb/in^3). Note that the high-strength D6AC steel (used for the F-111 wing carry-through structure discussed in Section 1.2.1) performs very well on this criterion, as does the 7178-T651 aluminum alloy selected for the KC-135A lower wing skin.

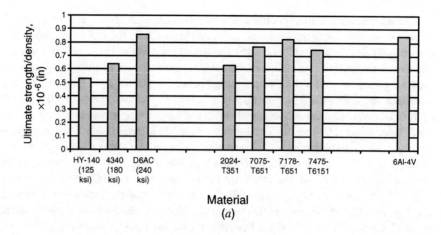

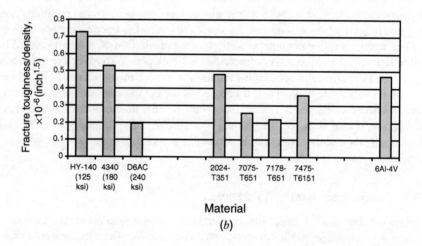

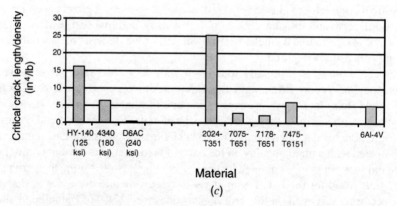

Figure 6.8. Comparison of (*a*) strength, (*b*) toughness, and (*c*) limit load critical crack size normalized with density for several aircraft alloys [11].

Comparing the same materials on the basis of plane strain fracture toughness K_{1c} (measured in psi-in$^{1/2}$) divided by density (lb/in^3) results in the second comparison shown in Figure 6.8. Note here that the steel and aluminum rankings reverse from the first case. In particular, the 2024-T351 material selected for the lower wing skin of the Boeing 707 commercial transport is now superior to the 7178-T651 alloy used on the KC-135A (the original military version of the Boeing 707). While the 7178 alloy was used to save weight in the KC-135A, severe cracking problems developed early in its life, and the lower wing skins were replaced with the 2024 material originally used for the 707. (The 707 fleet, by contrast, achieved many thousands of hours of commercial flight without serious cracking problems.)

The first two comparisons in Figure 6.8 can be combined in an interesting way that emphasizes the influence of material choice on critical crack size and its corresponding ramifications on inspection. While critical crack size is not a material property per se but depends on both structural configuration and applied load, one can make comparisons for a particular application. In this case the structure was assumed to be a wide center-cracked plate (crack length $2a$) loaded with a remote tensile stress σ, so that the stress intensity factor K is given by the familiar formula $K = \sigma\sqrt{\pi a}$. Now, selecting the applied stress to be 0.5 times the ultimate tensile stress (a common value used for subsonic aircraft design purposes), one can determine the critical crack size that would fracture the center-cracked panel. This value is divided by the material density as before and is also presented in Figure 6.8. Note that this plot compares the materials in the context of their impact on inspection requirements and further highlights the superiority of the 2024-T351 alloy over the 7178-T651 lower wing skin alloy and emphasizes the poor crack tolerance of the D6AC steel.

Although the emphasis on high-strength designs in the 1950s resulted in the selection of materials with high strength but poor toughness and led to some of the catastrophic fractures discussed in Chapter 1, materials producers have made significant strides in developing alloys with good combinations of yield strength and toughness [12]. Figure 6.9, for example, compares fracture toughness and yield strength

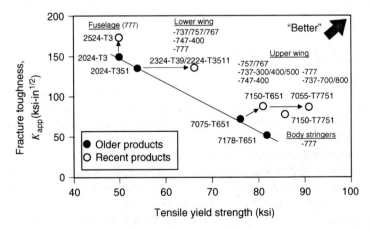

Figure 6.9. Trade-off in yield strength versus fracture toughness for aluminum aircraft alloys [12]. (Courtesy of Boeing.)

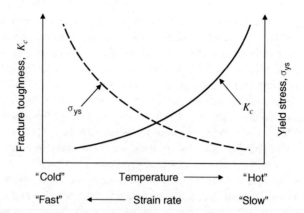

Figure 6.10. General effect of temperature and loading rate on yield strength and fracture toughness.

for various Boeing commercial aircraft. Note here that applications requiring resistance to crack-induced fracture (e.g., fuselage skins or the tension-dominated lower wing skins) require alloys with high toughness at the expense of yield strength. Compression-dominated components, such as the upper wing skin, emphasize materials with high yield strength and stiffness, however, since fracture is not as large of a concern for these applications. (Buckling and/or yielding would control failure in these areas.)

As a final comment regarding the influence of yield strength on fracture toughness, it is appropriate to note the general effect of temperature and loading rate (i.e., strain rate) shown schematically in Figure 6.10. As indicated, materials often demonstrate higher yield stresses when subjected to low temperatures and fast rates of loading (e.g., impact loads). The fact that low temperatures and high strain rates also promote "brittle" behavior (e.g., low fracture toughness) is again consistent with the crack tip plastic zone explanation for the inverse relationship between yield strength and K_c.

6.6   LEAK-BEFORE-BREAK

Leak-before-break is a design procedure intended to ensure that pressure vessels fail by the relatively benign mechanism of leaking rather than by explosive fracture. The concept is shown schematically in Figure 6.11, where surface cracks have partially penetrated the wall thickness of a pressure tank. If the pressure is repeatedly removed and reapplied or if the vessel contains a corrosive material, the surface cracks will grow by fatigue and/or stress corrosion cracking until one of two failure modes occurs. First, if the cracks extend "slowly" and pierce the wall thickness before the critical stress intensity factor K_c is reached, the vessel will leak and the hoop stresses will vanish as the pressure is relieved. If, however, K reaches the material fracture toughness K_c *before* the crack penetrates the wall thickness, the crack will extend suddenly, resulting in a catastrophic explosion (break).

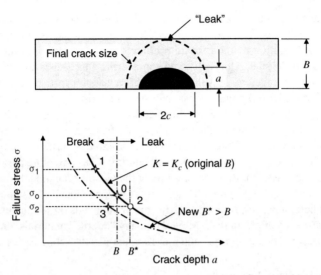

Figure 6.11. Schematic representation of leak-before-break conditions for a pressure vessel containing a surface crack that has partially penetrated the wall thickness.

Obviously the "leak" failure mode is preferred, and the designer's task is to ensure that the vessel will never "break" in service. This challenge is represented by the "failure" stress–crack depth plot shown by the solid line in Figure 6.11. (This curve represents a line of constant stress intensity factor $K = \sigma\sqrt{\pi a}\beta = K_c$.) For critical crack depths less than the wall thickness B, fracture (break) occurs when $K \geq K_c$. The largest stress σ_0 that can be applied without causing fracture occurs when the crack depth a equals the wall thickness B (point 0 in Figure 6.11). If the stress is less than σ_0, the crack penetrates the wall, causing the vessel to leak before the crack reaches a size that would fracture the tank.

Observe that if the hoop stress exceeds the "leak" value σ_0, as shown by point 1, the designer has several options to return the safer leak condition. The hoop stress may be reduced by decreasing the internal pressure or by increasing the wall thickness B (recall Equation 2.10), leading to point 2 in Figure 6.11. Note, however, that these are potentially conflicting approaches. Although the hoop stress (and K) will decrease for larger B (e.g., $B^* > B$ in Figure 6.11), it is possible that the fracture toughness could also decrease with the larger wall thickness (recall Section 6.5.1), and the fracture condition may actually be exacerbated (point 3). Thus, one needs to know the K_c-versus-thickness behavior for the material in question in order to make the correct design decision. Another design option would be to select a new material with a larger fracture toughness K_c for the same material thickness.

6.7 DUCTILE FRACTURE

The goal of this section is to summarize methods to predict fracture when plasticity invalidates the simple K_c (or G_c) criterion. This is an important issue for thin sections (e.g., fuselage skins) that are subjected to plane stress conditions that promote large crack tip plastic zones. Ductile fracture is, however, a complex subject that does not have a general solution method for all problems, and detailed discussion is beyond the scope of this volume. Although some engineering approaches are outlined below, the reader is encouraged to consult the cited references for further details.

Recall that the LEFM approach assumes that crack tip plasticity is small ($a \gg r_p$) and the material behaves in a "brittle" manner. In that case, fracture may be adequately characterized by assuming that the stress intensity factor $K = K_c = $ constant at fracture. For small amounts of plasticity, it may be possible to further assume an "effective" crack length $a_{\text{effective}} = a + r_p$ and still employ K_c as a fracture parameter based on the effective crack length. Here r_p is one of the plastic zone models described earlier in Chapter 5 (e.g., the Irwin or Dugdale plastic zone model). An iterative solution is required here, however, since the plastic zone r_p depends on K (and crack length). For larger amounts of plasticity exhibited in many "ductile" materials, yielding is not confined to a small crack tip zone, and more sophisticated fracture criteria are needed.

6.7.1 Fedderson Approach

Another issue associated with fracture of thin sheets is that uncontained plasticity in narrow specimens can cause the apparent fracture toughness to depend on the specimen width as well as thickness. The width phenomenon depends on material and is manifested in larger measured K_c's for wider specimens (i.e., K_c is no longer solely a material property). Although this apparent fracture toughness eventually achieves a constant value for large widths, there is a practical problem when very large panels (on the order of 5 ft = 1.5 m in width) must be tested to determine the appropriate K_c for large components. As discussed below, the problem stems from the fact that it is possible for net section yielding to control failure under certain conditions in narrow, cracked panels.

Figure 6.12 shows two competing failure criteria for a center-cracked panel subjected to a remote tensile stress. Here the failure stress σ_f is plotted versus the total crack length $2a$. The LEFM curve is the conventional fracture mechanics fracture criterion for the center-cracked sheet given by

$$K = K_c = \sigma_f \sqrt{\pi a} \left(\sec \frac{\pi a}{W} \right)^{1/2} \tag{6.23}$$

Here σ_f is the remotely applied stress at failure, a is the half crack length, and W is the panel width. The net section yield line in Figure 6.12 assumes failure when the net section stress equals the material yield strength σ_{YS}. In this case, the net section stress is the remote applied stress σ times the factor $W/(W - 2a)$, and the failure stress is given by

$$\sigma_f = \frac{W - 2a}{W} \sigma_{YS} \tag{6.24}$$

Note that LEFM gives the minimum failure stress for intermediate crack lengths, whereas the net section yield criterion controls large and small crack sizes. The

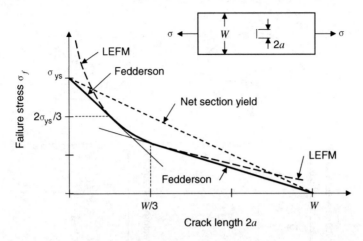

Figure 6.12. Schematic representation of Fedderson approach to combine LEFM and net section yield failure criteria for cracks in finite-width panels.

Fedderson [7, 13] approach shown in Figure 6.12 merges these two failure criteria by constructing tangent lines from the LEFM curve to the net section yield results for zero and maximum crack sizes. These two tangent lines, combined with the intermediate LEFM criterion (Equation 6.23), represent the failure stress for all crack sizes in finite-width panels. As shown in References 7 and 13, the Fedderson criterion agrees well with test data for various materials and specimen widths and can be easily constructed using the width, material yield strength σ_{YS}, and fracture toughness K_c for the given panel material and thickness. It is readily shown [7] that the failure stress where the LEFM curve first transitions into the Fedderson curve is $\frac{2}{3}\sigma_{YS}$ and the second tangent transition occurs at a crack length $2a = \frac{1}{3}W$. Thus, the original K_c tests used to establish the LEFM curve for the Fedderson approach must also fall within these limits for valid results.

6.7.2   Crack Growth Resistance Curves (R-Curves)

In brittle materials, cracks immediately become unstable when the applied strain energy release rate G (or stress intensity factor K) reaches the fracture toughness G_c (or K_c) for the given material. Consider, for example, the center-cracked panel shown in Figure 6.13. Here G and K are given by (for plane stress conditions)

$$G = \frac{K^2}{E} = \frac{\sigma^2 \pi a}{E} \qquad (6.25)$$

For a given initial crack size a_0 rapid fracture occurs when the applied stress σ causes G to reach G_c. In ductile materials, however, it is possible for the crack to extend a small amount in a stable manner and then *arrest*, requiring *additional* load for continued crack extension. In this case, the material "resistance" to fracture increases

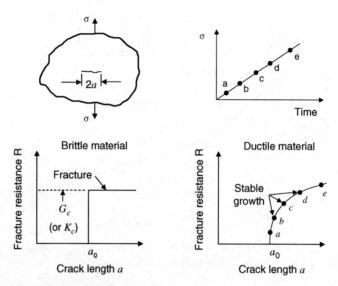

Figure 6.13. Comparison of fracture resistance R versus crack length for a wide center-cracked sheet in brittle and ductile materials.

due to plastic zone development and strain hardening. This stable crack extension and arrest may continue with further load application until at some point rapid fracture occurs.

The fracture resistance R for "brittle" and "ductile" materials are compared as a function of crack length in Figure 6.13. Note that for the brittle material R (i.e., G_c or K_c) remains constant as the crack rapidly extends beyond its initial size a_0. In the ductile material, however, the fracture resistance increases with crack growth. (Here the toughness is computed after each increment of stable crack growth using the applied stress corresponding to the arrested crack length.) Since R increases with initial crack extension, the issue here is to determine conditions for unstable fracture in ductile materials that resist initial crack extension in a stable manner. The R-curve approach described below provides one engineering approach to this problem.

Crack growth resistance (R-curve) concepts [7, 14, 15] are based on Irwin's interpretation of the Griffith energy formulation, which stated that unstable crack extension occurs when the change in strain energy release rate with crack growth $\partial G/\partial a$ equals or exceeds the change in material resistance to crack extension $\partial R/\partial a$. This situation is shown schematically in Figure 6.14 for the wide center-cracked panel considered previously. Here strain energy release rate and fracture resistance R are plotted versus crack length on the same set of axes for both the brittle and ductile materials. In this case, Equation 6.25 indicates that G is a linear function of crack length a, so that the G-curves are a family of straight lines with increasing slope as the applied stress is increased (i.e., $\sigma_1 < \sigma_2 < \sigma_3$ in Figure 6.14). Note that unstable fracture results when the applied G and material resistance R-curves are tangent. In the brittle material, this occurs at the original crack length a_0, whereas for the ductile material, unstable growth does not occur until the crack has extended to length a_c.

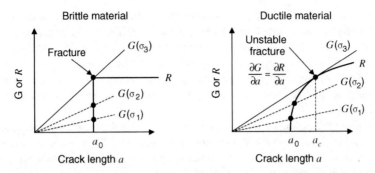

Figure 6.14. Strain energy release rate G- and R-curves for brittle and ductile materials showing onset of rapid fracture when G is tangent to R-curve.

(Note that this tangency criterion may also be followed for other specimen geometries, where the G–crack length relation is nonlinear.)

Thus, the R-curve characterizes a materials' ability to resist slow-stable crack growth, and fracture occurs at the crack length where the G- and R-curves are tangent ($\partial G/\partial a = \partial R/\partial a$). Although R-curves were originally developed in terms of G (units of force per length), subsequent work often treats them in terms of the stress intensity factor K (units of stress times length$^{1/2}$). The terms G_R and K_R are used to distinguish the presentation format. Although there may be a crack size effect (i.e., the R-curve may depend on the initial crack length in the specimen), R-curves are often expressed in the form K_R (or G_R) versus crack extension Δa ($\Delta a = a - a_0$). Experimental procedures for establishing the material R-curve are discussed in Reference 15.

6.7.3 *J*-Integral

The J-integral has been suggested as a parameter for characterizing crack growth under conditions of large-scale yielding. Rice [16] defined the J-integral as

$$J = \oint_\Gamma \left(W\,dy - T_i \frac{\partial u_i}{\partial x_i}\,ds \right) \tag{6.26}$$

where W = strain energy density, $= \int \sigma_{ij}\,d\varepsilon_{ij}$

  Γ = any closed contour surrounding the crack tip

  T_i = traction vector acting on contour Γ

  u_i = displacement in the direction of T_i

  ds = increment of arc length along path Γ

The J-integral has been shown to have several interesting features. First, J is defined for a *nonlinear* elastic body and as such is a valid parameter in the deformation theory of plasticity. The integral is *path independent*, indicating that the same value for J will be computed no matter what contour Γ is chosen. Thus, J calculated along a path Γ remote from the crack tip (where elastic conditions might be expected) would give the same value as for a contour chosen "near" the crack tip. For linear elastic behavior, J reduces to the strain energy release rate (i.e., $J = G = K^2/E$ for plane stress). The J-integral has also been interpreted as an "intensity parameter" that describes elastic–plastic stress fields at the tip of a crack, analogous to the role of K for LEFM [17, 18].

Experimental estimates of J may be obtained from compliance measurements (i.e., load–displacement records) on cracked specimens. Experimental evidence indicates J may provide a means for describing nonlinear crack growth. It has been suggested, for example, that J reaches a maximum value for a given material (i.e., $J = J_c$) and may be employed as a criterion for elastic–plastic fracture [7, 14, 19]. The cyclic J has been correlated with fatigue crack growth rates in situations involving considerable yielding [20]. Moreover, the time derivative of J has had some success in correlating "creep" crack growth [21].

Although J has shown promise for these nonlinear applications, considerable more research and a degree of caution is appropriate. Note that some mathematical rigor is lost when J is applied to crack growth and that analysis and test techniques can become quite complex for engineering applications.

6.7.4   Crack Tip Opening Displacement

The crack tip opening displacement (CTOD) method for analyzing ductile fracture is based on pioneering work by Wells [22], who suggested that the CTOD measured at the onset of rapid fracture may be treated as a material property. For linear elastic behavior, CTOD is related to the strain energy release rate and the stress intensity factor and is consistent with the K_c (or G_c) fracture criterion [7, 14]. In principle, the CTOD can also be measured and/or computed for cases involving large-scale yielding at the crack tip and used to characterize nonlinear fracture. The key point here is where to define and measure the crack tip opening in a consistent manner suitable for a general fracture criterion. Progress has been made with these issues, however, and versions of CTOD criteria have been used to successfully predict the onset of fracture for single and multiple cracks in stiffened and unstiffened panels [7, 14, 23–25]. Reference 26 provides a recent review of applications and limitations of the CTOD fracture criterion.

6.8   CONCLUDING REMARKS

This chapter has focused on the general problem of determining when structural cracking reaches a size that reduces residual strength to an unacceptable margin of safety. For small-scale yielding, the simple LEFM criterion based on the assumption that the stress intensity factor K is limited to a maximum value (i.e., the fracture toughness K_c) performs quite well. This measure of fracture toughness does, however, depend on specimen thickness, decreasing to a minimum value of K_{Ic} for thick plates. The variation of toughness with thickness and the associated transition in the fracture appearance are readily explained in terms of the plane stress versus plane strain plastic zone sizes determined by the von Mises plastic zone model described in Section 5.4.

Development of new materials with increased capacity for crack tip plasticity, however, leads to the perplexing dilemma that although more resistance to fracture is obtained, the simple LEFM analysis is no longer valid. For such situations that involve large-scale yielding, fracture prediction is much more complicated, and no single fracture criterion has been shown to apply for all cases. Nevertheless, reasonable engineering estimates of residual strength can often be obtained by the approaches outlined in Section 6.8. (Other nonlinear failure criteria are also described in References 7, 14, and 24.) Any structural design or analysis that employs one of those ductile fracture criteria should, however, be subjected to extensive experimental verification if at all possible. It is expected that continued research and development, along with increased computational power, will provide more efficient tools for analyzing ductile fracture.

PROBLEMS

6.1 Briefly discuss the validity of the following statements:

  (a) Fracture toughness is a unique material property.

  (b) Crack tip plasticity invalidates the practicality of the linear elastic stress intensity factor as a crack parameter.

  (c) There are two schools of thought in fracture mechanics: Those who believe in energy methods employ G as a crack parameter, whereas those who follow the stress approach use K.

  (d) Each material has its own critical crack size that can be considered a material property.

6.2 The fracture toughness data given in Figure 6.6 indicate that K_c varies with specimen thickness. Give a brief explanation for this behavior.

6.3 A large plate containing a central through-the-thickness crack of length $2a = 2$ in fails at a remote stress $\sigma = 50$ ksi. The plate thickness B is 1 in and the material yield strength is 200 ksi.

  (a) What is the fracture toughness for this material?

  (b) Is this the plane strain value for fracture toughness K_{Ic}? Why?

  (c) What is the maximum stress that can be applied to a 4-in-wide by 1-in-thick plate made from the same material if it contains a 1.5-in-long edge crack?

6.4 A large steel sheet (0.1 in thick) containing a center crack with length $2a = 4$-in fractures at a remote tensile stress $\sigma = 30$ksi. The yield strength of this steel is 100 ksi. Answer the following questions. If you feel you cannot answer a question with the data given, explain in detail.

(a) What is the fracture toughness for this sheet of steel?

(b) What is K_{1c} for this material?

(c) What remote stress σ will fracture an identical plate containing a 9-in center crack?

(d) What is the failure stress for a 4-in-long center crack in a 1-in-thick plate loaded as before?

(e) What is the fracture stress if the original 0.1-in-thick plate contains a 0.1-in-long center crack loaded as before?

6.5 It has been learned that many metals will strengthen when exposed to nuclear radiation. This phenomenon, known as radiation hardening, is characterized by an increase in the ultimate tensile stress. Would you expect the ability of these materials to withstand preexistent cracks to also change when subjected to radiation? Explain your answer.

6.6 The plane strain fracture toughness for a certain material is 100 ksi-in$^{1/2}$ and its tensile yield strength is 200 ksi. It is desired to predict the fracture behavior of large center-cracked panels (crack length $2a$) made of this material that are loaded with a remote tensile stress σ. Answer the following questions within the limitations (if any) of LEFM:

(a) Are there any constraints on the plate thicknesses you can analyze with the data provided? What plate thickness (or thicknesses) can you consider?

(b) What range in crack lengths (if any) can be safely analyzed?

(c) Would you expect your answers to parts a and b above to change if the center-cracked panels were tested at a different temperature (i.e., hotter or colder)? Explain your answer fully.

(d) What is the plane strain fracture toughness for a three-point bend specimen ($S/W = 4$, $a/W = 0.3$) made from the same material?

6.7 A *wide* panel is made from a steel that has a fracture toughness of 100 ksi-in$^{1/2}$, a tensile yield stress of 200 ksi, and fatigue crack growth properties that are given by the following equation:

$$\frac{da}{dN} = 4 \times 10^{-10} \Delta K^3$$

Here the units of da/dN are measured in inches per cycle and ΔK is expressed in ksi-in$^{1/2}$.

(a) If the panel used to measure the fracture toughness was 0.5 in thick, is the observed fracture toughness the plane strain value K_{1c}? Why?

(b) A 0.5-in-thick steel panel made from this material initially contains a 0.2-in-long edge crack. The *wide* panel is subjected to 75,000 cycles of a remotely applied tensile stress that varies between 0 and 20 ksi. The remotely applied stress is then increased so that it varies between 0 and 30 ksi, and 10,000 more cycles are applied. The load is then increased until the plate fractures. What is the magnitude of the fracture stress σ_f in this case?

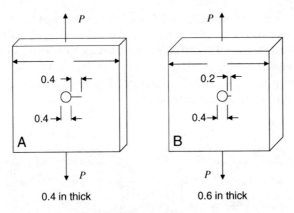

Figure P6.8

6.8 The 0.4-in-diameter holes in the 10-in-wide plates A (0.4 in thick) and B (0.6 in thick) contain radial through-the-thickness cracks of lengths 0.4 and 0.2 in, as shown in Figure P6.8. Answer the following questions using the material properties given in Figure 6.6:

(a) If both components are made of D6-AC steel, which one will withstand the largest remote load P? What is this value of P (in pounds)?

(b) Which member will take the largest load if both are made from 7075-T6 aluminum?

6.9 An engineer has been asked to design a large tension panel that contains a hole. The material in question has a plane strain fracture toughness of 50 ksi-in$^{1/2}$ and a tensile yield strength of 100 ksi. He has suggested two designs: plate A, which is 1.0 in thick, and plate B, which is a laminated design consisting of two 0.5-in-thick sheets loaded in parallel, as shown in Figure P6.9. The engineer claims that design B will be more resistant to cracks that could develop at the hole. Critique his statements regarding his preference for design B over design A.

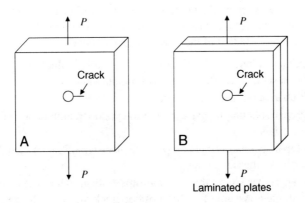

Figure P6.9

6.10   A 10-in-wide plate contains a 0.5-in-diameter hole that has a 0.5-in-through-thickness radial crack emanating from one side of the hole bore. The plate is to carry a remote force $P = 100{,}000 \, \text{lb} = 100$ kips without fracturing. Determine the minimum thickness B that the plate can have if it is made from 7075-T6 aluminum (i.e., design the *lightest* plate). Fracture toughness data for 7075-T6 aluminum are given in Figure 6.6.

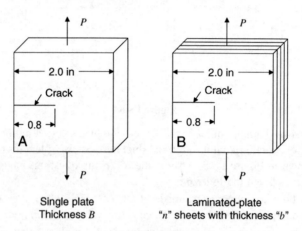

Single plate
Thickness B

Laminated-plate
"n" sheets with thickness "b"

Figure P6.11

6.11   **(a)** It is desired to construct a tensile member from the Ti–6Al–4V material given in Figure 6.6. The bar is to be 2.0 in wide and must resist fracture from a 0.8-in-long edge crack, as shown in Figure P6.11. Determine the minimum thickness B that the bar can have if it is to withstand 28 kips tensile force P without fracture.

(b) As an alternate to the design determined above, assume that the solid bar is replaced by a series of thin sheets with individual thickness b_i. Again assume that each sheet contains a 0.8-in-edge crack and that individual sheets carry the same remote stress (i.e., $\sigma = P/BW = P/nb_iW$, where n is the number of individual sheets and b_i is the thickness of each individual sheet. Determine the number of sheets n and their individual thickness b_i to give the *lightest* design that will not fracture at the 28-kip load. Compare the weight to that of the solid bar given above.

6.12   Consult a handbook and determine representative values of K_{Ic}, σ_{YS}, and density for several steel, aluminum, and titanium alloys.

(a) Prepare a table that summarizes those values.

(b) Prepare bar charts that compare the yield strength/density for all of these materials.

(c) Prepare bar charts that compare the plane strain fracture toughness/density for all of these materials.

(d) Prepare bar charts that compare the critical crack size/density for all of materials. Assume that the critical crack size is to be determined for a center crack of length $2a$ in a *wide sheet* subjected to a remote tensile stress equal to two-thirds of the tensile yield strength for that particular material.

(e) Discuss the results of these comparisons.

6.13 Compare Equation 6.22 with the 7075-T6 toughness data given in Figure 6.6. Assume that the tensile yield stress for aluminum alloy 7075-T6 is 69 ksi.

6.14 Verify that the tangent intercepts for the Fedderson approach given in Figure 6.12 have the $\frac{2}{3}\sigma_{YS}$ and $\frac{1}{3}W$ values shown.

REFERENCES

1. A. A. Griffith, "The Phenomena of Rupture and Flow in Solids," *Philosophical Transactions of the Royal Society of London*, Vol. A-221, pp. 163–197, 1920 (reprinted with commentary in *Transactions of the ASM*, Vol. 61, p. 871, 1968).

2. A. A. Griffith, "The Theory of Rupture," in *Proceedings of the 1st International Conference for Applied Mechanics*, 1925, pp. 55–63.

3. G. R. Irwin, "Fracture Dynamics," in *Fracturing of Metals*, American Society for Metals, Materials Park, Ohio, 1948, pp. 147–166.

4. E. Orowan, "Energy Criteria of Fracture," *Welding Journal Research Supplement*, Vol. 34, No. 3, March 1955, pp. 1575–1605.

5. G. R. Irwin and J. A. Kies, "Critical Energy Rate Analysis of Fracture Strength," *Welding Journal Research Supplement*, Vol. 33, No. 4, pp. 1935–1985, April 1954.

6. G. R. Irwin, "Analysis of Stresses and Strains Near the End of a Crack Traversing a Plate," *Journal of Applied Mechanics*, Vol. 24, No. 3, pp. 361–364, September 1957.

7. D. Broek, *Elementary Engineering Fracture Mechanics*, 3rd rev. ed., Martinus Nijhoff, The Hague, 1983.

8. D. P. Wilhem, *Fracture Mechanics Guidelines for Aircraft Structural Applications*, Technical Report AFFDL-TR-69–111, Air Force Flight Dynamics Laboratory, Wright-Patterson Air Force Base, Ohio, February 1970.

9. W. D. Pilkey, *Formulas for Stress, Strain, and Structural Matrices*, Wiley, New York, 1994, p. 299.

10. "Standard Test Method for Plain-Strain Fracture Toughness Testing of Metallic Materials," ASTM Test Specification E-399, *Annual Book of ASTM Standards, Section 3, Metals Test Methods and Analytical Procedures*, American Society for Testing and Materials, West Conshohocken, Pennsylvania, 2002.

11. R. J. Bucci, circa 1974, personal communication.

12. R. J. Bucci, C. J. Warren, and E. A. Starke, Jr., "Need for New Materials in Aging Aircraft Structures," *AIAA Journal of Aircraft*, Volume 37, No. 1, pp. 122–129, January/February 2000.

13. C. E. Fedderson, "Evaluation and Prediction of the Residual Strength of Center Cracked Tension Panels," ASTM STP 486, American Society for Testing and Materials, West Conshohocken, Pennsylvania, 1971, pp. 16–38.

14. T. L. Anderson, *Fracture Mechanics, Fundamentals and Applications*, 2nd ed., CRC Press, Boca Raton, Florida, 1995.

15. "Standard Practice for R-Curve Determination," Designation E 561–98, in *Annual Book of ASTM Standards, Section 3, Metals Test Methods and Analytical Procedures,* Vol. 03.01, American Society for Testing and Materials, West Conshohocken, Pennsylvania, 2002.

16. J. R. Rice, "A Path Independent Integral and the Approximate Analysis of Strain Concentration by Notches and Cracks," *Journal of Applied Mechanics,* Vol. 35, pp. 379–386, 1968.

17. J. R. Rice and G. G. Rosengren, "Plane Strain Deformation Near a Crack Tip in a Power-Law Hardening Material," *Journal of the Mechanics and Physics of Solids,* Vol. 16, pp. 1–12, 1968.

18. J. W. Hutchinson, "Singular Behavior at the End of a Tensile Crack Tip in a Hardening Material," *Journal of the Mechanics and Physics of Solids,* Vol. 16, pp. 13–31, 1968.

19. J. A. Begley and J. D. Landes, in *Fracture Toughness,* ASTM Special Technical Publication 514, American Society for Testing and Materials, West Conshohocken, Pennsylvania, 1972, pp. 1–20.

20. N. E. Dowling and J. A. Begley, "Fatigue Crack Growth During Gross Plasticity and the J-Integral," in *Mechanics of Crack Growth,* ASTM Special Technical Publication 590, American Society for Testing and Materials, West Conshohocken, Pennsylvania, 1976, pp. 82–103.

21. J. D. Landes and J. A. Begley, "A Fracture Mechanics Approach to Creep Crack Growth," in *Mechanics of Crack Growth,* ASTM Special Technical Publicaiton 590, American Society for Testing and Materials, West Conshohocken, Pennsylvania, 1976, pp. 128–148.

22. A. A. Wells, "Unstable Crack Propagation in Metals: Cleavage and Fast Fracture," in *Proceedings of the Crack Propagation Symposium,* Vol. 1, Paper 84, Cranfield, UK, 1961.

23. "Standard Test Method for Crack-Tip Opening Displacement (CTOD) Fracture Toughness Measurement, " Designation E 1290–99, in *Annual Book of ASTM Standards, Section 3, Metals Test Methods and Analytical Procedures,* Vol. 03.01, American Society for Testing and Materials, West Conshohocken, Pennsylvania, 2002.

24. D. S. Dawicke and J. C. Newman, Jr., "Evaluation of Various Fracture Parameters for Predictions of Residual Strength in Sheets with Multi-Site Damage," in *Proceedings of the First Joint DoD/FAA/NASA Conference on Aging Aircraft,* Vol. II, Ogden, Utah, July 1997, pp. 1307–1326.

25. B. R. Seshadri, J. C. Newman, Jr., D. S. Dawicke, and R. D. Young, "Fracture Analysis of the FAA/NASA Wide Stiffened Panels," in *Proceedings of the Second Joint NASA/FAA/ DoD Conference on Aging Aircraft,* NASA/CP-1999-208982, Part 2, Williamsburg, Virginia, August 1998, Langley Research Center, Hampton, Virginia, pp. 513–534.

26. J. C. Newman, Jr., M. A. James, and U. Zerbst, "A Review of the CTOA/CTOD Fracture Criterion," *Engineering Fracture Mechanics,* Vol. 70, pp. 371–385, 2003.

CHAPTER 7

FATIGUE CRACK GROWTH

7.1 OVERVIEW

This chapter extends the LEFM approach introduced in Chapter 3 for analyzing fatigue crack growth. It discusses techniques for treating stress ratio and sequence effects associated with variable-amplitude loading, the growth of small cracks, the influence of environment on crack growth, and the role of fatigue crack closure. Multi-degree-of-freedom analyses are also developed for the cyclic growth of surface cracks and the interaction between multiple flaws located in a single component.

Recall that fatigue is the failure mode associated with *cyclic* loading. As shown in Figure 7.1, the fatigue process consists of three general steps: crack nucleation, crack growth, and final fracture. The final fracture process was discussed in Chapter 6, and techniques for determining the crack nucleation life are summarized later in Section 9.2. Attention at this time is directed to the fatigue crack propagation phase. The precrack could have occurred from initial material or manufacturing defects or prior service (fatigue, environmental attack, foreign object damage) or have been specified by decree (i.e., design code requirements).

Many factors influence the rate of fatigue crack growth, including the material composition and heat treatment, microstructure, the type of product (plate, extrusions, forging, etc.), residual stresses, and batch-to-batch variations in material processing. Once a component enters service, the environment and loading have a significant influence on the crack growth rate. Loading variables include the magnitude and sign (tension or compression) of the applied load, the mean stress level (i.e., $R =$ min/max stress), the sequence of load applications, and the occurrence of overloads or underloads. Other important factors include the type of environment (e.g., salt water, humidity), temperature, and the frequency and shape of the loading cycle. This

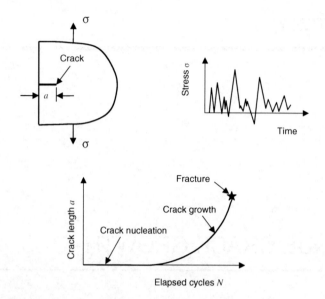

Figure 7.1. Schematic representation of fatigue crack formation, growth, and fracture.

chapter is mainly concerned with the influence of loading and environmental factors on fatigue crack growth rate.

7.2   STRESS RATIO EFFECTS

The fatigue crack growth rate da/dN was shown in Chapter 3 to be controlled by the cyclic range in stress intensity factor ΔK. Recall that the constant-amplitude da/dN–ΔK relation is sigmoidal in shape, as shown in Figure 7.2. Note that there is an upper asymptote when the maximum K during the cycle approaches the fracture toughness K_c. There may also be a lower asymptote for small ΔK, known as the threshold stress intensity factory ΔK_{th}. (There is no observable crack growth for $\Delta K < \Delta K_{th}$.) These three segments of the fatigue crack growth curve are commonly known as stage I (near threshold behavior), stage II (steady-state growth or Paris regime), and stage III (onset of final fracture). These three stages are not to be confused with the modes I, II, and III crack opening directions discussed in Section 3.2.

The cyclic stress intensity factor is not, however, the sole load parameter to control the fatigue crack growth rate. As shown in Figure 7.3, for example, da/dN at a given ΔK often increases with stress ratio. This influence of mean stress depends on the material in question, with some materials showing a large effect of R on fatigue crack growth rate and others having little dependence on mean stress (compare Figures 7.3a, b). As shown in Figure 7.4, complete specification of the load cycle requires at least *two* of the following quantities: σ_{min}, σ_{max}, $\Delta\sigma$, R, or σ_{mean}.

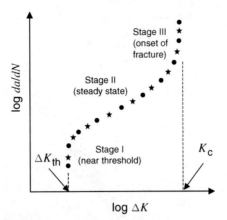

Figure 7.2. Schematic representation of sigmoidal relationship between fatigue crack growth rate and the cyclic stress intensity factor.

7.3 FATIGUE CRACK GROWTH RATE MODELS

Sophisticated fatigue crack growth models have been developed to correlate constant-amplitude fatigue crack growth rates with various loading parameters. These models are based on curve-fitting techniques and are primarily used with computer programs to interpolate between the experimental data obtained for various test conditions. A few of the many proposed fatigue crack growth models are listed below.

Paris Equation [2]

$$\frac{da}{dN} = C \, \Delta K^m \tag{7.1}$$

Here C and m are empirical constants and ΔK is the sole load parameter. This simple equation is quite limited in that it does not contain either the upper or lower asymptotes in the general sigmoidal da/dN–ΔK behavior and does not have a mean stress term. A series of "segmented" Paris laws may, however, be used to fit various regions of the crack growth curve, providing one method to treat the asymptote issue.

Forman Equation [3]

$$\frac{da}{dN} = \frac{C \, \Delta K^m}{(1 - R)K_c - \Delta K} \tag{7.2}$$

Here C, m, and K_c are empirical constants and the two load variables are ΔK ($\Delta K = K_{max} - K_{min}$) and the stress ratio R ($R = \sigma_{min}/\sigma_{max}$). This formula allows da/dN to depend on R and yields an upper asymptote as the crack growth rate gets very

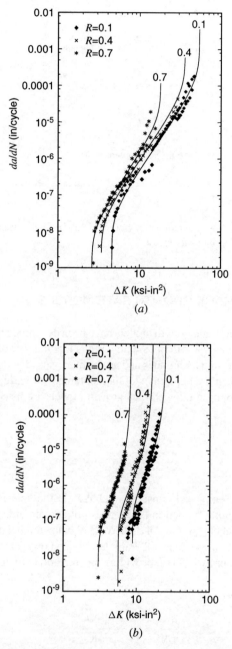

Figure 7.3. (*a*) Fatigue crack growth rate data for BT 14 titanium showing effect of stress ratio and NASGRO curve fit (from p. 6, Ref. 17). (*b*) Fatigue crack growth rate data for A357 cast aluminum showing effect of stress ratio and NASGRO curve fit (from p. 7, Ref. 1).

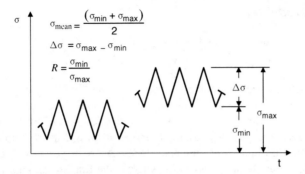

Figure 7.4. Schematic representation of two stress histories with the same stress range but different mean stresses.

large when K_{max} in ΔK approaches K_c. Note that K_{min} cannot be negative when determining ΔK since the stress intensity factor is zero for compressive loading. (The crack faces close in compression so that the crack tip singularity is eliminated and $K = 0$ by the formal definition discussed in Section 3.2.)

Walker Equation [4]

$$\frac{da}{dN} = C\left[K_{max}(1-R)^m\right]^n \tag{7.3}$$

Here the empirical constants are C, m, n and the two load variables are R and K_{max} (where K_{max} is the maximum stress intensity factor per cycle). Although this expression does account for mean stress through R, it does not provide the upper or lower asymptotic behavior.

Collipriest Equation [5]   As indicated by the following two models, the mathematical expressions used to correlate fatigue crack growth data can be quite complex. The Collipriest expression given in Equation 7.4, for example, is based on an inverse hyperbolic tangent function that provides the ability to model both upper and lower asymptotes and was used for damage tolerant analyses on the original design for the space shuttle:

$$\frac{da}{dN} = C(K_c\,\Delta K)^{n/2}\exp\left[\ln\left(\frac{K_c}{\Delta K_0}\right)^{n/2}\operatorname{arctanh}\left(\frac{\ln\left[\Delta K^2/(1-R)K_c\,\Delta K_0\right]}{\ln\left[(1-R)K_c/\Delta K_0\right]}\right)\right] \tag{7.4}$$

Here the two load variables are ΔK and R and the empirical fitting constants are $C, n, K_c,$ and ΔK_0. (Typical values for these constants are given for over 60 materials in Reference 5.)

NASGRO Model [1]

$$\frac{da}{dN} = C\left[\left(\frac{1-f}{1-R}\right)\Delta K\right]^n \frac{(1-\Delta K_{th}/\Delta K)^p}{(1-K_{max}/K_c)^q}$$ (7.5)

This model has been developed for use with the NASGRO crack growth analysis software [1] and also describes the full sigmoidal nature of the fatigue crack growth rate curve. The three load terms are the cyclic stress intensity factor ΔK, the maximum stress intensity K_{max}, and R, and C, n, p, and q are empirical material constants. The threshold stress intensity factor ΔK_{th} in Equation 7.5 is related to crack length, material, and stress ratio, whereas the fracture toughness K_c depends on yield strength, specimen thickness, and other empirical material constants. The crack opening function f also depends on R, K_{max}, and other material constants as described in Reference 1. The NASGRO equation is compared with test data for BT 14 titanium and A357 cast aluminum in Figures 7.3a, b [1].

Table Lookup   Many other models have been employed to correlate stress intensity factor terms with da/dN [6]. Some crack growth models yield the complete sigmoidal shaped curve, whereas others are limited to specific ranges of data. Note that these expressions are simply empirical relations between the fatigue crack growth rate da/dN and the applied stress intensity factor level, and their main function is to give a compact mathematical representation of experimental data that may be employed in a numerical analysis scheme. While they may also be useful for interpolation, great care must be taken when extrapolating a given model beyond the range of experimental test conditions.

The modeling task may also be readily accomplished by computerized table lookup procedures. In that case, it is not necessary to obtain an explicit mathematical representation for the $da/dN-\Delta K$ behavior. Instead, the crack growth data are input in a tabular format, and an appropriate interpolation scheme is employed to find crack growth rates for a particular value of K, R, and so on. This procedure is particularly effective for numerical life prediction schemes.

7.4   LOAD INTERACTION EFFECTS

As discussed in Chapter 5, crack tip plasticity often limits application of LEFM and leads to some peculiar crack growth behavior. In fracture problems, for example, through-the-thickness variation in crack tip plasticity led to a thickness dependence on fracture toughness K_c (recall Figure 6.5). As discussed below, crack tip plasticity also causes load interaction effects that must be accounted for in fatigue crack growth predictions.

Consider, for example, the "fatigue crack retardation" phenomenon shown schematically in Figure 7.5, where application of large tensile overloads can actually *increase* cyclic life. Note that this increase in life (or reduction in da/dN) is *not* predicted by any of the fatigue crack growth rate models discussed in the previous

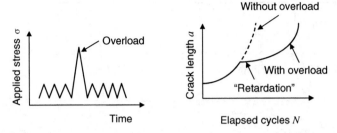

Figure 7.5. Schematic representation of fatigue crack retardation phenomenon.

section. Although the amount of retardation (increased life) depends on the particular test material, in general, more ductility leads to larger crack tip plastic zones and more retardation. Increasing the magnitude or number of the overloads can also increase the fatigue life, and in some cases, it may be possible to permanently stop subsequent crack growth. (It is assumed here, of course, that $K_{max} < K_c$ during the overload cycle, so that the overload does not cause fracture.)

Test data showing the influence of overloads on fatigue crack growth lives in 7075-T6 aluminum specimens are given in Figure 7.6 [7]. Note that in the absence of peak overloads the fatigue life for a center-cracked specimen was less than 10 blocks of 4001 constant-amplitude cycles. Here the constant-amplitude loading during each block varied between 2.7 and 8.0 ksi (18.3 and 55.2 MPa) remote stress. When *one cycle* per block was changed to vary between 2.7 and 14.4 ksi (18.3 and 99.3 MPa), however, the life *increased* to approximately 480 blocks (a 50-fold increase in life).

The sign and sequence of the overload can have a tremendous influence on life, as shown in Figure 7.7 [8]. Here fatigue crack growth curves are given for three center-cracked 2024-T3 aluminum specimens. Compare the crack growth behavior after

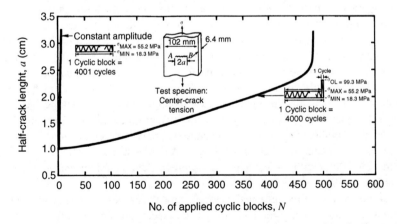

Figure 7.6. Fatigue crack retardation produced by a single peak overload in aluminum alloy 7075-T6. (Reprinted from Reference 7, copyright 1979, with permission from Elsevier.)

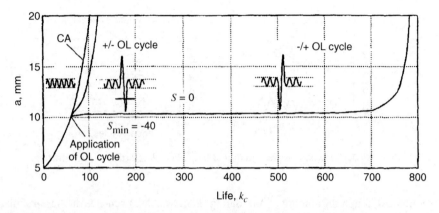

Figure 7.7. Test data showing the effect of two different overload cycles on fatigue crack growth in 2024-T3 aluminum. Mean stress is 80 MPa in all cases. Stress amplitude is 25 MPa for constant-amplitude cycles and 120 MPa for overload cycles. (From p. 117, Ref. 8.)

a 0.4-in (10-mm) crack length for these three cases. In the first instance constant-amplitude cycling continued between 8.0 and 15.2 ksi (55 and 105 MPa) remote stress, and failure occurred within 50,000 additional cycles (curve CA in Figure 7.7). Another specimen, however, was subjected to a single peak load of 29 ksi (200 MPa) preceded by a minimum peak of −5.8 ksi (−40 MPa) at that time and then returned to the original constant-amplitude cycling. Note that this specimen lasted nearly 700,000 more cycles, an order-of-magnitude life increase compared to the original specimen without the overload cycle. When a third specimen was subjected to an overload cycle that was identical to that applied to the second specimen, *except that the order* of the 29- and −5.8-ksi (200- and −40 MPa) peaks were *reversed*, the retardation effect was nearly eliminated, and the specimen had approximately the same life as for the original constant-amplitude loading. Thus, subtle changes in load sequencing can be very significant when the fatigue crack retardation phenomenon is present.

Appreciation of the retardation effect plays an important role in fatigue crack growth life analyses and test considerations. Ignoring "retardation," for example, could give life predictions that are too conservative and of little engineering use. The issue also has important implications for testing, since care must be taken when "precracking" specimens to generate valid baseline fatigue crack growth rate data. As shown schematically in Figure 7.8, for example, large precracking loads could delay subsequent crack growth at smaller stress amplitudes and give erroneously slow da/dN measurements for a given ΔK. Note that the high precrack loads would give *unconservative* material behavior in this case. ASTM test specification E647 [9] gives specific guidelines with respect to preparing precracked specimens that are free of this "retardation effect."

Careful consideration must also go to selecting the proper magnitude and sequence of loads that comprise a variable-amplitude test spectrum, since addition and/or truncation of certain loads can have a significant effect. Two examples of how improper load selection led to misleading full-scale fatigue test results are described below.

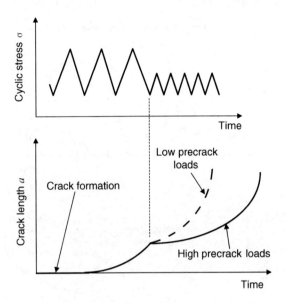

Figure 7.8. Large precracking loads can cause fatigue crack retardation upon subsequent data collection at smaller cyclic stresses and give artificially slow fatigue crack growth rates (nonconservative measurement).

A full-scale fatigue test was conducted on the KC-135 aircraft in the 1950s to verify that no fatigue failures would occur in service [10]. In order to make the test as "severe" as possible, it was decided to introduce additional large loads to those that were actually expected to occur in service (see Figure 7.9). This attempt to conduct a "conservative" fatigue test (by applying more large loads) gave an unconservative result, however, in that the actual service life was less than predicted by the test.

A similar misleading experience occurred during the full-scale testing conducted for the Comet aircraft in the 1950s [8] when a production fuselage was pressurized to 200% of the maximum load expected in service. After passing this static "overload" test unharmed, the same test article was then used for fatigue testing. Again, the structure successfully survived the fatigue test, and the aircraft was believed to have a satisfactory fatigue life. This mistaken assurance was quickly shattered, however, when two aircraft failed from fatigue at relatively short lives (1286 and 903 flight hours). Subsequent analysis of these disasters revealed the significant error in using the same test article for both the static overload and the fatigue tests. It was determined later, for example, that crack retardation resulting from the static overload test actually increased the subsequent fatigue life by a factor of 16. Had another, untested fuselage been employed for the original fatigue tests, it is possible that the fatigue problems would have been discovered and corrected before the aircraft entered service. (In fairness, it should be pointed out that the retardation effect was largely unknown at the time of the KC-135 and Comet fatigue tests, both of which were conducted well before development of the LEFM approach to fatigue crack growth.)

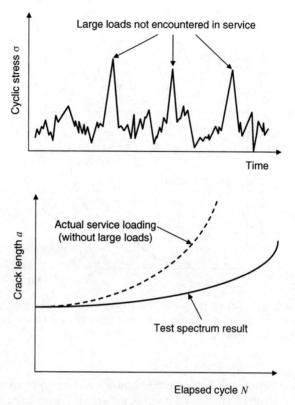

Figure 7.9. Schematic representation of "severe" load spectrum with extra overloads and subsequent effect on fatigue crack growth life.

7.5 CRACK RETARDATION MODELS

At least three mechanisms have been proposed to explain how crack tip plasticity causes fatigue crack retardation: blunting *at* the crack tip, residual stresses *ahead* of the crack tip, and residual deformations *behind* the crack tip

The crack tip blunting mechanism assumes that the effect of plastic flow is to "blunt" the crack and effectively turn it into a "notch" (Figure 7.10). Subsequent load cycles are then required to "reinitiate" a sharp crack tip at the blunted notch, effectively providing a delay in crack growth. Evidence for the blunting phenomenon is reported by the results of overload fatigue crack growth experiments conducted with polycarbonate, a transparent polymer [11]. Although this blunting mechanism has a plausible physical basis, it has not been used for predictive models and will not be considered further here. The other two retardation mechanisms lead to the yield stress and crack closure models described in the following sections.

Before describing specific models for predicting fatigue crack retardation, however, it is useful to consider the following argument for development of *compressive*

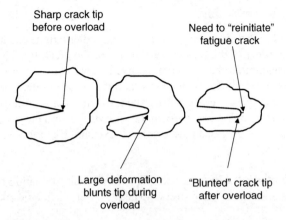

Figure 7.10. Schematic representation of crack tip blunting.

residual stresses inside the crack tip plastic zone following application of remote tensile loads. As shown schematically in Figure 7.11, consider the stress–strain behavior of point A (within the crack tip plastic zone) and an adjacent point B (just outside the plastic zone). Note that inside the plastic zone (point A) there is positive strain after returning to zero stress but outside the plastic zone (point B) one has zero strain after returning to zero stress. Since there cannot be a discontinuous jump in strain across the plastic zone boundary, the elastic material acts as a "clamp" that forces the plastic zone material to zero strain and, thus, to a state of compressive residual stress.

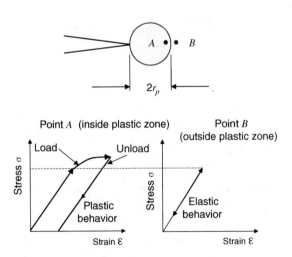

Figure 7.11. Schematic development of compressive residual stresses in crack tip plastic zone.

7.5.1 Yield Stress Models

Several overload models have been developed to predict fatigue crack retardation by considering the effect of the plastic zone *ahead* of the crack tip. These models are based on the assumption that the compressive residual stresses in the plastic zone reduce the stress intensity factor and slow the crack growth rate. Overloads cause a larger plastic zone that the crack must grow through upon returning to the smaller cyclic load. The subsequent crack growth rate is reduced until the crack tip grows out of the influence of the overload plastic zone. Two examples of these types of models are outlined below.

Wheeler Retardation Model   The original Wheeler retardation model provides one *example* of this type of yield stress model [12]. Let the fatigue crack growth rate be given by

$$\frac{da}{dN} = C_p F(\Delta K) \tag{7.6}$$

Here the $F(\Delta K)$ term is one of the fatigue crack growth rate models that relates the cyclic stress intensity factor with the fatigue crack growth rate da/dN (e.g., Equations 7.1–7.5) and C_p is a "retardation parameter" given by

$$C_p = \left(\frac{R_y}{a_p - a}\right)^n \tag{7.7}$$

As shown schematically in Figure 7.12, R_y is the current plastic zone size, a is the current crack length, $a_p - a$ is the distance from the current crack tip to the plastic zone boundary caused by the overload, and n is a "shaping" exponent determined

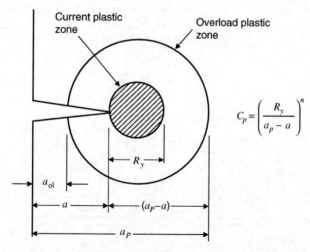

Figure 7.12. Schematic representation of Wheeler fatigue crack retardation model.

by experiment (typically $1 < n < 4$). Also in Figure 7.12, the crack length when the overload is applied is a_{ol}. The crack tip plastic zone size employed by Wheeler is the plane strain version of Irwin's plastic zone given by

$$R_y = \frac{1}{4\sqrt{2}\pi} \left(\frac{K_I}{\sigma_{YS}} \right)^2 \tag{7.8}$$

Immediately following the overload, the retardation parameter $C_p < 1$. As the crack grows through the overload plastic zone, C_p increases monotonically to a value of 1, at which time the crack is assumed to have grown out of the influence of the overload, retardation stops, and the crack growth rate (da/dN) has returned to normal. The retardation term C_p is not allowed to exceed 1 (i.e., no crack acceleration).

Although the Wheeler retardation model has been useful for aircraft life predictions, it has several limitations. First, the shaping exponent n has to be determined for each load history and is not generally applicable for arbitrary loading. Second, the model does not predict the reduction in retardation caused by negative loads (recall Figure 7.7). Finally, the Wheeler retardation scheme does not distinguish between single and multiple overloads.

Willenborg Retardation Model The Willenborg retardation model [13] reduces the maximum stress intensity factor per cycle during the retardation period through computation of an effective compressive stress in the plastic zone. As shown schematically in Figure 7.13, an overload plastic zone with diameter D_{ol} is caused by the tensile overload when the crack length is a_{ol}. The crack begins to grow through this overload plastic zone until reaching a current crack length a_i. The plastic zone size associated with the current crack length a_i and load is D_i. As with the Wheeler model, Willenborg also assumes that the fatigue crack growth rate is retarded by the tensile overload as long as the current plastic zone is contained in the shadow of the large-overload plastic zone.

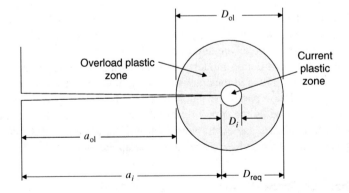

Figure 7.13. Schematic representation of Willenborg crack retardation model.

The Willenborg model computes an effective cyclic stress during the retardation period. This stress is obtained by determining the stress required (σ_{req}) to extend the current plastic zone to the end of the overload plastic zone. The following relations can be defined in connection with Figure 7.13:

$$D_{ol} = \frac{2}{\alpha\pi}\left(\frac{K_{ol}}{\sigma_{YS}}\right)^2 \tag{7.9}$$

$$K_{ol} = \sigma_{ol}\sqrt{\pi a_{ol}}\,\beta\,(a_{ol}) \tag{7.10}$$

$$D_{req} = a_{ol} + D_{ol} - a_i = \frac{2}{\pi\alpha}\left[\frac{\sigma_{req}\sqrt{\pi a_i}\,\beta(a_i)}{\sigma_{YS}}\right]^2 \tag{7.11}$$

Here the Irwin plastic zone model is used to estimate the plastic zone sizes with $\alpha = 2$ for plane stress and $\alpha = 6$ for plane strain conditions. Since all of the terms in Equations 7.9 and 7.10 may be found on a cycle-by-cycle basis, Equation 7.11 may be solved for σ_{req}.

It is now assumed that there is a compressive stress in the crack tip plastic zone (σ_{comp}) that must be subtracted from the actual applied maximum and minimum stresses (σ_{max_i} and σ_{min_i}). This compressive stress and the resulting effective maximum and minimum stresses are given by

$$\sigma_{comp} = \sigma_{req} - \sigma_{max_i} \tag{7.12}$$

$$\sigma_{max\ eff} = \sigma_{max_i} - \sigma_{comp} \tag{7.13}$$

$$\sigma_{min\ eff} = \sigma_{min_i} - \sigma_{comp} \tag{7.14}$$

The minimum effective stress given by Equation 7.14 is truncated to zero if the calculations indicate that it is negative. These effective maximum and minimum stresses are then used to compute the effective cyclic stress intensity factor and the subsequent fatigue crack growth rates during the retardation period.

Although the Willenborg plastic zone model eliminates the empirical constant of the Wheeler plastic zone and depends only on the tensile yield stress, it cannot distinguish between single and multiple overloads. Additional modifications to the Willenborg retardation model are described in Reference 14.

There are other retardation models based on similar arguments concerning the crack tip plastic zone *ahead* of the crack tip [8, 14]. In general, these models describe some type of retardation parameter that is based on the current plastic zone size and the plastic zone caused by the earlier overload. As the crack grows, the value of the retardation parameter changes. Although such models are quite useful for specific situations, none has been shown to have universal application.

7.5.2 Fatigue Crack Closure

Another explanation for fatigue crack retardation was proposed by Elber [15, 16], who suggested that retardation was caused by plasticity acting *behind* the crack tip,

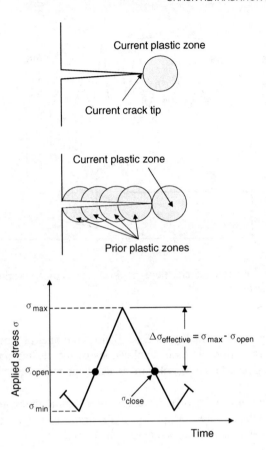

Figure 7.14. Plastic wake behind final crack tip and resulting crack opening load and effective stress.

as shown schematically in Figure 7.14. Since the crack tip continually grows through earlier plastic zones, there is a "wake" of prior plastic deformation that attempts to "close" the crack tip. This residual deformation must be overcome by the remote tensile load before the crack surfaces can separate and the crack can extend. Thus, there is an "opening" stress σ_{open} that must be exceeded before the crack fully opens, producing the "effective" cyclic stress $\Delta\sigma_{\text{eff}}$ (here $\Delta\sigma_{\text{eff}} = \sigma_{\text{max}} - \sigma_{\text{open}}$). Overloads cause more plasticity, result in greater residual deformations (and a larger plastic wake), and effectively increase the opening load. Thus, the effective cyclic stress (and ΔK) is reduced following the overload and the fatigue crack growth rate is decreased (see Figure 7.15).

The closure phenomenon has been observed by many investigators by a variety of experimental techniques, including microscopes, strain gages, interferometry, electric potential, displacement gages, and fractographic observations. The closure mechanism initially attracted many researchers by its explanation of fatigue crack retardation, and closure-based models have been developed to an extent where they

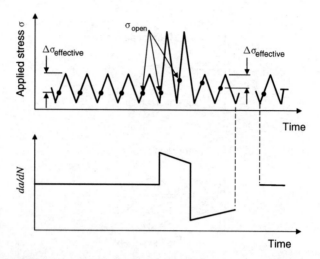

Figure 7.15. Change in fatigue crack opening load caused by an overload and subsequent effect of crack growth rate.

have been incorporated in crack growth prediction computer programs. Although the remainder of this section discusses applications of the closure mechanism to predict fatigue crack retardation, the closure concept has many other implications (see Section 7.12).

The strip yield model [1] is one example of a fatigue crack growth retardation scheme based on the closure concept. Elber [16] proposed modifying the Paris crack growth law with an effective stress intensity factor ΔK_{eff}:

$$\frac{da}{dN} = C(\Delta K_{\text{eff}})^m \tag{7.15}$$

where

$$\Delta K_{\text{eff}} = K_{\max} - K_{\text{open}} \tag{7.16}$$

Here $K_{\max}$ is the peak stress intensity factor during a given load cycle (computed with $\sigma_{\max}$) and K_{open} is the stress intensity factor when the crack tip first becomes completely open. (If the crack tip is already open at the minimum applied stress in a given cycle, K_{open} is the normal $K_{\min}$ computed for the minimum stress $\sigma_{\min}$.)

The key to applying the strip yield model is the ability to predict how K_{open} changes during a given load history and, in particular, to determine how it is affected by overloads. As described in Reference 1, the strip yield model employs the Dugdale plastic zone to represent the material left in the plastic wake by a series of bar elements. These elements break at maximum load but come in contact as the remote load is removed, effectively modeling the crack closure phenomenon. Although originally designed for thin sheets (i.e., plane stress conditions), the Dugdale model is modified by an empirical constraint factor to treat plane strain conditions. Further details of the strip yield model and its application to variable-amplitude life predictions are given in References 1, 8, and 14.

7.6 THICKNESS AND MATERIAL EFFECTS

Crack tip plasticity may also manifest itself in various other thickness and material effects on fatigue crack growth. Although constant-amplitude fatigue crack growth data do not usually exhibit a strong dependence on specimen thickness, variable-amplitude stress histories may lead to significantly longer crack growth lives for thin specimens than for thicker ones. Since the crack tip plastic zone size depends on the state of stress, with thinner sheets having larger plastic zones (plane stress) than thicker plates (plane strain), the thickness effect is associated with through-the-thickness crack tip plastic zone changes (recall Section 5.4).

Results of single peak-overload experiments [17] conducted with 7075-T6 aluminum specimens subjected to constant ΔK conditions are shown in Figure 7.16. Here load-shedding techniques were used to maintain the $R = 0.05$ baseline cycling at $\Delta K = 15$ ksi-in$^{1/2}$ (16.5 MPa-m$^{1/2}$), and periodic overloads of 30 ksi-in$^{1/2}$ (33 MPa-m$^{1/2}$) were applied. The measured delay cycles are shown as a function of specimen thickness in Figure 7.16. Note that delay caused by the peak overload decreased with specimen thickness for the uniform-thickness specimens, varying from over 100,000 cycles for a 0.025-in- (0.64-mm-) thick specimen to less than 25,000 cycles for 0.05-in- (1.27-mm-) thick specimens. The "side-groove" data shown in Figure 7.16 came from specimens of the same material but which contained side grooves added along the crack plane to artificially introduce plane strain in the

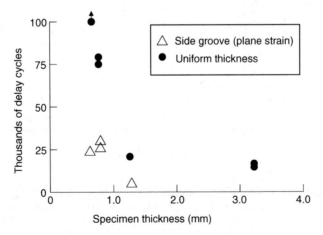

Figure 7.16. Summary of fatigue crack retardation experiments showing effect of specimen thickness on delay cycles in 7075-T6 aluminum specimens [17]. Side grooves simulate plane strain conditions in thin sections.

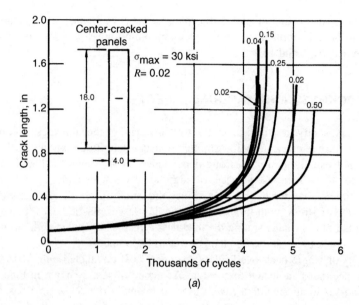

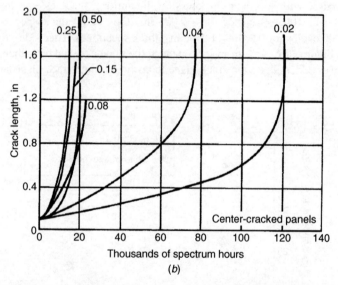

Figure 7.17. Effect of specimen thickness on (*a*) constant-amplitude and (*b*) variable-amplitude fatigue crack growth in 7075-T6 aluminum specimens [18]. All dimensions are in inches.

thin section [17]. Note that the side-groove specimens gave less retardation (due to plane strain conditions) than uniform-thickness (plane stress) members with the same thickness, confirming the premise that the thickness effect is associated with its influence on the crack tip plastic zone size.

A similar thickness result is demonstrated by fatigue crack growth tests conducted by Saff [18]. Constant-amplitude fatigue crack growth curves for center-cracked 7075-T6 aluminum specimens are shown in Figure 7.17a. The specimens varied in thickness from 0.02 to 0.5 in (0.5–12.7 mm). The constant-amplitude fatigue lives range from approximately 4200 to 5500 cycles. This variation is typical of the normal "scatter" found in fatigue behavior, and there is no systematic influence of specimen thickness on the constant-amplitude fatigue life. Figure 7.17b, however, shows a definite dependence on specimen thickness for similar specimens subjected to a *variable-amplitude* loading. In this case, the thin specimens [0.02 in (0.5 mm) and 0.04 in (1 mm) thickness] had much longer lives than the 0.08- to 0.5-in- (2.0- to 12.7-mm-) thick specimens. This difference in life is associated with the fact that the thin specimens exhibit plane stress conditions, whereas the thicker members more closely represent plane strain through their thickness. Thus, the plastic zones resulting from the peak overloads in the variable-amplitude fighter spectrum cause more retardation in the thin (plane stress) members than in the thicker (plane strain) specimens.

The retardation phenomenon may also cloud material rankings based on fatigue crack growth considerations. Bucci [7], for example, has shown that alloy rankings based on constant-amplitude fatigue crack growth tests may change when the same materials are subjected to variable-amplitude loading. He points out that retardation, cyclic hardening, alloy strength–toughness combinations, and microstructural factors may have a different influence under variable-amplitude than constant-amplitude loadings. Thus, if possible, materials selection decisions involving fatigue crack growth considerations should always be based on load histories that match the expected service conditions.

7.7 SMALL FATIGUE CRACKS

"Natural" fatigue cracks usually nucleate simultaneously at various material imperfections or slip planes located in high-stress areas that cause local yielding. These cracks grow independently until influenced by adjacent cracks and then coalesce into a few dominant cracks that control final failure. Although the "small-crack" phase of growth [i.e., crack lengths on the order or 0.001–0.10 in (0.025–2.5 mm)] can represent a substantial period of total fatigue life, several complex technical problems arise when one attempts to analyze them by conventional LEFM methods. The crack tip plastic zones may be comparable in size to the crack length, for example, and plasticity considerations take greater importance. Moreover, continuum assumptions regarding material homogeneity lose validity as crack lengths shrink to the size of characteristic material dimensions (e.g., grain sizes), casting further doubt on the validity of ΔK as a crack driving force.

Numerous investigations have shown that growth rates for small cracks can exceed those of large cracks tested at similar ΔK levels [19–21]. Small cracks may, in fact, continue to grow at applied ΔK's below the large-crack threshold. These faster growth rates are of particular concern because life predictions for "small" cracks that are based on data collected from conventional "large"-crack specimens could lead to nonconservative results. Figure 7.18, for example, compares small- and large-crack growth rates for three versions of 7050-T7451 aluminum plate [22]. Although there is only a minor difference in behavior for the three material pedigrees, the small-crack data [crack lengths < 0.01 in. (0.25 mm)] all lie significantly above the large-crack data in the low-ΔK regime.

References 19 and 21 distinguish between "short" cracks, which are small in length only (i.e., through-the-thickness crack) and "small" cracks, in which all dimensions are physically small (i.e., surface cracks). Small cracks are further characterized as microstructurally small, mechanically small, and chemically small [21].

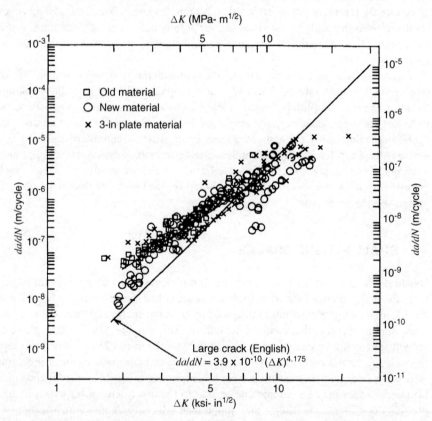

Figure 7.18. Fatigue crack growth rate data for 7050-T7451 aluminum plate specimens showing faster growth rates for small cracks. The data represented by symbols were obtained for crack lengths of less than 0.01 in (0.25 mm), and the line labeled "large crack" was fit through conventional large-crack (length > 0.5 in = 12.7 mm) specimen data [22].

Microstructurally small cracks are those whose dimensions cannot be assumed large relative to microstructural features such as grain size. Mechanically small cracks are those whose dimensions compare to mechanical features such as the crack tip plastic zone or the plastic region near a notch. Chemically small cracks may be observed in corrosion fatigue and result when factors that control the crack tip environment, such as convective mixing, ionic diffusion, or surface electrochemical reactions, are affected by crack length. Further discussion of the important consequences and causes for the small-crack problem are given in References 19–21.

7.8 ENVIRONMENTAL EFFECTS

Many components operate in the presence of aggressive environments that combine the fatigue and corrosion damage mechanisms into a process known as corrosion fatigue. Although corrosion fatigue is a very complex subject that defies simple description, some general observations are given below.

Speidel [23], for example, suggests that the major differences in fatigue crack growth rates observed for various materials are due primarily to their response to the environment. To support this conclusion, he has correlated da/dN measured in a vacuum with the elastic-modulus-corrected stress intensity factor range $\Delta K/E$. Based on the results of approximately 100 experiments, Speidel suggests that the following expression may be used to determine fatigue crack growth rates in a *vacuum* for a wide variety of different materials:

$$\frac{da}{dN} = 1.7 \times 10^6 \left(\frac{\Delta K}{E} \right)^{3.5} \quad \text{(m/cycle)} \tag{7.17}$$

Here, the stress intensity factor range and elastic modulus E are given in metric units. Speidel also proposes Equation 7.18 as an estimate for the stress intensity factor threshold in a vacuum:

$$\frac{\Delta K_{th}}{E} = (2.7 \pm 0.3) \times 10^{-5} \quad \left(\text{m}^{1/2} \right) \tag{7.18}$$

Although Equations 7.17 and 7.18 are proposed as master relationships for fatigue cracking in a vacuum, crack growth rates in other environments are known to differ significantly for various alloys and materials. Speidel suggests that much of this material dependence is associated with their responses to environmental effects, as "aggressive" environments often accelerate the rate of fatigue crack growth. The influence of environment on fatigue crack growth is highly dependent on material composition, chemical content of the environment, temperature, loading frequency, and shape of the load cycle. A given environment may have little influence on da/dN in one alloy but significantly increase fatigue crack growth rates in another material. Indeed, one of the metallurgist's goals is to develop materials with superior environmental resistance.

Fatigue crack growth rates measured in inert environments and at temperatures where creep does not act are usually independent of the cyclic frequency (i.e., number of applied load cycles per unit time). When an environment is present, however, frequency effects often occur, with faster crack growth rates (per cycle of loading) observed at slower cyclic frequencies than at higher rates of loading. In addition, increasing the temperature usually increases the environmental influence on fatigue crack growth.

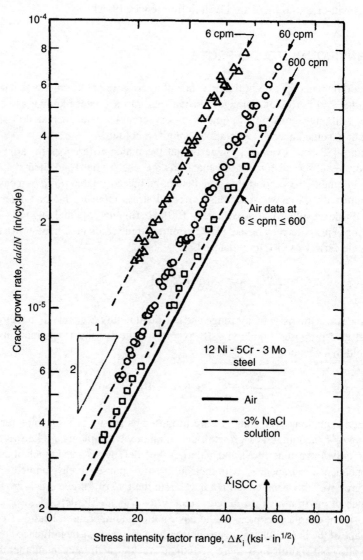

Figure 7.19. Effect of cyclic frequency on fatigue crack growth rate behavior of 12Ni–5Cr–3Mo steel tested in air and in salt water. (From [24], copyright ASTM International. Reprinted with permission.)

The frequency effect is shown in Figure 7.19 [24], where 12Ni–5Cr–3Mo steel has been tested at various cyclic frequencies in both "air" and salt water (3.5% NaCl solution). In this example, air is the reference "inert" environment and salt water is the "aggressive" medium. Note that the salt water has little influence on the fatigue crack growth rate (when compared with the air data) at high cyclic frequencies (600 cycles/min) but greatly increased da/dN at slower loading rates (6 cycles/min). Air was determined to be a benign environment in this case, and as shown in Figure 7.20, frequency changes had little effect on da/dN measured in specimens that were tested in the benign environment. Fatigue crack growth in other materials, such as aluminum alloys, however, can be quite sensitive to changes in relative humidity that occur in

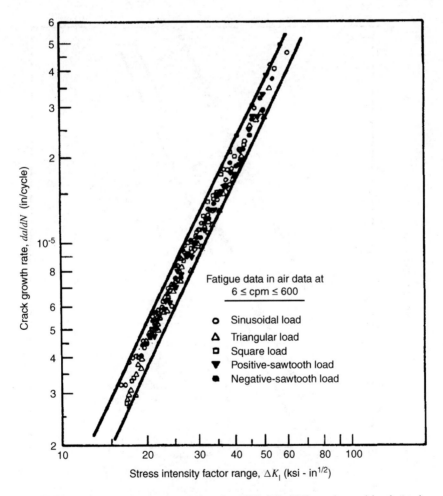

Figure 7.20. Fatigue crack growth rate data for 12Ni–5Cr–3Mo steel tested in air (an inert environment for this material) showing little effect of cyclic frequency or load wave shape. (From [24], copyright ASTM International. Reprinted with permission.)

normal laboratory air, so the test environment must be carefully controlled and/or reported when generating baseline fatigue crack growth data.

The shape of the applied load–time cycle (e.g., sine wave, square wave) can also play a significant role in corrosion fatigue crack growth. Figure 7.20, for example, gives fatigue crack growth rate data for 12Ni–5Cr–3Mo steel tested in the inert air environment under several different load wave shapes [24]. Note that neither the frequency nor the wave shape had much impact on fatigue crack growth rate in this case. As reported in Figure 7.21, similar tests were repeated in an aggressive environment (salt water) at a constant cyclic frequency. Note here that when the aggressive environment was present, there is a significant wave shape effect. The

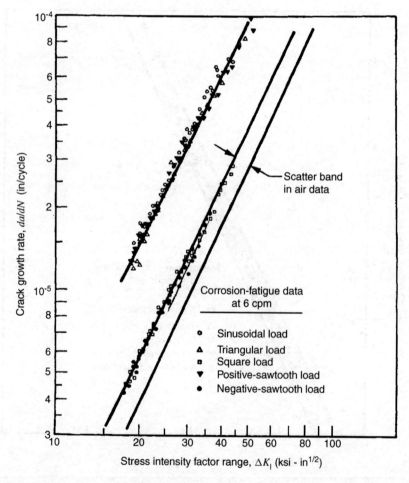

Figure 7.21. Effect of load wave shape on fatigue crack growth rate in 12Ni-5Cr-3Mo steel tested in salt water at a constant cyclic frequency. (From [24], copyright ASTM International. Reprinted with permission.)

sinusoidal, triangular, and positive sawtooth load forms yielded significantly faster crack growth rates than the square and negative sawtooth load waves. The square wave and negative sawtooth salt-water crack growth rates, however, differed little from the reference (air) data. Thus, the environment had little effect on da/dN for these load shapes. These data suggest that the corrosive process acts only during the time when the stresses are *increasing* and fresh material is being exposed to the environment as the crack tip extends. A protective film is formed during the corrosion process that slows subsequent rates of environmental attack. Thus, wave shapes that "slowly" increase load during the "fatigue" component of crack extension allow the environment more time to corrode fresh material at the crack tip during the cycle and as a result are more damaging during the total cycle.

Some investigators have suggested that as a *first approximation* it may be possible to estimate corrosion fatigue by a simple *linear summation* of the fatigue and stress corrosion cracking (SCC) processes. Here the fatigue crack growth rate in an inert reference environment is added to the environmental component of crack growth. The corrosive component of crack growth for a given cycle is computed for the load profile from the sustained load cracking data for the aggressive environment of interest (recall Section 3.5). This superposition of fatigue and stress corrosion cracking for a given cycle is given below:

$$\left(\frac{da}{dN}\right)_{\text{total}} = \left(\frac{da}{dN}\right)_{\text{fatigue}} + (da)_{\text{by scc}} \tag{7.19}$$

The crack growth da per cycle for the environmental load cracking is given by

$$da = \int_0^{t^*} \frac{da}{dt} \, dt \tag{7.20}$$

Here t^* is the period of the load cycle and da/dt is the sustained load crack growth rate which is a function of the varying load $K(t)$. This simple superposition argument is quite limited, however, and should be used with extreme caution. This superposition would predict no influence of environment for $\Delta K < K_{\text{ISCC}}$, which is in contrast to observed data. Nor does the superposition argument predict the waveform dependence discussed previously, since superposition would indicate that a "square-wave" load form gives the highest crack growth rate in corrosion fatigue.

7.9 VARIABLE-AMPLITUDE LIFE PREDICTION SCHEMES

Now turn your attention to the general problem of computing fatigue crack growth lives for components subjected to variable-amplitude load histories. As discussed previously, the fatigue crack growth life is given by the expression (recall Section 3.4)

$$N_f = \int_{a_0}^{a_f} \frac{da}{F(K)} \tag{7.21}$$

Here a_0 and a_f are the initial and final crack sizes, $F(K)$ is the fatigue crack growth rate model for the material of interest (e.g., Equations 7.1–7.5), and *crack length* is the *variable of integration* (da). Also recall that the stress intensity factor K in Equation 7.21 is a function of applied stress, crack length, and component geometry (i.e., β). Note that for constant-amplitude loading the stress may be moved outside the integral sign and the fatigue life N_f is readily calculated (either by exact integration as shown earlier in Section 3.4.1 or, more commonly, by a numerical integration procedure). When the stress history varies with time, however, stress is a function of crack length (since the crack also grows with time) and must remain inside the integral. Since one does not usually know how stress will vary with crack length at the onset, a different procedure must be used to integrate Equation 7.21 and compute life. Two such approaches are described in the following subsections.

7.9.1   Effective Stress Approach

If the problem of interest involves a fairly short, *repetitive* load sequence as shown in Figure 7.22, it may be possible to treat the loading as an "effective" constant-amplitude problem. First one would conduct baseline fatigue crack growth experiments with the test material for the given load history. Instead of measuring life in terms of elapsed cycles N, however, it is now expressed in terms of applied load blocks B (or flights), as shown in Figure 7.23. Next, the fatigue crack growth rate per block of loading da/dB is correlated with the effective stress intensity factor K_{eff} as shown in Figure 7.23. Now, the fatigue crack growth material properties *for this load history* may be fit with the equation

$$\frac{da}{dB} = F(K_{\mathrm{eff}}) \tag{7.22}$$

The crack growth life measured in the number of applied loading blocks B is then given by

$$B = \int_{a_0}^{a_f} \frac{da}{F(K_{\mathrm{eff}})} \tag{7.23}$$

Here the effective stress intensity factor K_{eff} is based on an "effective" stress σ_{eff} (determined for the particular loading block), the crack length for the current loading block, and the stress intensity factor coefficient β computed for the current crack length:

$$K_{\mathrm{eff}} = \sigma_{\mathrm{eff}}\sqrt{\pi a}\,\beta(a) \tag{7.24}$$

The key to this approach is determining an appropriate effective stress σ_{eff} for the given block of loading (see Figure 7.22). This effective stress could be the maximum stress level during the loading block, the average stress, or the root-mean-square value of stress per block. If a single value of effective stress can be found to characterize the loading block, it can be removed from the integral sign in Equation 7.23,

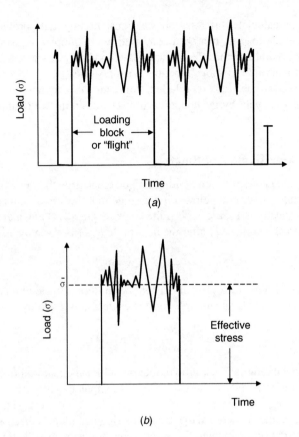

Figure 7.22. Schematic representation of a variable-amplitude loading history comprised of repeated blocks of similar loading or flights: (*a*) repeated loading blocks that are similar in content and history; (*b*) effective stress for given loading block or flight.

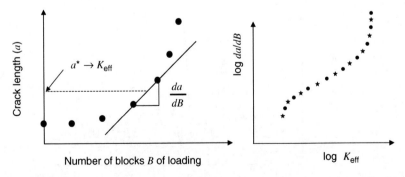

Figure 7.23. Schematic representation of fatigue crack length as a function of loading blocks showing crack growth rate per block of loading da/dB and subsequent correlation with effective stress intensity factor K_{eff}.

and life can be calculated as for a normal constant-amplitude problem. Thus, the variable-amplitude loading has been converted into a simpler, equivalent constant-amplitude case. Although the effective stress method has been successfully employed in practice, it is limited to short, repetitive load histories that do not exhibit significant amounts of fatigue crack retardation. For completely random or more complex load histories, the cycle-by-cycle approach discussed in the following section must be employed.

7.9.2   Cycle-by-Cycle Approach

A more general approach for computing fatigue crack growth lives is the "cycle-by-cycle" algorithim described below. First assume that the stress history is known as shown schematically in Figure 7.24. If the crack length a_n is known at a given cyclic life N_n, the crack length a_{n+1} after the next cycle N_{n+1} is given by

$$a_{n+1} = a_n + \Delta a \tag{7.25}$$

Now, the increment of crack growth Δa for one additional cycle of loading is simply the fatigue crack growth rate da/dN for that cycle and is given by

$$\Delta a = \frac{da}{dN} = F(K_{n+1}) C_r \tag{7.26}$$

Here $F(K)$ is the fatigue crack growth rate model for the material of interest, K_{n+1} is the stress intensity factor for cycle N_{n+1} (computed with stress σ_{n+1} and the known crack length a_n since $a_n \sim a_{n+1}$), and C_r is an appropriate "retardation parameter" (e.g., the Wheeler, Willenborg, or strip yield models discussed in Section 7.5). The retardation parameter C_r depends on the prior load history and can be readily computed from the known current and prior stresses and crack lengths.

Iterating Equations 7.25–7.26 on a cycle-by-cycle basis for the entire loading spectrum gives the final crack length at the end of the specified loading history. Note

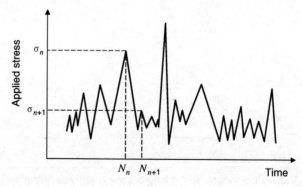

Figure 7.24. Variable-amplitude stress history showing stress peaks at successive cycles N and $N + 1$.

that this approach is quite flexible in that one can select from a variety of fatigue crack growth models (e.g., Equations 7.1–7.5), crack configurations (i.e., β terms), or crack retardation parameters. While this approach can be computationally intensive for long loading spectra, there are procedures to simplify calculations by truncating or lumping together certain portions of the loading spectrum. A variety of software packages are available to perform these cycle-by-cycle life calculations (e.g., the NASGRO program described in Reference 1 or the AFGROW software described in the Appendix of this volume. Problems 7.13–7.18 at the end of this chapter employ the AFGROW software discussed in the Appendix.)

7.10 MULTI-DEGREE-OF-FREEDOM ANALYSIS

So far it has been assumed that the component of interest contains a single crack that is characterized by one length dimension. Many practical problems, however, involve surface or corner cracks whose shape changes as they extend as well as components that contain multiple cracks that grow, interact, and coalesce into a dominant crack that leads to final fracture. The following two subsections describe how the cycle-by-cycle algorithm described previously may be modified to provide a multi-degree-of-freedom analysis for these more complex cracking situations.

7.10.1 Surface Cracks

Often cracks do not completely penetrate the component thickness but occur as surface or corner cracks, as shown in Figure 7.25. These cracks are frequently modeled as portions of ellipses with semimajor and semiminor axes c and a, and stress intensity factor solutions have been obtained for various flaw shapes and locations [25, 26]. A significant result of these analyses is the fact that K varies along the crack perimeter and often has different values at the points of maximum flaw penetration and at the intersection of the crack with the free surface (i.e., points A and C in Figure 7.25a). Thus, the crack will grow at different rates at these two locations, and the aspect ratio $(a/2c)$ will change with life. Since the stress intensity factor depends on both crack size and shape, these shape changes must be accounted for in the fatigue crack growth analysis.

Fortunately, this additional complication is readily treated with the cycle-by-cycle algorithm described in the preceding section. Begin with the assumption that the stress intensity factor solution is known for the surface or corner crack of interest and is expressed in the form

$$K = \sigma(\pi a)^{1/2}\beta \tag{7.27}$$

Here σ is the remotely applied stress, a is the crack depth, and β is the dimensionless stress intensity factor coefficient given by

$$\beta = F\left(a, c, \frac{a}{2c}, \varphi, \frac{a}{T}, \frac{c}{W}, \dots\right) \tag{7.28}$$

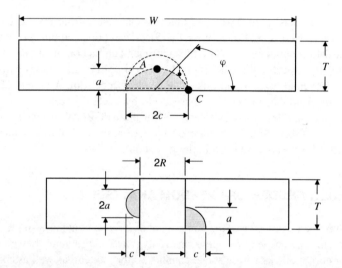

Figure 7.25. Schematic representation of typical surface and corner cracks: (*a*) semielliptical surface crack in a plate; (*b*) semielliptical and quarter-elliptical surface and corner cracks located at the bore of a hole in a plate.

Note that β depends on crack size (a, c, a/T, c/W), shape ($a/2c$), and position along the crack perimeter as defined by the angle φ. As discussed later in Section 8.2, several β solutions are available for various surface and corner crack configurations.

For the present time, assume that the known crack dimensions at cycle N_n are a_n and c_n. Now calculate the stress intensity factors K_A and K_C at the crack locations A and C in Figure 7.25a for the next load cycle N_{n+1} and compute the fatigue crack growth rates at these points by the fatigue crack growth model for the material of interest:

$$\frac{da}{dN} = F_A (\Delta K_A) \tag{7.29}$$

$$\frac{dc}{dN} = F_C (\Delta K_C) \tag{7.30}$$

Note that in general $K_A \neq K_C$, so the crack growth rates will differ at locations A and C, and the shape $a/2c$ will change as the crack grows. (If $K_A = K_C$, the rates are equal, and the crack maintains a "stable" shape throughout its life.) As per the earlier cycle-by-cycle analysis, crack dimensions a_{n+1} and c_{n+1} after the next load cycle N_{n+1} are

$$a_{n+1} = a_n + \frac{da}{dN}C_{ra} \tag{7.31}$$

$$c_{n+1} = c_n + \frac{dc}{dN}C_{rc} \tag{7.32}$$

Here C_{ra} and C_{rc} are fatigue crack retardation parameters computed at locations A and C by an appropriate retardation model as before. Now, iterating Equations 7.29–7.32 on a cycle-by-cycle basis allows the surface crack to grow into its natural shape as a function of elapsed cycles.

Again, note the flexibility in this approach. It is possible, for example, to account for potential material anisotropy by using different fatigue crack growth models $F_A(\Delta K)$ and $F_C(\Delta K)$ for the a and c crack growth directions. One can also employ separate fatigue crack retardation models C_{ra} and C_{rc} at locations A and C to account for potential differences in the state of stress at these locations. (The free surface point C, for example, is in plane stress and could be expected to undergo more plasticity and, thus, more retardation than interior point A, which, depending on its depth, may be in plane strain.)

This algorithm can be modified to speed computation time for *constant-amplitude* loading. Again, compute the stress intensity factors and fatigue crack growth rates at points A and C for the initial crack sizes a_n and c_n as before. Now assume a "small" increment of growth Δa at crack tip location A. The number of cycles ΔN required for this increment of crack extension is now given by

$$\Delta N = \frac{\Delta a}{da/dN} \tag{7.33}$$

The corresponding growth Δc at crack tip location C and the new values for crack lengths a and c after ΔN cycles are now given by

$$\Delta c = \left(\frac{dc}{dN}\right)\Delta N \qquad c_{\text{new}} = c_n + \Delta c$$

$$a_{\text{new}} = a_n + \Delta a \qquad N_{\text{new}} = N_n + \Delta N \tag{7.34}$$

Note here that the assumed value for Δa is the "step size" for the iteration and must be small enough for this approximate numerical solution to converge. (Of course, if Δa is chosen to be the fatigue crack growth rate da/dN, $\Delta N = 1$, and the algorithm reduces to the original cycle-by-cycle approach.) It should also be pointed out that the original assumed increment of crack growth could have been Δc rather than Δa and the above equations modified accordingly.

7.10.2   Multiple Cracks

It is common for more than one fatigue crack to develop at a single location (such as the bore of a fastener hole, as shown in Figure 7.26a) or at multiple locations (such as the row of fastener holes shown in Figure 7.26b). These cracks may grow independently for a period, then interact and coalesce into a dominant crack that grows to failure.

The multi-degree-of-freedom analysis scheme described above is readily extended to these simultaneous crack growth applications. One simply continues to compute the stress intensity factors, fatigue crack growth rates, and crack extensions Δa_i at all

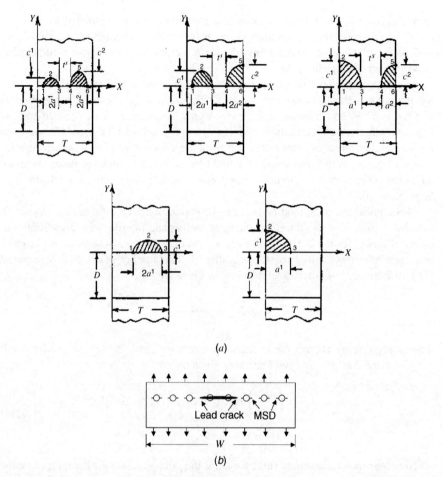

(a)

(b)

Figure 7.26. Two examples of problems involving the simultaneous growth of multiple fatigue cracks: (a) schematic representation of multiple cracks located along the bore of a hole; (b) simultaneous cracks along a row of fastener holes.

of the other i crack tips. Of course, appropriate stress intensity factor solutions must first be obtained for each individual crack, including interactions in K as separate tips approach each other [27, 28]. All of the crack lengths are then updated following the new ΔN cycles, and the process is repeated as before. When two cracks coalesce, the geometry is modified to reflect the new crack configuration, and the process is continued.

Figure 7.27 compares measured and predicted fatigue crack growth lives for approximately 80 constant-amplitude fatigue tests conducted with specimens that contained various combinations of surface and corner cracks located along the bore of open holes in remotely loaded tension members (i.e., Figure 7.26a). These experiments involved both metal (aluminum, titanium, and Waspalloy) and transparent

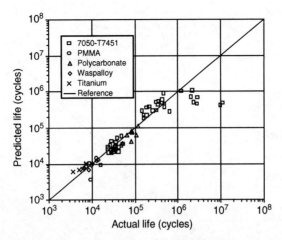

Figure 7.27. Comparison of experimental and measured fatigue crack growth lives for specimens with various combinations of multiple cracks located at an open hole. Data from References 29–32.

polymer (polycarbonate and PMMA) specimen materials and a wide variety of initial crack sizes, shapes, and locations [29–32]. Note that the constant-amplitude algorithm discussed above provides excellent life predictions for these sets of experiments.

A similar comparison for the multiple-site-damage (MSD) problem is shown in Figure 7.28. Here approximately 30 constant-amplitude MSD tests were conducted on 2024-T3 or 2524-T3 aluminum specimens loaded in remote tension. Each specimen contained a row of open holes or lap joint that had various combinations of large-lead and/or small-MSD cracks, as shown schematically in Figure 7.26*b* [33–36]. The lead cracks sizes were typically 2–9 in (5–23 cm) in length, and the MSD cracks emanating from the fastener holes were on the order of 0.04–0.14 in (1–3.6 mm). Specimen width varied between 9 and 16 in (23–41 cm) and specimen thickness ranged from 0.063 to 0.09 in (1.6–2.3 mm). In addition, some specimens had longitudinal strips riveted to them to simulate the stiffeners in aircraft structure, whereas a few were manufactured in a lap joint arrangement. Again, as shown in Figure 7.28, the multi-degree-of-freedom analysis gives excellent life predictions for this multi-cracked situation.

7.11 FATIGUE CRACK THRESHOLDS

This section briefly summarizes some issues associated with the threshold behavior of fatigue crack growth (i.e., the stage I portion of the $da/dN-\Delta K$ curve shown in Figure 7.2). This topic is of practical interest in that if the applied ΔK can be kept below the threshold stress intensity factor ΔK_{th}, cracks will not grow, suggesting the potential for "infinite" fatigue life. A similar situation occurs for smooth, uncracked fatigue specimens where extremely long lives are obtained when the stress amplitude

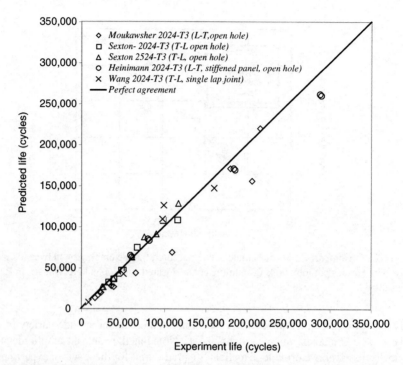

Figure 7.28. Comparison of measured and predicted fatigue crack growth lives for stiffened and unstiffened panels that contain various combinations of large lead cracks and small multiple-site-damage cracks located along a row of open holes [33–36].

stays below a certain value known as the endurance limit (S_e). The smooth specimen endurance limit is discussed in more detail in Chapter 9 (Section 9.2.2). Whereas there is obvious interest in avoiding fatigue failures through use of threshold concepts, many issues complicate this goal. (See Reference 37 for an in-depth review of this important subject.)

In theory, the maximum cyclic stress $\Delta\sigma_{th}$ that could be applied without fatigue failure can be determined from the threshold stress intensity factor as given by

$$\Delta\sigma_{th} = \frac{\Delta K_{th}}{\sqrt{\pi a}\beta(a)} \tag{7.35}$$

The "Kitagawa" diagram shown in Figure 7.29a plots this threshold stress $\Delta\sigma_{th}$ as a function of crack size (line labeled LEFM). Note that the threshold stress is also bounded by the smooth specimen endurance limit S_e for zero crack length and that Equation 7.35 overestimates the threshold stress for small crack lengths (recall the "small-crack" problem discussed in Section 7.7). El Haddad et al. [38] proposed a small-crack threshold correction obtained by adding a small value a_0 to the crack length of interest. The fictitious crack length a_0 is found by solving Equation 7.36

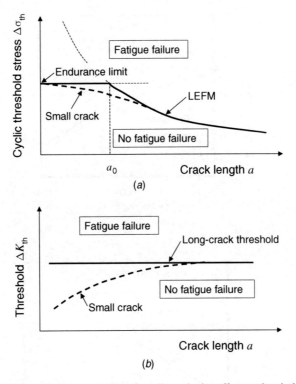

Figure 7.29. (*a*) Schematic representation of small-crack-size effect on threshold fatigue stress $\Delta\sigma_{th}$. (*b*) Influence of small crack size on threshold stress intensity factor ΔK_{th}.

when the applied cyclic stress is S_e, the smooth stress endurance limit, and ΔK_{th} is the conventional "long-crack" threshold stress intensity factor:

$$\Delta K_{th} = S_e \sqrt{\pi a_0} \beta(a_0) \qquad (7.36)$$

Now once a_0 is determined, the small-crack corrected threshold stress $\Delta\sigma_{th}^{sc}$ and the small-crack threshold intensity factor ΔK_{th}^{sc} are given by

$$\Delta\sigma_{th}^{sc} = \frac{\Delta K_{th}}{\sqrt{\pi(a + a_0)}\beta(a + a_0)} \qquad (7.37)$$

$$\Delta K_{th}^{sc} = \Delta\sigma_{th}^{sc}\sqrt{\pi(a)}\beta(a) = \Delta K_{th}\sqrt{\frac{a}{a + a_0}}\frac{\beta(a)}{\beta(a + a_0)} \qquad (7.38)$$

Note in Equation 7.37 that the effect of the a_0 correction decreases as the actual crack length $a > a_0$, so that one obtains the conventional long-crack stress intensity factor threshold ΔK_{th} when $a \gg a_0$. This dependence of ΔK_{th} with crack size is shown schematically in Figure 7.29b. Figure 7.30a compares Equation 7.37 with small-crack threshold stress data obtained for titanium alloy Ti-6-4, whereas Figure 7.30b shows a

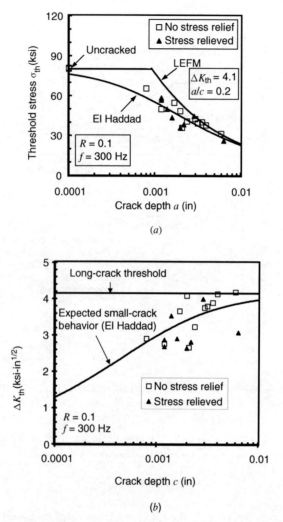

Figure 7.30. (*a*) Comparison of El Haddad small-crack-size correction with threshold stress behavior in Ti-6-4 specimens. (*b*) Comparison of small-crack-corrected stress intensity factor threshold with Ti-6-4 data [39].

similar comparison between Equation 7.38 and measured small-crack threshold stress intensity factor. Note that the El Haddad small-crack correction gives a reasonable estimate for the effect of crack size on fatigue threshold behavior in this alloy [39].

A number of issues lead to the small-crack threshold behavior described above, including changes in crack plane as naturally formed cracks adjust from maximum shear stress directions to maximum principal stress (i.e., mode I) directions, the influence of microstructural features on crack growth, and the influence of crack tip plasticity. Crack closure also plays a significant role, as discussed in the following section. A very important practical concern deals with precracking the test specimens used to measure ΔK_{th}, where it is essential that loads be gradually reduced during the precrack stage to avoid the crack retardation phenomenon and artificially high measurements of ΔK_{th}. Additional details of threshold fatigue behavior are discussed in Reference 37.

7.12 FATIGUE CRACK CLOSURE

Crack closure concepts, first reported by Elber in 1970 to explain fatigue crack retardation (recall Section 7.5.2), have subsequently been found to clarify many other aspects of fatigue crack growth. Thus, it is appropriate to conclude this chapter by describing the broad impact of crack closure on the growth of fatigue cracks. It is hoped this discussion will demonstrate the general importance of crack closure as well as provide a means to summarize some particular aspects of fatigue crack growth discussed previously.

Recall that crack closure is the physical phenomenon where fatigue crack surfaces do not immediately separate upon remotely applied loading. As shown in Figure 7.14, there is an initial "opening" load that must be overcome before the crack faces completely separate and additional damage can occur at the crack tip. Upon unloading, the crack faces again come in contact at a stress level known as the "closure" load. Although the opening and closure loads may differ in a given cycle, they are often similar in magnitude, and no further distinction will be made here between them. The key point here is that only the portion of a load cycle that exceeds the opening (or closure) load is effective in causing further crack growth. This effective cyclic stress $\Delta\sigma_{effective}$ is given by

$$\Delta\sigma_{effective} = \sigma_{max} - \sigma_{open} \tag{7.39}$$

In the event that the minimum remotely applied stress σ_{min} exceeds the opening stress, so that the crack faces are separated during the entire cycle, the effective stress is the same as the nominal cyclic stress:

$$\Delta\sigma_{nominal} = \sigma_{max} - \sigma_{min} \tag{7.40}$$

Note that the cyclic stress intensity factor may be based on either the effective stress range or the nominal stress ranges as defined above. When the cyclic stress intensity factor is computed with the effective stress range $[\Delta K_{eff} = \Delta\sigma_{eff}\,\beta(\pi a)^{1/2}]$, however, it is often possible to obtain a truer picture of the actual crack driving force than when considering the nominal stress intensity factor range $[\Delta K_{nominal} = \Delta\sigma_{nominal}\,\beta(\pi a)^{1/2}]$. This point will be more apparent in the following discussion.

7.12.1 Closure Mechanisms

At least five *mechanisms* have been identified [41] as causing the crack closure effect:

- *Plasticity-induced* closure is caused when a fatigue crack grows through the wake of plastically deformed material left by prior crack tip plastic zones created when the crack was smaller. As discussed in Section 7.5.2, residual deformations in the prior plastic zone wake act to hold the crack surfaces shut during initial loading.

- *Asperity-induced* closure results from surface roughness that prevents mating crack faces from completely closing upon unloading. The crack surfaces are propped open at zero load and require additional force to cause further opening (and damage) at the crack tip.

- *Oxide-induced* closure results from corrosion products that build up along the crack surfaces. These corrosion products again prop the crack faces open, requiring an initial applied load before further crack separation (and flaw extension) occurs.

- *Viscous fluid–induced* closure occurs from hydrodynamic wedging caused by a chemically inert fluid (such as oil) that penetrates the crack surfaces.

- *Phase transformation–induced* closure can occur in materials that undergo metallurgical phase transformations in the plastic zone that cause large volume increases in the wake of the crack path.

7.12.2 Closure Measurements

Fatigue crack closure is a well-established phenomenon that has been observed by a variety of methods in many different materials. (See References 40 and 41 for extensive reviews of published literature dealing with the closure phenomenon.) Perhaps the most popular method to quantify crack opening is to obtain a load–displacement record during the fatigue cycle, as shown schematically in Figure 7.31. Here the remotely applied force P is increased while displacement is measured at some location that is sensitive to crack opening. The displacement δ may be measured with an extensometer placed across the crack mouth, for example, or with strain gages placed at various points near the crack tip. The sensitivity of the load–displacement record depends of the displacement sensor as well as the measurement location.

When the force is first applied, the crack surfaces are completely closed, and there is no observed displacement at the measurement point. As the force is increased, part of the crack surfaces separate, leading to a measured displacement. Additional increase in load causes more crack opening, and the crack appears to lengthen. This apparent crack extension alters the specimen compliance until the crack surfaces are completely open, at which time the force–displacement slope remains constant. The opening load is then defined as the force that first results in constant compliance.

Crack closure has also been measured by a variety of other techniques, including direct microscopic observation, optical interferometry, ultrasonics, and electric potential methods. Evidence of closure has also been observed from fractographic observations of the fatigue striation spacing or by examination of wear debris left when the crack surfaces come in contact during the closure period. Numerical simulations have also demonstrated the occurrence of crack closure. As discussed later in Section 7.12.4, fatigue crack closure is a complex three-dimensional phenomenon,

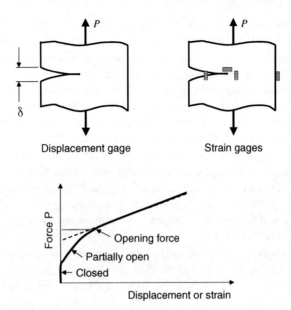

Figure 7.31. Schematic representation of load–displacement record showing change in specimen compliance as crack tip opens.

however, and precise measurement of the opening load can be complicated by the type of measurement employed and by the measurement location [42–44].

7.12.3 Consequences of Crack Closure

Crack closure concepts provide the potential for explaining and predicting several distinct aspects of fatigue crack growth:

- *Variable-amplitude load interaction* effects (i.e., fatigue crack retardation) may be explained by changes in the closure load caused by tensile overloads. As discussed previously in Section 7.5.2, overloads increase the closure stress, reducing the effective stress amplitude and the effective ΔK, and cause slower (retarded) crack growth upon subsequent loading.
- *The short-crack problem* described in Section 7.7 occurs when small cracks grow faster than predicted by crack growth rate data collected from larger cracks. Closure concepts provide one explanation for this small-crack effect. Since short cracks have not had the opportunity to develop a large plastic wake, they exhibit less crack closure. Thus, for a small crack, a larger portion of the stress cycle exceeds the "opening load," resulting in larger "effective" ΔK's and faster growth rates at the same "nominal" ΔK.
- *Mean stress effects* on fatigue crack growth can also be explained by closure. Recall from Section 7.3 that large R's (where R is the minimum/maximum stress) cause faster crack growth than the same cyclic stress amplitude applied with a

lower mean stress. At larger R, more of the cyclic stress is above the closure load, and the effective cyclic stress is larger, leading to faster crack growth.

- *Threshold fatigue crack growth* studies are especially susceptible to crack closure, since small changes in stress amplitude can cause large changes in the corresponding crack growth rate. Measurement of ΔK_{th} is quite sensitive to precycling loads, mean stress, and crack size. Moreover, significant changes in surface roughness (which influence asperity-induced closure) can occur as small cracks transition from stage I to stage II growth.

- *Automated measurements of crack length* (compliance, electric potential, etc.) employed for computer-controlled crack growth testing can be significantly affected by crack closure. Recall Figure 7.31, for example, where the apparent compliance of the test specimen changes as the crack surfaces open. Failure to account for closure could lead to inaccurate crack length measurements and to invalid test results.

- *Environmental effects* that result when an aggressive environment is present at the crack tip (recall Section 7.8) may be affected by crack closure. The environment is drawn to the tip by capillary action as the crack faces open and close during a fatigue cycle. Corrosion products may form on the crack surfaces and influence subsequent crack opening and closing (i.e., cause oxide-induced closure). Since the amount of environment present at the crack tip depends on the fatigue cycle and crack opening depends on the resulting corrosion products, a complex interaction may occur between the fatigue and corrosion damage mechanisms.

- *Thickness effects* evident when fatigue crack growth in thin specimens is slower under variable-amplitude stress histories than in thicker specimens also have a closure explanation (recall Section 7.6). The overload plane stress crack tip plastic zones found in thin specimens would have more of an influence on the closure load and the effective stress amplitude than the smaller plane strain plastic zones in the thicker members.

- *Crack tunneling,* where the crack front grows faster in the specimen interior than at the surface, is related to three-dimensional variations in the crack closure load. Again, since crack tip plasticity depends on the state of stress (plane stress plastic zones are larger than plane strain zones), plasticity-induced closure is more significant at the plane stress specimen surface than in the plane strain interior (if the specimen is sufficiently thick).

- *Nondestructive inspection* methods for detecting fatigue cracks are frequently complicated by crack closure, since cracks whose faces are tightly held together by the closure mechanism are more difficult to detect. Thus, application of a remote tensile load to open the crack faces can aid detection.

7.12.4 Three-Dimensional Aspects of Crack Closure

Complete understanding of the role closure plays on fatigue crack growth requires characterization of through-the-thickness crack opening to determine how surface measurements of closure are related to internal crack behavior. Most techniques used

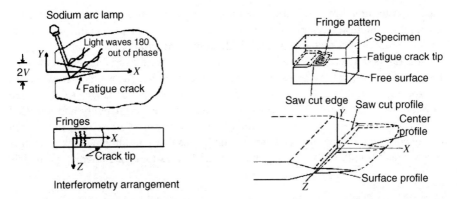

Figure 7.32. Schematic representation of interference mechanisms which give three-dimensional crack surface displacement profiles in transparent specimens [44].

to study crack closure either employ point measurements on the specimen surface or average bulk specimen measurements and cannot resolve internal crack opening at particular points. Thus, three-dimensional aspects of crack closure, including thickness effects, remain relatively unstudied.

Perhaps the first direct study of how crack closure varies through the thickness of a specimen is described in Reference 44. There relatively simple interferometric measurements were made possible by the use of transparent polymer [polymethylmethacrylate (PMMA)] specimens. As shown schematically in Figure 7.32, illumination of a crack in a transparent material with a monochromatic light source creates a set of interference fringes along the crack plane that completely describe three-dimensional crack surface displacements. Analysis of the interference fringes gives crack opening profiles as a function of applied load and provide the means to determine when the crack surfaces separate at the crack tip.

Similar measurements [42] of crack opening profiles for a fatigue crack grown under constant ΔK conditions in a polycarbonate (a ductile, transparent polymer) specimen are shown in Figure 7.33. Figure 7.33a presents crack profiles as a function of applied load as measured at the free surface. Note that a K of 76.18 psi-in$^{1/2}$ (83.9 kPa-m$^{1/2}$) is required to completely open the crack tip. Figure 7.33b presents crack opening shapes measured in the specimen interior (plane strain conditions) corresponding to the plane stress results obtained on the free surface. Note that the crack opens at a much smaller load in the specimen interior than at the specimen surface. Similar results reported in Reference 43 for experiments on 2024-T351 aluminum confirm the general contention that fatigue crack closure is, indeed, a three-dimensional phenomenon.

7.13   CONCLUDING REMARKS

As first introduced in Chapter 3, the LEFM approach assumes that the fatigue crack growth rate is controlled by the cyclic stress intensity factor range. Fatigue life may then be determined by integrating the fatigue crack growth rate relationship for the material of interest. This chapter extends that key concept to account for various complicating factors that arise in practical engineering problems.

Several loading effects must be treated when considering variable-amplitude load histories. Increasing the mean stress, for example, results in faster crack growth rates for specimens tested at the same cyclic stress intensity factor range. As described in Sections 7.2 and 7.3, this effect may be measured experimentally and modeled by various equations that account for changes in mean stress. Crack retardation can occur when tensile overloads are applied. This delay in fatigue crack growth is due to crack tip plasticity and may be estimated by various load interaction models

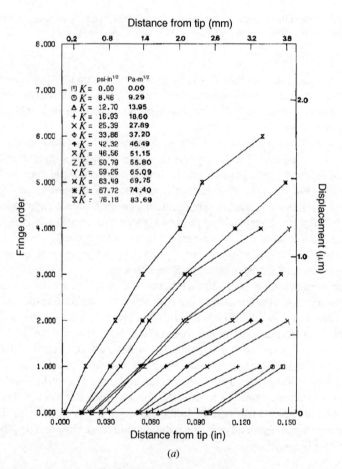

(a)

Figure 7.33. Crack opening profiles as a function of applied load determined by optical interference measurements in transparent polycarbonate specimens. Fringe order is directly related to distance between crack surfaces [42]. (*a*) Profiles measured at free surface where plane stress conditions dominate.

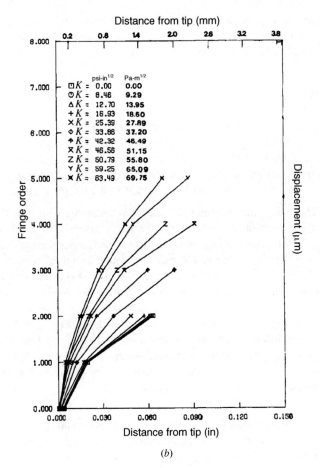

Figure 7.33. (*b*) Profiles measured at specimen midplane where plane strain conditions dominate.

described in Section 7.5. Although these load history issues prevent direct integration of the fatigue crack growth law for variable-amplitude loading, fatigue life may be calculated instead on a cycle-by-cycle basis. As described in Sections 7.9 and 7.10, this general approach is readily extended to consider the growth of surface cracks as well as multiple cracks that exist in a single component.

Other complications arise when considering fatigue of small cracks or structures that operate in corrosive environments. These matters are discussed in Sections 7.7 and 7.8. Finally, fatigue crack closure was shown to be a key mechanism that explains several aspects of fatigue crack growth.

Thus, although the simplistic view of fatigue crack growth described in Chapter 3 is clouded by several practical concerns, fracture mechanics techniques have been developed to allow reasonable life estimates for many realistic fatigue crack problems.

Sophisticated software packages are available to perform these calculations with a minimum of difficulty, providing the engineer with the basic tools needed for detailed damage tolerant analyses and designs.

PROBLEMS

7.1 A newspaper account of an aircraft accident reported that investigators discovered metal fatigue and fractures on the mountings that connect a missing propeller blade to the aircraft. A metallurgist also found evidence of a preexisting crack and a fatigue zone. Briefly discuss the appearance of a fatigue failure and discuss what would be evidence of a fatigue zone.

7.2 A *wide* center-cracked panel is made from a material with $R = 0.333$ fatigue crack growth properties given by a Paris law where $C = 6.75 \times 10^{-10}$ and $m = 3.89$ (corresponding units of da/dN are inches per cycle and ΔK are in ksi-in$^{1/2}$). The panel is assumed to contain an initial center crack whose length $2a = 0.5$ in. The plate is subjected to 50,000 cycles of remote tensile stress that varies between 5 and 15 ksi. The applied load is then reduced, and the plate is cycled to fracture at a remote stress that varies between 4 and 12 ksi. Determine the remaining fatigue life of the cracked panel after the load is reduced following the first 50,000 cycles. (Ignore any load history effects that could possibly result from the fatigue crack retardation phenomenon.) For this calculation, use a fracture toughness value K_c that corresponds to the maximum stress intensity factor per cycle which causes a crack growth rate $da/dN = 0.001$ in/cycle.

7.3 The ASTM standard test procedure [9] for measuring fatigue crack growth rates requires a *precracking* procedure to introduce a fatigue crack at a notch machined into the specimen. The section of the standard test method that describes the precracking procedure contains the following paragraph: "*The final K_{max} during the precracking shall not exceed the initial K_{max} for which test data are to be obtained. If necessary, loads corresponding to higher K_{max} values may be used to initiate cracking at the machined notch. In this event, the load range shall be stepped down to meet the above requirement. Furthermore, it is suggested that the reduction in maximum load for any of these steps be no greater than 20% and that measurable crack extension occur before proceeding to the next step. To avert transient effects in the test data, apply the load range in each step over a crack length increment of at least $3\pi(K'_{max}/\sigma_{YS})^2$, where K'_{max} is the terminal value of K_{max} from the previous loadstep.*"

(a) Briefly explain what you believe to be the purpose for this paragraph in the standard test method.

(b) What are the "transient effects" referred to and what do you think is the basis for the term $3\pi(K'_{max}/\sigma_{YS})^2$?

7.4 The occurrence of *crack tunneling* is often observed in fatigue crack growth tests and is characterized by a curved crack front that grows faster in the specimen interior than at the free surface. Give a brief explanation for the tunneling phenomenon and why it might be more pronounced for variable-amplitude than for constant-amplitude loading.

7.5 A *wide* edge-cracked strip (original edge-cracked length $a = 1.0$ in) is made from a material with 200 ksi yield stress and fatigue crack growth properties given by a Paris law with $C = 10^{-10}$ and $m = 4$ (units of da/dN are ksi-in$^{1/2}$ and ΔK are ksi-in$^{1/2}$). The specimen is subjected to *K-controlled* loading where 10,000 cycles at $K_{min} = 0$ and $K_{max} = 40$ ksi-in$^{1/2}$ are applied. Next, a single peak load to $K_{max} = 80$ ksi-in$^{1/2}$ is applied, followed by more cycles at the original $R = 0$, $\Delta K = 40$ ksi-in$^{1/2}$ loading.

(a) How would one obtain the constant ΔK loading for this experiment?

(b) Determine the crack length in the specimen immediately before the overload is applied.

(c) The 80-ksi-in$^{1/2}$ overload causes a "crack retardation" period that lasts for a certain number of cycles, after which the crack resumes its normal growth behavior. Explain what is meant by crack retardation.

(d) Sketch the expected shape of the crack length–cycles curve for this experiment.

(e) Estimate the edge-cracked length immediately after the retardation period has ceased.

7.6 A 3-in-wide tension member contains a single edge crack of length a located perpendicular to the direction of the remotely applied tensile stress. It is desired to conduct a fatigue test with this specimen so that the crack grows at a *constant rate da/dN* throughout its *entire* life. The initial edge-cracked length $a = 0.6$ in and the final crack length after 1,000,000 cycles is to be $a = 1.6$ in. The desired constant crack growth rate is to be achieved by periodically adjusting the remotely applied stress during the test. If the specimen is made from the material considered in problem 7.5, specify the magnitude of the cyclic stress range ($R = 0$) that must be applied as a function of elapsed cycles. Report the required value of the cyclic stress at the start of the test and at cyclic lives N of 100,000, 250,000, 500,000, 750,000, and 1,000,000 cycles.

7.7 A wide center-cracked panel is made from the material considered in problem 7.5. The panel has a thickness $B = 0.1$ in and is loaded with a concentrated force P that varies between zero and 10 kips $= 10,000$ lb, as shown in Figure P7.7. Note that this is the wedge-loaded configuration considered in Figure 3.2c.

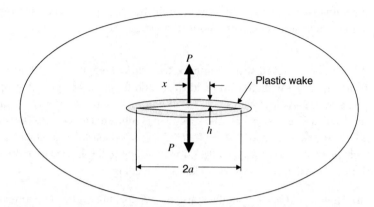

Figure P7.7 Wedge-loaded crack.

(a) How many cycles will it take for the crack to grow from a size $2a = 0.5$ in to a length $2a = 5$ in?

(b) Note that a plastic "wake" develops as the crack grows by cyclic loading, as shown schematically in Figure P7.7 (the wake is shown with a greatly expanded scale). Briefly define what is meant by the term *plastic wake* and discuss its significance for fatigue crack growth.

(c) Determine an approximate expression for the height h of the plastic wake as a function of the distance x from the centerline of the specimen. Assume that the wake is measured on the *surface* of the specimen.

7.8 A *wide* center-cracked specimen is made from a material whose fatigue crack growth properties are given by a Paris law with $C = 10^{-9}$ and $m = 4$ (here the units of da/dN are inches per cycle and ΔK is measured is ksi-in$^{1/2}$.) The specimen has an initial center-cracked length $2a = 2$ in, and is subjected to an $R = 0$ cyclic stress whose maximum value per cycle is *decreased as a function of crack length* during the test. In this case the maximum stress per cycle is given by the expression $\sigma = 3.4a^{-1}$ ksi, where a is the current value of crack length. When the crack length $2a = 8$ in, the crack stops growing (i.e., $da/dN = 0$). At this point in time, the stress s is increased until the plate fractures. The value of σ when fracture occurs is $\sigma = 20$ ksi.

(a) What is the fracture toughness K_c of the material?

(b) What is the threshold stress intensity factor ΔK_{th} of the material?

(c) How many cycles did it take for the crack to grow from its initial size $2a = 2$ in to its final size $2a = 8$ in?

7.9 Briefly explain the phenomenon of fatigue crack closure and discuss its significance with respect to fatigue crack growth.

7.10 A material has a threshold stress intensity factor $\Delta K_{th} = 1$ ksi-in$^{1/2}$, a fracture toughness $K_c = 100$ ksi-in$^{1/2}$, and fatigue crack growth properties that are represented by a "segmented" Paris law. In this case, when $1 < \Delta K < 10$ ksi-in$^{1/2}$, the Paris constants $C = 10^{-8}$ and $m = 2$. When $10 < \Delta K < 100$ ksi-in$^{1/2}$, the Paris constants are $C = 10^{-10}$ and $m = 4$ (the units of da/dN are inches per cycle in both cases). A *wide* center-cracked panel is made from this material and subjected to a constant-amplitude cyclic stress $\Delta\sigma$. The initial crack length $2a = 0.2$ in.

(a) Determine the fatigue life of the specimen if the $R = 0$ constant-amplitude cyclic stress $\Delta\sigma = 1.5$ ksi.

(b) Repeat part a if $\Delta\sigma = 3$ ksi.

(c) Repeat part a if $\Delta\sigma = 20$ ksi.

7.11 The following statement was found in a recent journal article: "Several different values of fracture toughness were assumed for the fatigue crack growth analysis, but it was found that the value of the material fracture toughness had little effect on the fatigue crack growth life for the specimens examined." Briefly discuss the context in which this statement could be true and when it would not be true.

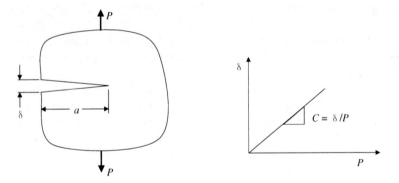

Figure P7.12 Specimen compliance.

7.12 Joe Graduate Student has been learning to conduct fatigue crack growth tests with a computer-controlled test machine that allows him to "automatically" determine crack length from compliance measurements that are recorded and analyzed by the computer-controlled test system (see Figure P7.12). This technique employs LEFM theory to relate the compliance of the specimen with crack length, where compliance is the slope obtained from real-time applied load versus crack opening displacement measurements (recall Sections 6.3 and 6.4). Assuming the cracked specimen behaves as a linear spring, there is a one-to-one relationship between compliance and crack length that can be programmed into the computer and used to determine crack length. Unfortunately Joe's compliance crack lengths do not agree with visual measurements that he is also taking. (The visual observations give consistently longer crack lengths than the compliance measurements.) Joe knows that his colleague, Suzy Ph.D. Student, has successfully used this apparatus in the past and has obtained excellent agreement between both visual and automated compliance crack length measurements. Joe has also verified that his equipment is in proper calibration and working condition. Briefly discuss what might be causing Joe's problems and how he might solve them.

AFGROW Problems

The following problems employ the AFGROW software described in the Appendix.

7.13 Read the Appendix and obtain a copy of the AFGROW software. Examine the help menus and familiarize yourself with this fatigue crack growth analysis software.
(a) What types of cracks can be analyzed by AFGROW?
(b) What $da/dN-\Delta K$ models are incorporated in AFGROW?
(c) What fatigue crack retardation models are incorporated in AFGROW?

7.14 Consider a wide sheet that contains a through-thickness center crack of length $2a$. Assume that the sheet is subjected to a constant-amplitude ($R = 0$) tensile stress of 20 ksi and that the sheet is made from a material whose fatigue crack growth properties are given by a Paris law with $C = 10^{-9}$ and $m = 4$ (here the units of da/dN are inches per cycle and ΔK is measured is ksi-in$^{1/2}$.)
(a) Determine the number of cycles required to grow the crack from an initial size $2a = 2$ in to $2a = 4$ in. Obtain the "exact" result by direct integration of the fatigue crack growth model.
(b) Repeat this calculation using the AFGROW software and compare your result with the "exact" solution obtained in part a. In this case, assume the "wide" panel has a width of 200 in. (*Hint:* The Walker equation option in AFGROW can be used to simulate the Paris fatigue crack growth model.)
(c) Repeat part b assuming that the "wide" sheet now has a finite width of 10 in. How does the answer change from the wide-panel solution?
(d) Repeat part c for a 6-in-wide panel.

7.15 Consider a surface-cracked plate loaded in remote tension, as shown in Figure P7.15. Assume that the plate is 1 in thick by 10 in wide, is made from 7075-T6 aluminum (L-T direction), and is subjected to a constant-amplitude ($R = 0$) stress of 15 ksi. Use the NASGRO database option in AFGROW to select the 7075-T6 material properties.

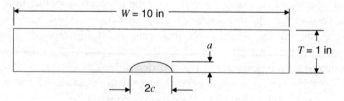

Figure P7.15   Surface-cracked plate.

(a) Determine the fatigue life if the initial surface crack is semicircular in shape with crack dimensions $a = c = 0.4$ in (i.e., the total surface crack length $2c = 0.8$ in). When does the crack depth penetrate the plate thickness? (Use the AFGROW default criterion of 95% thickness penetration.) What is the corresponding surface crack dimension $2c$ at this time?

(b) Repeat part a if the initial surface-cracked length $2c = 0.4$ in and the initial crack depth $a = 0.4$ as before.

(c) Repeat part a if the initial surface-cracked length $2c = 1.6$ in and the initial crack depth $a = 0.4$ as before.

7.16 Repeat problem 7.15 for a stress ratio $R = \frac{1}{3}$.

7.17 The purpose of this problem is to develop a "maintenance plan" to maximize the cyclic life of an edge-cracked strip loaded in tension, as shown in Figure P7.17. The basic idea is to "repair" the strip by machining off a surface layer of depth d at various times to "shorten" the crack and, thus, extend life. Although removing a layer of material reduces the strip width W and increases the stress, shortening the crack length by an amount d can increase the life in some cases. Repeating this repair process results in the series of crack growth curves shown. Since one cannot be positive that the total crack is removed (due to inspection limitations), assume a small residual crack remains after the repair. For this problem assume that the strip is made from 2024-T3 clad aluminum (L-T direction) and use the NASGRO data base option in AFGROW to obtain the material properties. Assume an initial crack length $a = 0.1$ in, $W = 1.5$ in, and thickness $B = 0.5$ in. Let the applied force P vary between 4000 and 15,000 lb.

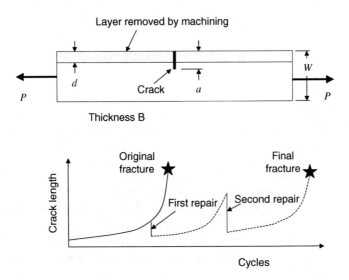

Figure P7.17   Edge-cracked strip repair.

(a) Determine the original fatigue life (to fracture) for the cracked beam.

(b) Determine the *maximum fatigue* life that can be achieved if a layer of depth *d* is removed *two times* during the life of the component. Specify how much material is to be removed (depth *d*) and when (number of elapsed cycles) this repair should be made to prevent fracture. Conservatively assume that a crack length $a = 0.02$ in remains after each machining operation. What increase (if any) has been obtained in component life by this repair procedure? Plot the fatigue crack growth curves for parts *a* and *b*.

7.18   Consider a 4-in-wide by 0.25-in-thick panel with a 0.25-in-diameter hole as shown in Figure P7.18. The panel is made from titanium alloy Ti-6-4 alpha–beta forging and is subjected to a remotely applied constant-amplitude stress that varies between 3 and 30 ksi. Assume that the hole contains an initial 0.05-in-radius quarter-circular hole as shown. (Use the Harter T-method model option in AFGROW to obtain the fatigue crack growth properties for the material.)

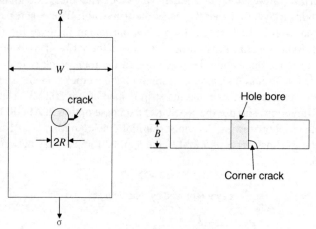

Figure P7.18   Corner-cracked hole.

(a) What is the original fatigue life of the panel if no repairs are made?

(b) Repeat the maintenance procedure described in P7.17. In this case the repair is accomplished by redrilling the hole to a larger diameter in order to remove the crack. Assume that a residual 0.02-in-radius corner crack remains after each drilling operation and that you are allowed to make two repairs (i.e., oversize drilling) to remove the crack during the life of the component. When should these repairs be made to obtain the longest possible fatigue life and what should the new hole diameter be in each case? Plot the original crack growth curves along with the "repaired" curves. How much additional life, if any, were you able to obtain?

(c) What practical restrictions might be associated with this hole oversizing method of repair?

REFERENCES

1. R. G. Forman, V. Shivakuman, S. R. Mettu, and J. C. Newman, *Fatigue Crack Growth Computer Program "NASGRO" Version 3.00,* Reference Manual, NASA Johnson Space Center, Houston, Texas, JSC-22267B, November 1998.

2. P. Paris and F. Erdogan, "A Critical Analysis of Crack Propagation Laws," *Journal of Basic Engineering,* Vol. 85, pp. 528–534, December 1963.

3. R. G. Forman, V. E. Kearney, and R. M. Engle, "Numerical Analysis of Crack Propagation in Cyclic-Loaded Structures," *Journal of Basic Engineering,* Vol. 89, pp. 459–464, September 1967.

4. K. Walker, "The Effect of Stress Ratio during Crack Propagation and Fatigue for 2024-T3 and 7075-T6 Aluminum," in *Effects of Environment and Complex Load History on Fatigue Life,* ASTM STP 462, American Society for Testing and Materials, West Conshohocken, Pennsylvania, 1970.

5. R. G. Forman and T. Hu, "Application of Fracture Mechanics on the Space Shuttle," in J. B. Chang and J. L. Rudd (Eds.), *Damage Tolerance of Metallic Structures: Analysis Methods and Applications,* ASTM STP 482, American Society for Testing and Materials, West Conhohocken, Pennsylvania, 1984, pp. 108–133.

6. A. F. Liu, *Structural Life Assessment Methods,* ASM International, Materials Park, Ohio, 1998.

7. R. J. Bucci, "Selecting Aluminum Alloys to Resist Failure by Fracture Mechanisms," *Engineering Fracture Mechanics*, Vol. 12, No. 3, pp. 407–441, 1979.

8. J. Schijve, "Fatigue Crack Growth Under Variable-Amplitude Loading," in *ASM Handbook,* Vol. 19: *Fatigue and Fracture*, ASM International, Materials Park, Ohio, 1996, pp. 110–133.

9. "Standard Test Method for Measurement of Fatigue Crack Growth Rates," ASTM Standard E 647, in *Annual Book of ASTM Standards*, Vol. 03.01, American Society for Testing and Materials, West Conshohocken, Pennsylvania, 2002.

10. J. W. Lincoln, circa 2000, personal correspondence.

11. D. H. Banasiak, A. F. Grandt, Jr., and L. T. Montulli, "Fatigue Crack Retardation in Polycarbonate," *Journal of Applied Polymer Science*, Vol. 21, pp. 1297–1309, 1977.

12. O. E. Wheeler, "Spectrum Loading and Crack Growth," *Journal of Basic Engineering*, Vol. 94, pp. 181–186, March 1972.

13. J. Willenborg, R. M. Engle, and H. A. Wood, "A Crack Growth Retardation Model Using an Effective Stress Concept," Technical Report AFFDL-TR-71-1, Air Force Flight Dynamics Laboratory, Wright-Patterson AFB, Ohio, January 1971.

14. C. R. Saff, "Crack Growth Retardation and Acceleration Models," in J. B. Chang and J. L. Rudd (Eds.), *Damage Tolerance of Metallic Structures: Analysis Methods and Applications,* ASTM STP 842, American Society for Testing and Materials, West Conshohocken, Pennsylvania, 1984, pp. 36–49.

15. W. Elber, "Fatigue Crack Closure under Cyclic Tension," *Engineering Fracture Mechanics*, Vol. I, No. 4, pp. 705–718, 1970.

16. W. Elber, "The Significance of Fatigue Crack Closure," *Damage Tolerance in Aircraft Structures*, ASTM STP 486, American Society for Testing and Materials, West Conshohocken, Pennsylvania, 1971, pp. 230–242.

17. J. P. Hess, A. F. Grandt, Jr., and A. Dumanis, "Effect of Side-Grooves on Fatigue Crack Retardation," *Fatigue of Engineering Materials & Structures*, Vol. 6, No. 2, pp. 189–199, 1983.

18. C. R. Saff, "F-4 Service Life Tracking Program (Crack Growth Gages)," Technical Report AFFDL-TR-79–3148, Air Force Flight Dynamics Laboratory, Wright-Patterson Air Force Base, Ohio, December 1979.

19. S. Suresh and R. O. Ritchie, *International Metals Review,* Vol. 29, pp. 445–476, 1984.

20. K. J. Miller, *Materials Science Technology,* Vol. 9, pp. 453–462, 1993.

21. R. C. McClung, K. S. Chan, S. J. Hudak, and D. L. Davidson, "Behavior of Small Fatigue Cracks," in *ASM Handbook,* Vol. 19: *Fatigue and Fracture*, ASM International, Materials Park, Ohio, 1996, pp. 153–158.

22. A. F. Grandt, Jr., A. J. Hinkle, C. E. Zezula, and J. N. Elsner, "The Influence of Initial Quality on Durability of 7050-T7451 Aluminum Plate," paper presented at the 1993 USAF Structural Integrity Program Conference, San Antonio, Texas, November 30–December 2, 1993.

23. M. O. Speidel, "Fatigue Crack Growth at High Temperatures," in P. R. Sahm and M. O. Speidel (Eds.), *High Temperature Materials in Gas Turbines*, Elsevier Scientific, Amsterdam, 1974, pp. 207–251.

24. J. M. Barsom and S. T. Rolfe, *Fracture and Fatigue Control in Structures: Applications of Fracture Mechanics*, Third Edition, ASTM International, West Conshohocken, Pennsylvania, 1999.

25. H. Tada, P. C. Paris, and G. R. Irwin, *The Stress Analysis Cracks Handbook*, Paris Productions (and Del Research Corporation), St. Louis, Missouri, 1985.

26. Y. Murakami (Editor-in-Chief), *Stress Intensity Factors Handbook*, Pergamon, New York, 1987.

27. A. Kamei and T. Yokoburi, "Two Colinear Asymmetrical Elastic Cracks," Report of the Research Institute of Strength of Materials, Vol. 10, Tohoku University, Japan, December 1974.

28. B. J. Heath and A. F. Grandt, Jr., "Stress Intensity Factors for Coalescing and Single Corner Flaws along a Hole Bore in a Plate," *Engineering Fracture Mechanics* Vol. 19, No. 4, pp. 665–673, 1984.

29. A. F. Grandt, Jr., R. Perez, and D. E. Tritsch, "Cyclic Growth and Coalescence of Multiple Fatigue Cracks," *Advances in Fracture Research*, Vol. 3, pp. 1571–1578, December 1984.

30. A. F. Grandt, Jr., A. B. Thakker, and D. E. Tritsch, "An Experimental and Numerical Investigation of the Growth and Coalescence of Multiple Fatigue Cracks at Notches," *Fracture Mechanics: Seventeenth Symposium*, ASTM Special Technical Publication 905, American Society for Testing and Materials, West Conshohocken, Pennsylvania, 1986, pp. 239–252.

31. T. H. McComb, J. E. Pope, and A. F. Grandt, Jr., "Growth and Coalescence of Multiple Fatigue Cracks in Polycarbonate Test Specimens," *Engineering Fracture Mechanics*, Vol. 24, No. 4, pp. 601–608, 1986.

32. A. F. Grandt, Jr., A. J. Hinkle, T. D. Scheumann, and R. E. Todd, "Modeling the Influence of Initial Material Inhomogeneities on the Fatigue Life of Notched Components," *Fatigue & Fracture of Engineering Materials and Structures*, Vol. 16, No. 2, pp. 199–213, 1993.

33. E. J. Moukawsher, M. A. Neussl, and A. F. Grandt, Jr., "Fatigue Life of Panels with Multiple Site Damage," *Journal of Aircraft*, Vol. 33, No. 5, pp. 1003–1013, September/October 1996.

34. D. G. Sexton, "A Comparison of the Fatigue Damage Resistance and Residual Strength of 2024-T3 and 2524-T3 Panels Containing Multiple Site Damage," M.S. Thesis, School of Aeronautics and Astronautics, Purdue University, West Lafayette, Indiana, August 1997.

35. M. B. Heinimann and A. F. Grandt, Jr., "Fatigue Analysis of Stiffened Panels with Multiple Site Damage," in *Proceedings of the First Joint DoD/FAA/NASA Conference on Aging Aircraft*, Vol. II, Ogden, Utah, July 8–10, 1997, pp. 1263–1305.

36. H. L. Wang and A. F. Grandt, Jr., "Fatigue Analysis of Multiple Site Damage in Lap Joint Specimens," ASTM STP 1300, in *Fatigue and Fracture Mechanics, Vol. 30* American Society for Testing and Materials, West Conshohocken, Pennsylvania, 2000, pp. 214–226.

37. L. Lawson, E. Y. Chen, and M. Meshii, "Near-Threshold Fatigue: A Review," *International Journal of Fatigue*, Vol. 21, pp. S15–S34, 1999.

38. M. H. El Haddad, K. N. Smith, and T. H. Topper, "Fatigue Crack Propagation of Short Cracks," *ASME Journal of Engineering Materials and Technology*, Vol. 101, pp. 42–46, 1979.

39. P. J. Golden, B. Bartha, A. F. Grandt, Jr., and T. Nicholas, "Fatigue Crack Growth Threshold of Fretting Induced Cracks in Ti-6Al-4V," paper presented at the Sixth National Turbine Engine HCF Conference, Jacksonville, Florida, March 5–8, 2001.

40. S. Banerjee, "A Review of Crack Closure," Technical Report AFWAL-TR-84-4031, AFWAL Materials Laboratory, Wright-Patterson AFB, Ohio, April 1984.

41. S. Suresh and R. O. Ritchie, "Near-Threshold Fatigue Crack Propagation: A Perspective of the Role of Crack Closure," in D. L. Davidson and S. Suresh (Eds.), *Fatigue Crack Growth Threshold Concepts*, TMS-AIME, Warrendale, Pennsylvania, 1984, pp. 227–261.

42. S. Ray, A. F. Grandt, Jr., and S. P. Andrew, "Three-Dimensional Measurements of Fatigue Crack Opening and Closure," ASTM STP 924, in R. P. Wei and R. P. Gangloff (Eds.), *Basic Questions in Fatigue*, Vol. 2, American Society for Testing and Materials, West Conshohocken, Pennsylvania, 1988, pp. 275–393.

43. D. S. Dawicke, A. F. Grandt, Jr., and J. C. Newman, Jr., "Three-Dimensional Crack Closure Behavior," *Engineering Fracture Mechanics*, Vol. 36, No. 1, pp. 111–121, 1990.

44. F. J. Pitoniak, A. F. Grandt, Jr., L. T. Montulli, and P. F. Packman, "Fatigue Crack Retardation and Closure in Polymethylmethacrylate," *Engineering Fracture Mechanics*, Vol. 6, pp. 663–670, 1974.

CHAPTER 8

STRESS INTENSITY FACTOR ANALYSIS

8.1 OVERVIEW

As discussed previously, the stress intensity factor K is the LEFM quantity that characterizes the crack tip stress field and is the key parameter that controls fracture and subcritical crack growth rates. The objective of this chapter is to overview methods to obtain stress intensity factor calibrations for new crack configurations. Engineering estimates and experimental and analytical/numerical techniques for determining this important factor are briefly described.

Recall that the modes I, II, and III stress intensity factors are based on the elastic stresses σ_i at a point (r, θ) near the crack tip and were formally defined in Section 3.2 as follows:

$$K_\mathrm{I} = \lim_{r \to 0} \sqrt{2\pi r} \sigma_y(\theta = 0) \quad \text{(mode I)}$$

$$K_\mathrm{II} = \lim_{r \to 0} \sqrt{2\pi r} \sigma_{xy}(\theta = 0) \quad \text{(mode II)} \tag{8.1}$$

$$K_\mathrm{III} = \lim_{r \to 0} \sqrt{2\pi r} \sigma_{yz}(\theta = 0) \quad \text{(mode III)}$$

These stress intensity factors characterize the square-root singular crack tip stress field. They relate load, crack length, and specimen geometry and are often expressed in the form

$$K = \sigma \sqrt{\pi a} \beta \tag{8.2}$$

where σ = remotely applied stress

    a = crack length

    β = β(a) = dimensionless function of crack length that depends on given specimen geometry

The LEFM approach assumes that the stress intensity factor controls crack growth. As discussed previously, fracture occurs when $K = K_c$, a material constant, and the rates of fatigue crack growth and stress corrosion cracking are also controlled by the applied K level. Several handbooks catalog existing K solutions into a form convenient for use by the engineer [1–4]. The specific objective of this chapter is to describe methods for obtaining the geometry dependent factor β in Equation 8.2 for crack configurations that have not been previously solved. Once β is established, the LEFM methods described previously may then be used to evaluate the significance of a given crack.

8.2 SURFACE CRACKS

Before discussing various K analysis methods, it is useful to comment here on the three-dimensional nature of stress intensity factor solutions for cracks that do not completely penetrate the specimen thickness in a uniform manner. As shown schematically in Figure 8.1, such surface and embedded cracks are quite common in service. Fatigue cracks often initiate at component surfaces, for example, where they are exposed to mechanical and/or environmental damage. Cracks may also start from internal material defects or extend nonuniformly through the specimen thickness. Three-dimensional analyses indicate that K varies along the crack perimeter for these cases, so that the crack may grow faster at certain locations than others, leading to a change in crack shape. Thus, complete characterization of crack growth requires description of both crack shape (length and width) and size. If the variation of K is known around the crack border, however, one may readily predict both crack shape and length changes, as described previously in Section 7.10.1.

In order to introduce the nomenclature often employed for surface crack problems, consider the stress intensity factor solution for a flat elliptical crack embedded in a large body, as shown schematically in Figure 8.2. Here the major and minor axes

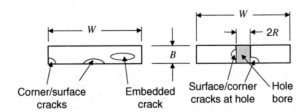

Figure 8.1. Schematic representation of crack plane showing various embedded, surface, and corner cracks in plate of thickness B (loading is perpendicular to the crack plane).

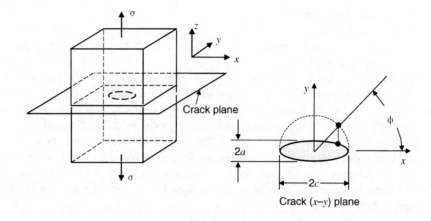

$$K = \frac{\sigma\sqrt{\pi a}}{E(K)}\left[\sin^2\phi + \left(\frac{a}{c}\right)^2\cos^2\phi\right]^{1/4}$$

Figure 8.2. Embedded elliptical crack located in a large body loaded in remote tension.

of the elliptical crack are $2c$ and $2a$, respectively, and positions along the crack perimeter are specified by the *elliptic angle* ϕ, defined by a projection from a circle circumscribed about the elliptical crack. Coordinates (x, y) of points along crack perimeter are given by $x = c\cos\phi$ and $y = a\sin\phi$. The uniform tensile stress σ is applied remotely in a direction perpendicular to the crack plane.

The stress solution for this embedded crack was obtained by Green and Sneddon [5] and used by Irwin [6, 7] to compute the following mode I stress intensity factor K_1:

$$K_1 = \frac{\sigma\sqrt{\pi a}}{E(k)}\left[\sin^2\phi + \left(\frac{a}{c}\right)^2\cos^2\phi\right]^{1/4} \tag{8.3}$$

Here

$$E(k) = \int_0^{\pi/2}\left(1 - k^2\sin^2\theta\right)^{1/2}d\theta \tag{8.4}$$

and

$$k^2 = 1 - \left(\frac{a}{c}\right)^2 \tag{8.5}$$

Here crack size and component load are specified by the $\sigma\sqrt{\pi a}$ term in Equation 8.3. Crack shape, defined by the aspect ratio a/c, is reflected by $[\sin^2\phi + (a/c)^2\cos^2\phi]^{1/4}$ and by the *elliptic integral* $E(k)$. Note that $E(k)$ depends only on crack shape a/c, and limiting values are $E(k) = \pi/2$ for $a/c = 1$ and $E(k) = 1$ for $a/c = 0$.

Although values for $E(K)$ are reported in standard mathematical tables, an empirical expression for estimating $E(k)$ to within 0.13% is given by [8]

$$E(k) = \left[1 + 1.464 \left(\frac{a}{c} \right)^{1.65} \right]^{1/2} \quad \text{for } \frac{a}{c} \le 1 \qquad (8.6)$$

The position along the crack border is defined by the elliptic angle ϕ in the $[\cdot]^{1/4}$ term in Equation 8.3. The effect of the crack tip perimeter location on the stress intensity factor is indicated below.

Circular Cracks $(a = c)$ Here $[\sin^2 \phi + (a/c)^2 \cos^2 \phi]^{1/4} = 1$, and since $E(k) = \pi/2$ when $a/c = 1$,

$$K_{\mathrm{I}} = \frac{\sigma \sqrt{\pi a}}{\pi/2} [1]^{1/4} = 2\sigma \sqrt{\frac{a}{\pi}} \qquad (8.7)$$

Note that K is constant for all ϕ in this case, so circular cracks are a "stable" shape for embedded cracks loaded in remote tension.

Noncircular Cracks $(a < c)$ At the free surface location $(\phi = 0)$, the ϕ term in Equation 8.3 becomes

$$\left[\sin^2 \phi + \left(\frac{a}{c} \right)^2 \cos^2 \phi \right]^{1/4} = \sqrt{\frac{a}{c}}$$

and

$$K_{\mathrm{I}} = \frac{\sigma \sqrt{\pi a}}{E(k)} \sqrt{\frac{a}{c}} \quad \text{for } \phi = 0 \qquad (8.8)$$

At the crack perimeter location $\phi = \pi/2$, the position term becomes

$$\{\cdot\}^{1/4} = 1$$

and

$$K_{\mathrm{I}} = \frac{\sigma \sqrt{\pi a}}{E(k)} \quad \text{for } \phi = \frac{\pi}{2} \qquad (8.9)$$

Since it is assumed that $a \le c$ in Figure 8.2 (and in Equation 8.3), K_{I} is greatest at the $\phi = \pi/2$ location, and a fatigue crack would tend to grow faster at that point until eventually stabilizing at the circular shape $a = c$.

Irwin [6, 7] modified the embedded-flaw solution to estimate stress intensity factors for semielliptical *surface* cracks in plates with a finite thickness t. Following the format of the embedded-flaw solution, Irwin expressed K_{I} for the surface crack in the form

$$K_1 = \sigma \sqrt{\pi \left(\frac{a}{Q}\right)} M_f \tag{8.10}$$

The dimensionless "Q" term is a modification of the earlier elliptic integral $E(k)$ and represents an attempt to provide a "plasticity correction." Here

$$Q = [E(k)]^2 - 0.212 \left(\frac{\sigma}{\sigma_{YS}}\right)^2 \tag{8.11}$$

where σ_{YS} is the material yield stress. The factor Q depends on crack shape [reflected by the a/c dependence of $E(k)$] and load level σ/σ_{YS}. Note that when $\sigma \ll \sigma_{YS}$, $Q^{1/2} = E(k)$.

The dimensionless magnification factor M_f in Equation 8.10 represents a modification to the embedded-flaw solution to account for the finite plate thickness. This term is sometimes separated further into front- and back-face "correction" terms. Note that, in general, M_f will depend on crack shape a/c, crack size expressed by a/t or c/W (see Figure 8.1), position along the crack perimeter (angle ϕ), and the type of loading (tension, bending, etc.). A stress intensity solution reported by Newman and Raju [8] for a surface cracked plate loaded in remote tension is given in Figure 8.3 (results for bending loads are also reported in Reference 8). Other typical stress intensity factor solutions for surface and corner crack problems are given in References 1–4.

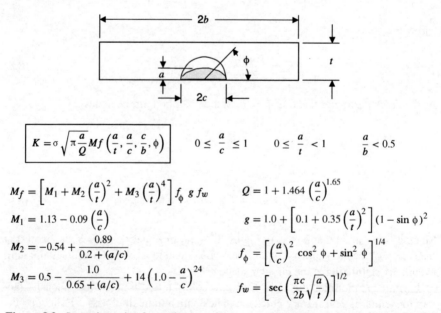

$$K = \sigma \sqrt{\pi \frac{a}{Q}} Mf\left(\frac{a}{t}, \frac{a}{c}, \frac{c}{b}, \phi\right) \qquad 0 \le \frac{a}{c} \le 1 \qquad 0 \le \frac{a}{t} < 1 \qquad \frac{a}{b} < 0.5$$

$$M_f = \left[M_1 + M_2 \left(\frac{a}{t}\right)^2 + M_3 \left(\frac{a}{t}\right)^4\right] f_\phi \, g \, f_w \qquad\qquad Q = 1 + 1.464 \left(\frac{a}{c}\right)^{1.65}$$

$$M_1 = 1.13 - 0.09 \left(\frac{a}{c}\right) \qquad\qquad g = 1.0 + \left[0.1 + 0.35 \left(\frac{a}{t}\right)^2\right](1 - \sin \phi)^2$$

$$M_2 = -0.54 + \frac{0.89}{0.2 + (a/c)} \qquad\qquad f_\phi = \left[\left(\frac{a}{c}\right)^2 \cos^2 \phi + \sin^2 \phi\right]^{1/4}$$

$$M_3 = 0.5 - \frac{1.0}{0.65 + (a/c)} + 14 \left(1.0 - \frac{a}{c}\right)^{24} \qquad\qquad f_w = \left[\sec \left(\frac{\pi c}{2b} \sqrt{\frac{a}{t}}\right)\right]^{1/2}$$

Figure 8.3. Stress intensity factor for a surface crack in a finite plate subjected to remote tensile stress σ applied perpendicular to the crack plane shown [8].

Although the nomenclature described here for the surface crack problem is common, it is by no means standard. Thus, one should be cautious when interpreting another author's results and pay particular attention to how crack shape and position are defined, the format for presentation of K, and so on. It should also be emphasized that determination of K for surface cracks is a very difficult three-dimensional stress analysis problem, and one should be cautious when using results from the literature that have not been subjected to experimental and/or analytical validation.

8.3 ENGINEERING ESTIMATES OF K

Now return to the general problem of determining stress intensity factors for new crack configurations. As a first approximation, it is often possible to estimate or bound K with existing solutions for similar problems. Since K is a linear elastic parameter, one may add individual solutions, for example, to account for combined loading effects. Figure 8.4, for example, shows schematic summation of the solutions for an edge-cracked strip loaded in tension and bending (i.e., see Figures 3.2*b, g*). It may also be possible to "correct" other solutions obtained for infinite bodies by use of finite-width correction factors.

Some examples of engineering estimates for K solutions are given below.

Example 8.1. K Bounds for Cracked Hole The goal here is to estimate K for radially cracked holes in tension plates shown in Figure 8.5. The "exact" solution for a *wide* plate was obtained by Bowie [9], and the dimensionless factor β is given in Table 8.1 (as reported in Reference 10). First, consider an estimate for *short* crack

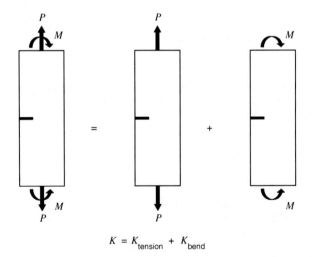

$$K = K_{tension} + K_{bend}$$

Figure 8.4. Example of superposition of stress intensity factor solutions for combined tension and bending.

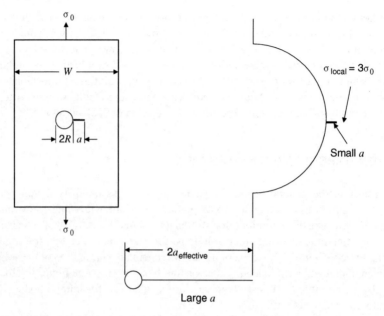

Figure 8.5. Cracked hole in a finite plate showing small- and large-crack estimates.

lengths. In this case, K_I may be approximated by the solution for an edge crack in a semi-infinite plate:

$$K_I = 1.12\sigma\sqrt{\pi a}$$

Now, next to the bore of the hole, the local stress σ is three times the remote stress σ_0 due to the stress concentration (i.e., $K_t = 3$). Thus,

TABLE 8.1 Comparison of Exact and Estimated K Solutions for Through-Cracked Hole in Plate Loaded in Remote Tension

a/R	$\beta(a/R)_{\text{exact}}$	$\beta(a/R)_{\text{short}}$	$\beta(a/R)_{\text{long}}$
0.0	3.39	3.36	—
0.1	2.73	—	—
0.2	2.30	—	—
0.4	1.86	—	1.73
0.8	1.47	—	1.32
1.5	1.18	—	1.03
3.0	0.94	—	0.91
5.0	0.85	—	0.84
—	0.707	—	0.707

$$K_I \approx 1.12(3\sigma_0)\sqrt{\pi a}$$

and

$$\beta\left(\frac{a}{R}\right)_{\text{short}} = 3.36 = \frac{K_I}{\sigma_0\sqrt{\pi a}}$$

Note that this result would only be expected to be valid for very small crack lengths, however, since it does not take into consideration the actual stress gradient at the hole.

Now, estimate K_I for *long* crack lengths (i.e., $a \gg R$) by considering the hole as part of the crack. This assumption gives the following "effective" crack length:

$$2a_{\text{effective}} = a + 2R$$

Using the well-known center-cracked solution yields the following results for this case:

$$K_I = \sigma_0\sqrt{\pi a_{\text{effective}}} = \sigma_0\sqrt{\tfrac{1}{2}\pi(a + 2R)} \qquad \beta\left(\frac{a}{R}\right)_{\text{long}} = \sqrt{0.5 + \frac{R}{a}}$$

Both the short- and long-crack estimates for $\beta(a/R)$ are compared with Bowie's exact solution in Table 8.1. Note that the short and long estimates effectively bound the exact solution and provide a reasonable estimate for K_I.

Example 8.2. Cracked Hole in Finite-Width Plate   Now estimate the stress intensity factor for the radially cracked hole in the *finite-width* plate shown in Figure 8.5. This estimate may be obtained by combining the Bowie solution for a *wide* plate with a "width" correction factor as follows:

$$K = \sigma\sqrt{\pi a}\beta\left(\frac{a}{R}\right)\beta\left(\frac{a}{W}\right)$$

Here $\beta(a/R)$ is the Bowie solution for infinite plates as before. When $a/R \leq 10$, the Bowie solution $\beta(a/R)$ may be represented by the expression [11]

$$\beta\left(\frac{a}{R}\right) = \frac{0.8733}{0.3245 + a/R} + 0.6762$$

$$\beta\left(\frac{a}{W}\right) = \text{width correction for a center-cracked panel [12]}$$

$$= \left[\sec\left(\frac{\pi a^*}{W}\right)\right]^{1/2}$$

where the effective crack length $a^* = a/2 + R$.

8.4   LINEAR SUPERPOSITION

Linear superposition provides another relatively simple means for using *crack face pressure* solutions to compute K from "uncracked" stress analyses. The procedure is demonstrated in Figure 8.6, where cracked member A (the problem of interest) is resolved into components B and C as shown. Here *cracked* member A and *uncracked* member B are both subjected to the desired external loads. Now, the desired crack plane in member B is made stress free by applying the negative of the normal stress that occurs for the unflawed configuration $[-p(x)$ in Figure 8.5]. Since this line in member B is now stress free, a crack may be introduced with no change. Member C is the cracked component loaded with the positive pressure $+p(x)$ applied along the crack faces but no external loading.

Note that member A = member B + member C, and, also by superposition, $K^A = K^B + K^C$. Now $K^B = 0$, since the crack plane is stress free and there are no singular crack tip stresses. Thus, the stress intensity factor $K^A = K^C$, where K^A is the stress intensity factor for the problem of interest, and K^C is the K for crack face pressure loading $p(x)$ chosen to be equal to the stresses in the uncracked member. The pressure $p(x)$ is determined by a conventional stress analysis of the uncracked member, but since the crack does not have to be analyzed directly, computational effort is greatly reduced. Now, solutions for the crack face pressure problem (member C) give the desired K for member A. Thus, the crack face pressure solution is a "general" solution (i.e., a Green's function for the flaw geometry) that can be used to determine K for many different loading cases.

Crack face pressure solutions are available for several crack geometries. The solution for a center crack in an infinite plate (see Figure 8.7) is given below [10]:

$$K_1 = \frac{1}{\sqrt{\pi a}} \int_{-a}^{+a} p(x) \left[\frac{a+x}{a-x}\right]^{1/2} dx \tag{8.12}$$

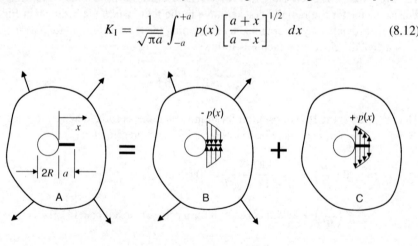

$$K^A = K^B + K^C$$

Figure 8.6. Linear superposition of a cracked member showing crack face pressure loading.

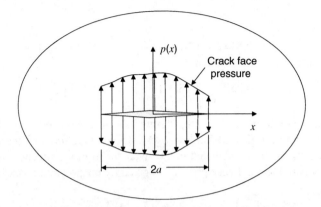

Figure 8.7. Center crack in a large plate subjected to crack face pressure loading.

Equation 8.12 may also be used to obtain the mode II stress intensity factor K_{II} by letting $p(x)$ be defined as the shear stress along the crack plane. (Note that x is measured from the crack center in Equation 8.12.)

Emery et al. [13] and Bueckner [14] have obtained stress intensity factors for edge-cracked strips loaded with pressure $p(t)$ as shown in Figure 8.8. Bueckner's solution is given as

$$K_1 = \sqrt{\frac{2}{\pi}} \int_0^a p(t)M(t)\, dt \qquad (8.13)$$

where a = crack length ($a \le 0.5W$)
  W = width of strip
  t = distance measured from crack tip ($t = a - x$)
$M(t) = t^{-1/2}\left(1 + m_1(t/a) + m_2(t/a)^2\right)$

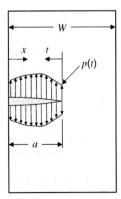

Figure 8.8. Edge-cracked strip subjected to crack face pressure loading considered by Bueckner.

and

$$m_1 = 0.6147 + 17.1844 \left(\frac{a}{W}\right)^2 + 8.7822 \left(\frac{a}{W}\right)^6$$

$$m_2 = 0.2502 + 3.2889 \left(\frac{a}{W}\right)^2 + 70.0444 \left(\frac{a}{W}\right)^6$$

Crack face pressure solutions for through-the-thickness, corner- and surface-cracked holes are described in References 11, 13–19. One further superposition allows these solutions to be tabulated in a convenient form for general use. Consider Figure 8.9, where the crack face pressure is given by the polynomial expansion

$$p\left(\frac{x}{R}\right) = A_0 + A_1\left(\frac{x}{R}\right) + A_2\left(\frac{x}{R}\right)^2 + \cdots \tag{8.14}$$

Here the polynomial coefficients A_i have units of stress and x/R is a dimensionless position term (x is measured from the edge of the hole in Figure 8.9). Now let K_n be the stress intensity factor for each individual term $A_n(x/R)^n$ in the pressure expansion. Then, by superposition,

$$K = K_0 + K_1 + K_2 + \cdots = \sum K_n \tag{8.15}$$

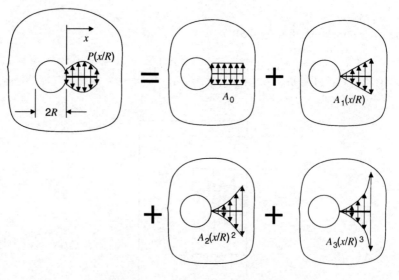

$$P(x/R) = A_0 + A_1(x/R) + A_2(x/R)^2 + A_3(x/R)^3$$

Figure 8.9. Crack face pressure loading of radial cracked hole broken into polynomial components.

Defining

$$M_n = \frac{K_n}{A_n\sqrt{\pi a}} \tag{8.16}$$

gives

$$K = \sum K_n = \sqrt{\pi a} \sum \frac{K_n}{\sqrt{\pi a}} = \sqrt{\pi a} \sum A_n M_n \tag{8.17}$$

References 15–19 present tabulated M_n for several two- and three-dimensional crack geometries. (Table 8.2 gives typical results for an open hole of radius R that contains a single radial through-thickness crack of length a.)

Example 8.3. Radially Cracked Hole   Use the crack face pressure solutions given in Table 8.2 to compute K for the radially cracked hole problem considered previously in Figure 8.5. Assume the through-crack length $a = 0.5R$ for this calculation and the plate width W is very large relative to the hole diameter. First it is necessary to determine the hoop stress distribution (σ_H versus x/R) for the uncracked member subjected to the given load. A closed form analysis for the stress distribution at the edge of a hole loaded with remote tensile stress σ is given in Reference 20. A least squares fit of that analysis gives the following expression for σ_H to within 3% of the exact solution for $0 \le x/R \le 1.0$:

$$\sigma_H = \sigma\left[2.9190 - 5.2853\left(\frac{x}{R}\right) + 6.2574\left(\frac{x}{R}\right)^2 - 2.6910\left(\frac{x}{R}\right)^2\right] \tag{8.18}$$

Now Table 8.2 and the least squares fit for σ_H give the following A_n and M_n (for $a/R = 0.5$):

$$A_0 = 2.9190\sigma \qquad M_0 = 0.9119$$
$$A_1 = -5.2853\sigma \qquad M_1 = 0.3020$$
$$A_2 = 6.2574\sigma \qquad M_2 = 0.1207$$
$$A_3 = -2.6910\sigma \qquad M_3 = 0.05167$$

The desired stress intensity factor is now given by

$$K_I = \sum_{n=0}^{3} K_n = \sqrt{\pi a}\,\frac{A_n K_n}{A_n\sqrt{\pi a}} = \sqrt{\pi a}\sum A_n M_n$$

$$= \sigma\sqrt{\pi a}\big[(2.9190)(0.9119) - (5.2853)(0.3020) + (6.2574)(0.1207)$$
$$- (2.6910)(0.05167)\big]$$

$$= 1.6819\sigma\sqrt{\pi a} = K_I$$

TABLE 8.2 Dimensionless Stress Intensity Factors for Single-Cracked Hole Loaded with Crack Face Pressure $p(x) = A_n(x/R)^n$

$$M \equiv K_n/A_n\sqrt{\pi a}$$

a/R	n = 0	n = 1	n = 2	n = 3	n = 4	n = 5	n = 6
0.10	0.9994	0.6348×10^{-1}	0.4994×10^{-2}	0.4244×10^{-3}	0.3756×10^{-4}	0.3405×10^{-5}	0.3137×10^{-6}
0.20	0.9817	0.1255	0.1980×10^{-1}	0.3369×10^{-2}	0.5968×10^{-3}	0.1083×10^{-3}	0.1996×10^{-4}
0.30	0.9592	0.1859	0.4416×10^{-1}	0.1130×10^{-1}	0.3004×10^{-2}	0.8181×10^{-3}	0.2263×10^{-3}
0.40	0.9351	0.2447	0.7785×10^{-1}	0.2661×10^{-1}	0.9449×10^{-2}	0.3434×10^{-2}	0.1267×10^{-2}
0.50	0.9119	0.3020	0.1207	0.5167×10^{-1}	0.2297×10^{-1}	0.1044×10^{-1}	0.4820×10^{-2}
0.60	0.8914	0.3583	0.1725	0.8882×10^{-1}	0.4744×10^{-1}	0.2590×10^{-1}	0.1436×10^{-1}
0.80	0.8610	0.4693	0.3030	0.2089	0.1490	0.1086	0.8036×10^{-1}
1.00	0.8442	0.5805	0.4702	0.4060	0.3625	0.3306	0.3059
1.25	0.8361	0.7214	0.7318	0.7907	0.8833	0.1008×10	0.1166×10
1.50	0.8337	0.8639	0.1052×10	0.1365×10	0.1831×10	0.2506×10	0.3480×10
1.75	0.8308	0.1006×10	0.1431×10	0.2167×10	0.3391×10	0.5418×10	0.8778×10
2.00	0.8248	0.1147×10	0.1866×10	0.3232×10	0.5783×10	0.1056×10^{2}	0.1956×10^{2}
2.50	0.8037	0.1418×10	0.2900×10	0.6292×10	0.1409×10^{2}	0.3220×10^{2}	0.7458×10^{2}

The hole radius is R and the radial through-crack length is a. From Reference 18.

The "exact" solution found by Bowie for $a/R = 0.5$ is $K = 1.73\sigma\sqrt{\pi a}$, which agrees within 3% of the superposition calculation.

Example 8.4. Coldworked Hole   Now compute K_I for a cracked fastener hole that is coldworked by an oversized mandrel to introduce compressive residual stresses that improve the fatigue life (see Section 17.6.1). Considering a particular example reported in Reference 21, assume the remote tensile stress is 40 ksi (276 MPa), the hole diameter $2R = 0.261$ in (6.63 mm), and the through crack is 0.1044 in (2.65 mm) long, giving $a/R = 0.8$. The residual hoop stresses left by the coldworking process are described in Reference 21 and can be fit for $0 \le x/R \le 1.0$ by

$$\sigma_H = \sum_{n=0}^{3} B_n \left(\frac{x}{R}\right)^n$$

$$= -130.53 + 296.25 \left(\frac{x}{R}\right) - 226.67 \left(\frac{x}{R}\right)^2 + 72.936 \left(\frac{x}{R}\right)^3$$

Here the coefficients B_n have units of ksi.

The final hoop stress distribution with the 40-ksi remote load can be estimated by adding the residual stresses to the remote stress distribution discussed in Example 8.3:

$$\sigma_{total} = \sigma_{remote} + \sigma_{coldwork}$$

$$= \sum_{n=0}^{3} (A_n + B_n) \left(\frac{x}{R}\right)^n = \sum_{n=0}^{3} C_n \left(\frac{x}{R}\right)^n$$

$$= -13.770 + 84.838 \left(\frac{x}{R}\right) + 23.626 \left(\frac{x}{R}\right)^2 - 34.704 \left(\frac{x}{R}\right)^3$$

Now, for a crack length $a = 0.104$ in ($a/R = 0.8$), the magnification factors M_n from Table 8.2 and the polynomial stress coefficients C_n are

$$M_0 = 0.8610 \qquad C_0 = -13.770 \text{ ksi}$$
$$M_1 = 0.4693 \qquad C_1 = 84.838 \text{ ksi}$$
$$M_2 = 0.3030 \qquad C_2 = 23.626 \text{ ksi}$$
$$M_3 = 0.2089 \qquad C_3 = -34.704 \text{ ksi}$$

Finally, by superposition, the desired stress intensity factor is given by

$$K_I = \sum K_n = \sqrt{\pi a} \sum M_n C_n$$

$$= \sqrt{(\pi)(0.1044)} \big[(0.8610)(-13.770) + (0.4693)(84.838) + (0.3030)(23.626)$$

$$+ (0.2089)(-34.704) \big]$$

$$= 15.96 \text{ ksi-in}^{1/2} = K_I$$

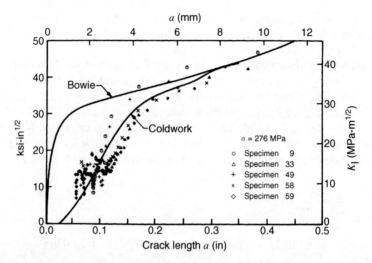

Figure 8.10. Example 8.4 showing stress intensity factors obtained for a coldworked fastener hole loaded with a remote tensile stress of 40 ksi (276 MPa).

This value for K compares well with experimental measurements on coldworked holes reported in Reference 21 and reproduced here in Figure 8.10. The calculation described here is designated by the large "star" on the figure. The line labeled "coldwork" was obtained by a similar superposition [21] to that described here, except that a slightly different stress solution was employed. Other superposition calculations for surface- and corner-cracked holes are reported in References 16–18, and details of the experimental measurements shown in Figure 8.10 are given below in Section 8.5.4.

8.5   EXPERIMENTAL METHODS

This section describes procedures for relating measurable quantities to the stress intensity factor K.

8.5.1   Compliance Calibration

As discussed in Reference 22, compliance calibrations are one of the oldest experimental crack analysis methods. Recall from Section 6.4 that the strain energy release rate G for a cracked member is related to the rate of change in compliance with crack length and, in turn, to the stress intensity factor. From Equation 6.6,

$$G = \frac{P^2}{2B}\frac{\partial C}{\partial a} \tag{8.19}$$

where G = strain energy release rate

P = applied load (units of force)

a = crack length

B = specimen thickness (assume constant)

C = compliance, inverse slope of load point–deflection curve

The compliance of a cracked member may be measured for different crack lengths $(a_1, a_2, \ldots, a_n)$, and the rate of change in compliance with crack length $\partial C/\partial a$ can be computed. Now calculate G for a given crack length (using measured $\partial C/\partial a$ at that crack length) by Equation 8.19. Also recall from Section 6.4 that G is related to the stress intensity factor K by

$$G = \frac{K_I^2}{E} \qquad \text{for plane stress}$$

$$ \tag{8.20}$$

$$G = \frac{1 - v^2}{E} K_I^2 \quad \text{for plane strain}$$

where E = elastic modulus

v = Poisson's ratio

Note that in order to obtain G (and K) from compliance, two derivatives are computed from the measured load P–displacement e data—the compliance $C = de/dP$ and $G = (P^2/2B)(\partial C/\partial a)$. Since derivatives may magnify experimental error, care must be taken in obtaining the original load–displacement information.

8.5.2 Crack Tip Opening Measurement

Recalling the elasticity solutions relating stresses and displacements near the crack tip to K discussed in Section 3.2, the mode I displacements near the crack tip are given by Equations 8.21 for plane stress conditions and by Equations 8.22 for plane strain:

$$v = \frac{K_1}{G}\sqrt{\frac{r}{2\pi}} \sin\frac{\theta}{2}\left(\frac{2}{1+v} - \cos^2\frac{\theta}{2}\right)$$

$$u = \frac{K_1}{G}\sqrt{\frac{r}{2\pi}} \cos\frac{\theta}{2}\left(\frac{1-v}{1+v} + \sin^2\frac{\theta}{2}\right) \tag{8.21}$$

$$u = \frac{K_1}{G}\sqrt{\frac{r}{2\pi}} \cos\frac{\theta}{2}\left(1 - 2v + \sin^2\frac{\theta}{2}\right)$$

$$v = \frac{K_1}{G}\sqrt{\frac{r}{2\pi}} \sin\frac{\theta}{2}\left(2 - 2v + \cos^2\frac{\theta}{2}\right) \tag{8.22}$$

where K_I = mode I stress intensity factor
   G = shear modulus
   v = Poisson's ratio
   (r, θ) = polar coordinates of point near crack tip $(r < a/10)$
   $u = x$ displacement of point (r, θ)
   $v = y$ displacement of point (r, θ)

   Any experimental method (e.g., interferometry, holography, Moiré, strain gages) capable of resolving crack tip displacements could be used to compute K_I by these relations. References 23 and 24, for example, describe use of a laser interferometry method to measure crack tip opening in metal specimens, and References 25 and 26 describe interferometric K-calibrations with transparent glass specimens. The key point here is the requirement to measure displacements near the crack tip where Equations 8.21 and 8.22 are valid.

8.5.3 Photoelastic Techniques

Photoelastic measurements [27, 28] of shear stresses may be used to compute K from the crack tip stress solutions given in Section 3.2 (recall Equations 3.6). Photoelastic data reduction often employs "higher order" terms in the crack tip stress expansion in order to improve experimental accuracy [27, 28]. Recall, for example, that close to the crack tip stresses are very large (infinite for elastic cases) and difficult to measure accurately. Further away from the crack tip, where stresses are easier to measure, the K-versus-stress relations given by Equation 3.6 are no longer accurate $(r > a/10)$ and higher order terms are required in the solution relating K with stress.

8.5.4 Fatigue Crack Growth Rate Calibrations

The LEFM fatigue crack growth law may be used to compute stress intensity factors from measured crack growth rates by the James and Anderson back-calculation method [21, 29–32]. First the relation between the cyclic range in stress intensity factor and fatigue crack growth rate is established for the test material by conventional baseline testing (e.g., center-cracked panels, compact tension specimens). Measured fatigue crack growth rates for the new specimen geometry then give ΔK from the fatigue crack growth law, as shown schematically in Figure 8.11 [i.e., find ΔK from the $da/dN = F(\Delta K)$ law for the test material]. Sample results from this approach are shown in Figure 8.10, where stress intensity factor measurements for coldworked fastener holes are compared with superposition predictions [21]. Here the individual data points were obtained by conducting constant-amplitude $(R = 0)$ fatigue crack growth tests with several coldworked hole test specimens and then converting the measured crack growth rates for a given crack length to ΔK (i.e., $\Delta K = K_{max}$ for the $R = 0$ tests).

   Although the fatigue crack growth method is relatively easy to apply to complex problems, care must be exercised since the fatigue crack growth behavior of most materials exhibits considerable scatter. Moreover, computing da/dN requires differentiation of experimental data, contributing to additional experimental error. Since

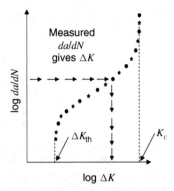

Figure 8.11. Schematic representation of fatigue crack growth rate method used to back calculate ΔK.

many data points can be obtained from a few experiments, however, statistical interpretation of data may often be accomplished.

8.5.5 Fracture Toughness Calibrations

Since the stress intensity factor reaches a critical value at fracture ($K = K_c =$ constant), fracture tests may also be used to estimate K-calibrations. First the fracture toughness K_c for the test material is measured in standard specimens with known stress intensity factor solutions (compact specimen, three-point bend, etc.). Next, cracked specimens of the desired geometry are loaded to fracture, recording critical crack size a_f and the fracture load σ_f. Now, since at fracture $K = K_c$, the desired dimensionless K calibration is given by

$$\beta(a) = \frac{K_c}{\sigma_f \sqrt{\pi a}} \tag{8.23}$$

Corner-cracked holes in plexiglass specimens were studied by this procedure in Reference 33. Note the following points of caution, however, regarding the fracture toughness method. First, fracture toughness values (K_c) often show considerable scatter for a given material. Next, since measurements are made at fracture loads where crack tip plasticity may be large, accuracy may be limited. Finally, since only a single measurement is obtained from each specimen, the testing costs can be extensive.

8.6 ANALYTICAL AND NUMERICAL METHODS

This section deals with analytical and numerical approaches for computing stress intensity factors. In principle, any technique capable of resolving the *singular* crack tip stress field with the detail needed to perform the limiting process in Equation 8.1

can be used to obtain K. While a detailed discussion of the various analytical methods available from the theory of elasticity is beyond the present scope, brief mention is made of several procedures that have had particular success with LEFM applications.

Williams [34, 35], for example, used the theory of elasticity to find an Airy stress function based on a series expansion in polar coordinates that provided stresses at the corner of a wedge subjected to arbitrary far-field loads. When the wedge angle was extended to 2π, closing on itself, the solution for the special case of a sharp crack in a wide plate was obtained. Westergaard [36] also found an Airy stress function to solve the inverse elasticity problem but employed complex variables to treat cracks in the complex domain. Although originally limited to infinite bodies, the Westergaard approach was subsequently generalized to analyze finite members as well [37, 38].

Other traditional crack analysis methods include boundary collocation, boundary integrals, and alternating methods. Boundary collocation finds stress functions (e.g., Williams or Westergarrd solutions) that satisfy boundary conditions at discrete points along external geometric boundaries. Since only a finite number of node points are considered, computational effort is significantly reduced. Numerical convergence is also improved by overconstraining the boundary conditions at the collocation points and then using statistical methods to solve for the unknown coefficients. The boundary integral method also relates unknown tractions and displacements on the surface of a member through a series of integral equations that are then discretized by algebraic equations. Since the boundary elements have one less dimension than the body being studied, the technique can be particularly efficient. The alternating method superimposes two related solutions to solve a third configuration. The first summation will ordinarily leave undesired stresses at free boundaries, but repeated superpositions can reduce the residual errors to an acceptable level. Although the two original solutions are usually exact, the alternating approach can also be conducted with numerical (e.g., finite element) solutions.

Additional details of the various analysis methods highlighted above are provided in References 36–39. The following subsections discuss the powerful weight function and finite element methods for obtaining stress intensity factors.

8.6.1   Weight Function Method

Bueckner [14, 40] and Rice [41] describe weight function methods that permit calculation of stress intensity factors for any loading of a given body provided K and crack opening displacement information are known for one loading (case 1). Here, the known stress intensity factors and crack opening data for the case 1 loading are used to define a universal weight function that allows calculation of K for any other (case 2) loading of the particular crack geometry.

Following the format of Rice [41], the desired case 2 stress intensity factor solution K_2 is given by

$$K_2 = \int_\Gamma \mathbf{T} \cdot \mathbf{h} \, d\Gamma + \int_A \mathbf{f} \cdot \mathbf{h} \, dA \qquad (8.24)$$

where Γ = boundary chosen around crack tip

A = region defined by Γ

$\mathbf{T}$ = stress vector acting on boundary

$\mathbf{f}$ = body forces acting in region A

$\mathbf{h}$ = universal weight function for given cracked body

The weight function is given by

$$h = h(x, y, a) = \frac{H}{2K_1} \frac{\partial \mathbf{u}_1}{\partial a} \tag{8.25}$$

where (x, y) = position coordinates

a = crack length

K_1 = stress intensity factor solution for case 1 loading

$\mathbf{u}_1$ = corresponding displacement field for case 1 loading

and the material constant is given by H

$$H = \begin{cases} E & \text{for plane stress} \\ \dfrac{E}{1 - v^2} & \text{for plane strain} \end{cases} \tag{8.26}$$

Note that the weight function $\mathbf{h}$ is completely defined by the case 1 loading parameters and material constants.

Example 8.5. Weight Function for Center-Cracked Sheet [41]   Consider the center-cracked sheet subjected to the two loadings shown in Figure 8.12. Here the goal is to use the K and crack opening displacement date for the remote stress loading (case 1) to compute K for the crack face pressure loads $p(z)$. As shown in Figure 8.13, the known stress intensity factor and crack opening displacement solutions for the remote stress (case 1 problem) are given as

$$K_1 = \sigma \sqrt{\pi \frac{c}{2}}$$

$$u_{y1} = \pm \frac{2\sigma}{H} z^{1/2} (c - z)^{1/2} \quad \text{along crack faces}$$

Now, assuming that boundary Γ consists of the crack surfaces, the weight function can be computed from the case 1 data:

$$h_y = \frac{H}{2K_1} \frac{\partial u_{y1}}{\partial c}$$

From the given case 1 solution,

$$\frac{\partial u_{y1}}{\partial c} = \frac{\sigma}{H} \frac{z^{1/2}}{(c - z)^{1/2}}$$

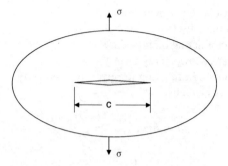

Case 1 loading: known K and displacements

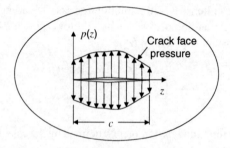

Case 2 loading: desired K

Figure 8.12. Schematic representation of weight function that allows K and displacement information for case 1 loading to be used to obtain K for case 2 loading.

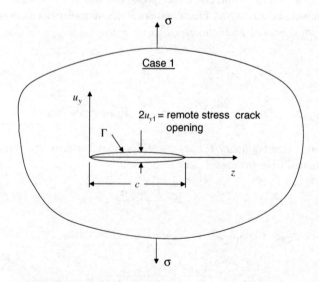

Figure 8.13. Weight function for center-cracked member.

which gives the following y component of the weight function:

$$h_y = \frac{H}{2\sigma\sqrt{\pi c/2}} \frac{\sigma}{H} \frac{z^{1/2}}{(c-z)^{1/2}} = \frac{z^{1/2}}{\sqrt{2\pi c}(c-z)^{1/2}}$$

Now, the desired case 2 stress intensity factor for the crack face pressure loading $p(z)$ is obtained from Equation 8.24:

$$K_2 = \int_\Gamma \mathbf{T} \cdot \mathbf{h} \, d\Gamma + \int_A \mathbf{f} \cdot \mathbf{h} \, dA$$

The remaining terms in Equation 8.24 for the case 2 crack face pressure loading are given as

$$f_Z = f_y = 0 \quad \text{(no body forces)}$$

$$T_y = p(z) \qquad T_Z = 0 \quad \text{along crack faces}$$

and

$$K_2 = \begin{cases} 2\int_0^c p(z)h_y \, dz \\ \dfrac{2}{\sqrt{2\pi c}} \displaystyle\int_0^c p(z)\left(\dfrac{z}{c-z}\right)^{1/2} dz \end{cases}$$

Note that this is the same solution given earlier in Equation 8.12 for a center crack loaded with a crack face pressure (substitute $c = 2a$ and $z = a + x$ in the earlier expression).

Although most authors have used analytical/numerical methods (e.g., finite elements) to obtain the original case 1 information, the weight function may also be constructed from experimental measurements of K and displacement. An experimental weight function, for example, was used in Reference 24 to convert stress intensity factor and crack opening measurements for edge-cracked strips loaded in four-point bending to a crack face pressure K solution. Both the initial case 1 K and crack tip displacement data were obtained experimentally in this case by a laser interferometric method, whereas a compliance gage gave the crack mouth opening.

In summary, the weight function technique is a relatively simple method for generalizing stress intensity factor and crack displacement data for one loading to a K solution for other loadings applied to that body. Additional details and examples are provided in References 40–45. Three-dimensional weight functions for a surface-cracked plate are, for example, given in Reference 45.

8.6.2 Finite Element Methods

The finite element method is perhaps the most common technique currently used to determine stress intensity factors for new crack configurations. This popularity is due, in part, to the wide availability of well-documented and supported software suitable for operation on many types of computers. In particular, there are several general-purpose finite element software packages developed specifically for stress intensity factor calculations.

In the finite element method, the structural member is subdivided into many smaller "elements" connected at "node" points. The displacements within each element are interpolated by certain polynomial "shape functions," and displacement compatibility is enforced at the node points. The given displacement functions allow computations for strains and, with an appropriate constitutive relation, stresses. The result is a large system of matrix equations with well-behaved properties that can be readily solved subject to various applied loads and boundary conditions. When modeling cracks, there must be a sufficient number of elements to resolve the stress singularity known to occur at the crack tip. This task is simplified, however, by using finite elements that have the square-root crack singularity built into them. These can be special crack tip elements or, by placing the midpoint nodes at special locations (i.e., the quarter points), conventional isoparametric elements may be used [46, 47].

Solving the equations associated with the finite element model gives displacements, strains, and stresses at discrete points throughout the member. This information may then be used to obtain the stress intensity factor by one of several procedures. Although, in principle, Equation 8.1 could be used to compute K from the crack tip stress distribution, this procedure is not very accurate due to the difficulty in resolving the singular stresses at the crack tip. Stress calculations can be employed further away from the crack tip by adding additional "higher order" terms to reflect the fact that stresses are being resolved away from the tip. Since the crack tip displacements are not singular, however, a better method would be to compute K from the crack tip displacement equations 3.7, 3.9, or 3.11. Other procedures for calculating K from finite element results involve computing the strain energy release rate G (and then K) from the change in compliance of the finite element model associated with crack extension (recall Equation 8.19), computing the J-integral (Equation 6.26), or nodal force methods. (See References 37 and 38 for more details on this topic.)

8.7   CONCLUDING REMARKS

Although stress intensity factors are tabulated in various handbooks [1–4] for many crack geometries, there is frequent need to obtain K for new configurations that may arise in a particular component design or failure investigation. This chapter has summarized methods to obtain stress intensity factor solutions for such occasions. As indicated, there are many practical tools to estimate, experimentally measure, or numerically compute K.

It should be noted, however, that it is usually not possible to obtain a closed form solution for K. Instead, the stress intensity factor is usually resolved for distinct crack lengths by one or more of the experimental or numerical methods discussed here. These discrete results are then fit with an appropriate mathematical expression or used with a computerized table lookup scheme to allow interpolation between crack lengths and for incorporation in life prediction software.

PROBLEMS

8.1 Use the Internet to locate the home page of a vendor that supplies fracture mechanics software for computing stress intensity factors and record this company's address for future reference. Examine the products and applications described on the web site and prepare a short report that describes the various analysis methods and capabilities of the software sold by this company.

8.2 A long strip has a 2-in-wide by 0.5-in-thick cross section. The strip contains a single edge crack and is subjected to a remote tensile force of 10,000 lb combined with a bending moment $M = 2000$ in-lb. (Here M is oriented so as to cause mode I loading of the edge crack.) If the strip is made from a material that has a fracture toughness $K_c = 60,000$ psi-in$^{1/2}$, determine the crack size that fractures the strip.

8.3 Use the Bueckner crack face pressure solution given in Equation 8.13 to compute the stress intensity factor coefficient $\beta = K/\sigma\sqrt{\pi a}$ for an edge-cracked strip loaded in remote tension. Consider dimensionless edge-cracked lengths $a/W = 0.1, 0.2, 0.4$ and compare your results with the edge-cracked solution given in Figure 3.2b.

8.4 Use the Bueckner crack face pressure solution given in Equation 8.13 to compute the stress intensity factor coefficient $\beta = K/\sigma\sqrt{\pi a}$ for an edge-cracked strip loaded in pure bending. Consider dimensionless edge-cracked lengths $a/W = 0.1, 0.2, 0.4$ and compare your results with the edge-cracked solution given in Figure 3.2g (pure bending).

8.5 Repeat example 8.3 (radially cracked hole) for dimensionless crack lengths $a/R = 0.2, 1.0, 2.0, 2.5$. Compare your results with the solution given in Figure 3.2d.

8.6 Repeat example 8.4 (coldworked hole) for crack lengths $a/R = 0.1, 0.5, 2.0$. Compare your results with Figure 8.10.

8.7 Repeat example 8.4 (coldworked hole) for crack lengths $a/R = 0.1, 0.5, 2.0$, except this time let the remotely applied stress be 25 ksi. Compare your results with those reported in Reference 21.

8.8 Repeat example 8.4 (coldworked hole) for crack lengths $a/R = 0.1, 0.5, 2.0$, except this time let the remotely applied stress be 16 ksi. What is the significance of your results for small crack lengths? Compare your results with those reported in Reference 21.

8.9 A long strip fails from an edge crack when subjected to a remote constant-amplitude cyclic tensile stress ($R = 0$). The cross section is 0.2 in thick by 4.0 in wide, and the member is made from an alloy with a fracture toughness of

60 ksi-in$^{1/2}$ and whose fatigue crack growth properties are given by $da/dN = 5 \times 10^{-10} \Delta K^{3.6}$ (units of da/dN are inches per cycle and ΔK are ksi-in$^{1/2}$). Microscopic examination of the fracture surface indicate that the striation spacing was 0.00005 in when the edge-cracked length is 1 in.

(a) What was the value of the constant-amplitude $R = 0$ cyclic stress applied to the component?

(b) How long was the edge crack when the component fractured?

8.10 Discuss how the fatigue crack closure phenomenon would influence the crack tip opening measurement technique described in Section 8.5.2.

REFERENCES

1. H. Tada, P. C. Paris, and G. R. Irwin, *The Stress Analysis Cracks Handbook*, 2nd ed., Paris Productions Incorporated and Del Research Corporation, St. Louis, Missouri, 1985.

2. G. C. Sih, *Handbook of Stress-Intensity Factors*, Institute of Fracture and Solid Mechanics, Lehigh University, Bethlemem, Pennsylvania, 1973.

3. D. P. Rooke and D. J. Cartwright, *Compendium of Stress Intensity Factors*, Her Majesty's Stationary Office, London, 1976.

4. Y. Murakami (Editor-in-Chief), *Stress Intensity Factors Handbook*, Pergamon Press, New York, 1987.

5. A. E. Green and I. N. Sneddon, "The Distribution of Stress in the Neighborhood of a Flat Elliptical Crack in an Elastic Solid," *Proceedings of the Cambridge Philosophical Society,* Vol. 46, pp. 159–164, 1950.

6. G. R. Irwin, "Crack Extension Force for a Part-Through Crack in a Plate," *Journal of Applied Mechanics*, December 1962, pp. 651–654.

7. P. N. Randall, *Plane Strain Crack Toughness Testing of High Strength Metallic Materials,* ASTM STP 410, American Society for Testing and Materials, West Conshohocken, Pennsylvania, 1966, pp. 88–125.

8. J. C. Newman and I. S. Raju, "An Empirical Stress Intensity Factor Equation for the Surface Crack," *Engineering Fracture Mechanics*, Vol. 15, No. 1/2, pp. 185–192, 1981.

9. O. L. Bowie, "Analysis of an Infinite Plate Containing Radial Cracks Originating at the Boundary of an Internal Circular Hole," *Journal of Mathematics and Physics*, Vol. 35, pp. 60–71, 1956.

10. P. C. Paris and G. C. Sih, "Stress Analysis of Cracks," in *Fracture Toughness Testing and Its Applications*, ASTM Special Technical Publication 381, American Society for Testing and Materials, West Conshohocken, Pennsylvania, 1965, pp. 30–81.

11. A. F. Grandt, Jr., "Stress Intensity Factors for Some Thru-Cracked Fastener Holes," *International Journal of Fracture*, Vol. 11, No. 2, pp. 283–294, April 1975.

12. C. F. Fedderson, *Plane Strain Crack Toughness Testing of High Strength Metallic Materials*, ASTM STP 410, American Society for Testing and Materials, West Conshohocken, Pennsylvania, 1966, pp. 77–79.

13. A. F. Emery, G. E. Walker, and J. A. Williams, "A Green's Function for the Stress Intensity Factors of Edge Cracks and Its Application to Thermal Stresses," *Journal of Basic Engineering*, Vol. 91, No. 4, pp. 618–624, December 1969.

14. H. F. Bueckner, "Weight Functions for Notched Bar," *Zeitschrift fur Angewandte Mathematik and Mechanik*, Vol. 51, pp. 97–109, 1971.

15. A. F. Grandt, Jr., "Stress Intensity Factors for Cracked Holes and Rings Loaded with Polynomial Crack Face Pressure Distributions," *International Journal of Fracture*, Vol. 14, No. 4, pp. R221–R229, August 1978.

16. A. F. Grandt, Jr. and T. E. Kullgren, " Stress Intensity Factors for Corner Cracked Holes Under General Loading Conditions," *Journal of Engineering Materials and Technology*, Vol. 103, No. 2, pp. 171–176, April 1981.

17. A. F. Grandt, Jr., "Crack Face Pressure Loading of Semielliptical Cracks Located Along the Bore of a Hole," *Engineering Fracture Mechanics*, Vol. 14, No. 4, pp. 843–852, 1981.

18. A. F. Grandt, Jr. and T. E. Kullgren, "Tabulated Stress Intensity Factor Solutions for Flawed Fastener Holes," *Engineering Fracture Mechanics*, Vol. 18, No. 2, pp. 435–451, 1983.

19. R. Perez, A. F. Grandt, Jr., and C. R. Saff, "Tabulated Stress Intensity Factors for Corner Cracks at Holes Under Stress Gradients," in *Surface-Crack Growth: Models, Experiments, and Structures,* ASTM STP 1060, ASTM, American Society for Testing and Materials, West Conshohocken, Pennsylvania, 1990, pp. 49–62.

20. S. Timoshenko and J. N. Goodier, *Theory of Elasticity*, 2nd ed., McGraw-Hill, New York, 1951, p. 80.

21. W. H. Cathey and A. F. Grandt, Jr., "Fracture Mechanics Consideration of Residual Stresses and Introduced by Coldworking Fastener Holes," *Journal of Engineering Materials and Technology*, Vol. 102, No. 2, pp. 85–91, January 1980.

22. R. T. Bubsey, D. M. Fisher, M. H. Jones, and J. E. Strawley, "Compliance Measurements," in A. S. Kobayashi (Ed.), *Experimental Techniques in Fracture Mechanics*, SESA Monograph Series, Society for Experimental Stress Analysis, Westport, Connecticut, 1973, pp. 76–95.

23. D. E. Macha, W. N. Sharpe, Jr., and A. F. Grandt, Jr., "A Laser Interferometry Method for Experimental Stress Intensity Factory Calibration," in *Cracks and Fracture*, ASTM Special Technical Publication 601, American Society for Testing and Materials, West Conshohocken, Pennsylvania, 1976, pp. 490–505.

24. D. Bar-Tikva, A. F. Grandt, Jr., and A. N. Palazotto, "An Experimental Weight Function for Stress Intensity Factor Calibrations," *Experimental Mechanics*, Vol. 21, No. 10, pp. 371–378, October 1981.

25. E. Sommer, *Engineering Fracture Mechanics*, Vol. 1, pp. 705–718, 1970.

26. P. B. Crosley, S. Motovoy, and E. J. Ripling, *Engineering Fracture Mechanics*, Vol. 4, pp. 421–433, 1971.

27. A. S. Kobayashi, "Photoelastic Techniques," in *Experimental Techniques in Fracture Mechanics*, SESA Monograph No. 1, Society for Experimental Stress Analysis, Westport, Connecticut, pp. 126–145, 1973.

28. C. W. Smith, "Use of Three-Dimensional Photoelasticity and Progress in Related Areas," in *Experimental Techniques in Fracture Mechanics*, SESA Monograph No. 2, Society for Experimental Stress Analysis, Westport, Connecticut, 1975, pp. 3–58.

29. L. A. James and W. E. Anderson, *Journal of Engineering Fracture Mechanics*, Vol. 1, pp. 565–568, April 1969.

30. A. F. Grandt, Jr. and G. M. Sinclair, "Stress Intensity Factors for Surface Cracks in Bending," in *Stress Analysis and Growth of Cracks*, Part I, ASTM Special Technical Publication 513, American Society for Testing and Materials, West Conshohocken, Pennsylvania, 1972, pp. 37–58.

31. M. B. Heinimann, M. T. Doerfler, and A. F. Grandt, Jr., "Analysis of Cracks at Deep Notches," *Engineering Fracture Mechanics,* Vol. 55, No. 4, pp 605–616, 1996.

32. A. I. Rifani and A. F. Grandt, Jr., "A Fracture Mechanics Analysis of Fatigue Crack Growth in a Complex Cross Section," *Engineering Failure Analysis,* Vol. 3, No. 4, pp 249–265, 1996.

33. T. E. Kullgren and F. W. Smith, "Static Fracture Testing of PMMA Plates Having Flawed Fastener Holes," *Experimental Mechanics,* Vol. 20, No. 3, pp. 95–102, March 1980.

34. M. L. Williams, "Stress Singularities Resulting from Various Boundary Conditions in Angular Corners of Plates in Extension," *Journal of Applied Mechanics,* Vol. 19, pp. 526–528, 1952.

35. M. L. Williams, "On the Stress Distribution at the Base of a Stationary Crack," *Journal of Applied Mechanics,* Vol. 24, pp. 109–114, 1957.

36. H. M. Westergaard, "Bearing Pressures and Cracks," *Journal of Applied Mechanics,* Vol. 6, pp. A49–A53, 1939.

37. R. J. Sanford, *Principles of Fracture Mechanics,* Prentice-Hall, Englewood Cliffs, New Jersey, 2003.

38. T. L. Anderson, *Fracture Mechanics—Fundamentals and Applications,* 2nd ed., CRC Press, Boca Raton, Florida, 1995.

39. A. S. Kobayashi, "Numerical Analysis in Fracture Mechanics," in *Experimental Techniques in Fracture Mechanics,* SESA Monograph No. 2, Society for Experimental Stress Analysis, Westport, Connecticut, 1975, pp. 166–199.

40. H. F. Bueckner, "A Novel Principle for the Computation of Stress Intensity Factors," *Zeitschrift fur Angewandte Mathematik und Mechanik,* Vol. 50, pp. 529–546, 1970.

41. J. R. Rice, "Some Remarks on Elastic Crack Tip Stress Fields," *International Journal of Solids and Structures,* Vol. 8, No. 6, pp. 751–758, June 1972.

42. H. J. Petroski and J. D. Achenbach, "Computation of the Weight Function from a Stress Intensity Factor," *Engineering Fracture Mechanics,* Vol. 10, No. 2, pp. 257–266, 1978.

43. R. Labbens, A. Pellissier-Tanon, and J. Heliot, "Practical Method for Calculating Stress Intensity Factors through Weight Functions," in *Mechanics of Crack Growth,* ASTM STP 590, American Society for Testing and Materials, West Conshohocken, Pennsylvania, 1976, pp. 368–384.

44. P. C. Paris, R. M. McMeeking, and H. Tada, "The Weight Function Method for Determining Stress Intensity Factors," in *Cracks and Fracture,* ASTM STP 601, American Society for Testing and Materials, West Conshohocken, Pennsylvania, 1976, pp. 471–489.

45. G. Shen and G. Glinka, "Weight Functions for a Surface Semi-elliptical Crack in a Finite Thickness Plate," *Theoretical and Applied Fracture Mechanics,* Vol. 15, pp. 247–255, 1991.

46. R. D. Henshell and K. G. Shah, "Crack Tip Finite Elements Are Unnecessary," *International Journal for Numerical Methods in Engineering,* Vol. 9, pp. 495–507, 1975.

47. R. D. Barsoum, "On the Use of Isoparametric Finite Elements in Linear Fracture Mechanics," *International Journal for Numerical Methods in Engineering,* Vol. 10, pp. 25–37, 1976.

CHAPTER 9

SERVICE-INDUCED DAMAGE

9.1 OVERVIEW

This chapter deals with damage mechanisms that can cause loss of structural integrity during service. In this instance, distinction is made between initial manufacturing and/or material damage and that which can develop with normal service usage (e.g., fatigue cracking or corrosion). First, stress–life and strain–life techniques are developed to determine the period required for fatigue cracks to form in smooth or notched components. Then, the various forms of corrosion are described, along with techniques for protecting structures from this form of material degradation. Next, total life calculations that include both a crack nucleation and propagation period are briefly discussed followed by description of an "equivalent initial flaw size" approach for treating fatigue crack formation and/or corrosion in the context of LEFM. Finally, brief consideration is given to supplying protection from accidental discrete source damage that may occur in service.

9.2 FATIGUE CRACK NUCLEATION

The preceding chapters in Part II deal with predicting the subcritical growth and final fracture of *preexistent cracks* in components subjected to cyclic loading and/or environmental attack. Whereas the initial crack supposition is key to damage tolerance analyses and results, in part, from limitations of nondestructive inspection, the first engineering approaches to fatigue employed methodology to predict the fatigue life of *pristine* structure and did not consider initial damage per se. The original stress–life analysis and the subsequent strain–life approach to fatigue are discussed in the following subsections. Although they do not consider cracks directly, they are often considered as methods for calculating the fatigue crack *nucleation* period.

9.2.1 Fatigue Testing

Before overviewing the stress–life and strain–life methods, however, a few comments about fatigue testing are appropriate. Recall that the goal of a fatigue test is to characterize material response under an applied cyclic load. As discussed in the following two subsections, it is important to distinguish between *force-controlled* and *displacement-controlled* loading, for unless the specimen behaves in an elastic manner, the material response will be different for these cases. Indeed, the recognition of this difference plus the ability to conduct highly precise force-controlled and/or displacement-controlled tests has led to significant advances in the understanding of fatigue.

Modern electrohydraulic fatigue machines such as those shown in Figure 9.1 employ hydraulic pressure to actuate a cylinder that pulls and/or twists the test specimen (see Figure 9.2). The applied force is sensed through a load cell placed in series with the test specimen and may be controlled in one of three ways: force control, displacement control, or external transducer control. Under force control, the test circuitry uses the output from the load cell to apply a specified *force* sequence to the specimen. A built-in function generator or other external source [e.g., a personal computer (PC)] specifies the maximum and minimum force as well as the frequency and shape of the load–time signal. The loading may be constant amplitude or may be set to represent a complex variable-amplitude service load history. The key point here is that the *force* applied to the specimen as measured by the load cell is the *controlling parameter* and is independent of specimen deformation. This type of loading is used for the stress-controlled testing described in Section 9.2.2.

Under displacement control, however, the hydraulic cylinder is set to expand and contract (or rotate in a torsion machine) between fixed *displacement limits.* Here the position of the loading grip is sensed by a linear variable-displacement transducer (LVDT), and that signal is used to control cycling of the test specimen. The force required to maintain the desired displacement limits may be measured by the load cell and recorded but is *not* the controlling parameter. In the event that the stiffness of the specimen changes, due to plastic yielding or crack formation, the force required to maintain the specified displacement will adjust as needed. Again, the function generator (or external PC) controls the frequency and shape of the desired displacement–time profile, which can also be constant or variable amplitude in nature. The key point here is that one end of the test specimen moves between specified limits of displacement and/or rotation.

The third method of load control uses the output from an external displacement transducer mounted on the specimen as the controlling parameter. The extensometer shown in Figure 9.3a, for example, measures specimen displacement over a set gage length and essentially determines the axial *strain* over that distance. The clip gage shown in Figure 9.3b also determines the displacement between two fixed points on the test specimen and, when mounted with legs on opposite sides of a crack,

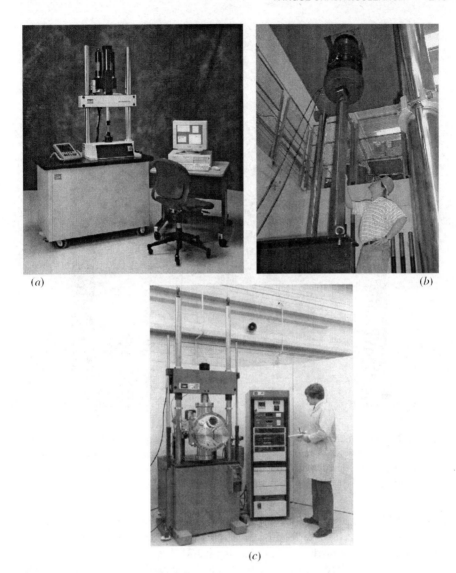

Figure 9.1. Typical electrohydraulic fatigue machines: (*a*) table-top system showing load frame, control panel, and PC; (*b*) large-capacity test machine; (*c*) test machine with vacuum chamber to control specimen environment. (Photographs courtesy of MTS Systems Corporation.)

measures crack opening displacement (COD) as a function of applied load. In this case, the output from the extensometer is used to control the test machine. Again, the force necessary to maintain the desired extensometer output (e.g., strain or COD) is measured and recorded but is *not* the controlling parameter. The key difference here from the preceding paragraph is that the *displacement between two local points* on the specimen is specified, rather than total displacement, allowing one to measure

Figure 9.2. Electrohydraulic test machine with panel removed to show hydraulic apparatus for loading fatigue specimen. The machine may be actuated in a manner that applies a specified force, displacement, or externally supplied displacement signal to the specimen grip. (Photograph courtesy of MTS Systems Corporation.)

and control localized strain (or crack opening) independent of deformations in other portions of the specimen or loading train. This type of arrangement is used for the strain-controlled testing discussed in Section 9.2.3.

Note that these three loading methods only give the same force history when the ends of the specimen displace in a linear elastic manner. The three force–time responses will differ if there is plastic deformation in the specimen, if the specimen geometry changes (e.g., due to crack growth), or if there is movement in the specimen grips. Many early fatigue machines were often displacement-controlled devices, and failure to recognize that the force applied to the specimen could change during the test led to inaccurate reporting and/or interpretation of results.

Another important matter in fatigue testing deals with how to report fatigue life. This issue can be explained by examining Figure 9.4 where the applied load is plotted versus time for both constant- and variable-amplitude fatigue testing. Here the load could be given in terms of force, displacement, or strain as discussed above.

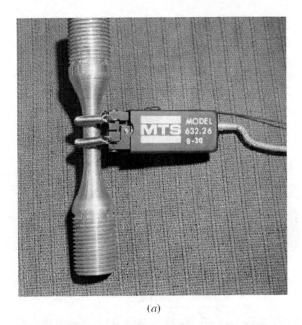

(a)

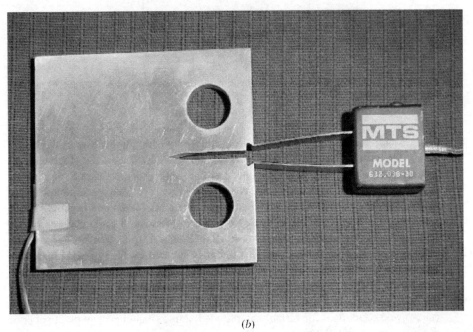

(b)

Figure 9.3. Displacement transducers mounted to test specimen used to measure and/or control specimen displacements: (a) extensometer used to measure axial displacement across a fixed gage length; (b) clip-gage used to measure crack opening displacement.

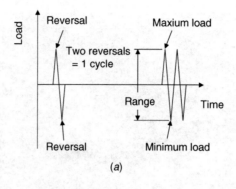

(a)

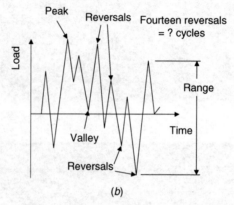

(b)

Figure 9.4. Definition of load cycle and reversals: (*a*) constant-amplitude loading; (*b*) variable-amplitude loading.

For constant-amplitude loading there is no ambiguity in defining a load cycle (Figure 9.4*a*). The initial force–time trace increases (or decreases) until the slope changes sign at a "reversal." The slope remains constant until changing sign at another reversal, and then the load returns to its original value. The load cycle characterized in this manner consists of two reversals that define a maximum and minimum peak, a unique load range, and mean load. Note that in this case a cycle consists of two load reversals. For variable-amplitude loading, however, it is common to have successive reversals that do not lead to a uniquely defined cycle, and specification of fatigue life in terms of cycles is problematic (see Figure 9.4*b*). To avoid this problem, fatigue life is often expressed in terms of load reversals, which are uniquely defined for both constant- and variable-amplitude loading. Relating cycles to load reversals (i.e., "cycle counting") for variable-amplitude loading is an important practical issue discussed in Section 9.2.6.

9.2.2 Stress–Life Approach

The stress–life method is the original approach to fatigue developed by Wohler and his colleagues during the latter portion of the nineteenth century to analyze fatigue failures of railroad components. The key to the stress–life approach is the *S–N* curve shown in Figure 9.5. Here the fatigue life is measured for "smooth" specimens that are subjected to a *constant-amplitude stress* (i.e., under force control). In this case, "life" could be the number of cycles to form a "detectable" crack in the small test specimen but is often simply the total number of cycles to specimen fracture.

The results from several individual fatigue tests shown schematically in Figure 9.5 indicate that fatigue life depends inversely on cyclic stress. This plot of applied stress amplitude $\Delta S/2$ versus specimen fatigue life N_f is known as the stress–life (or *S–N*) curve and is considered a material property that characterizes resistance to fatigue failure. Note that there is considerable variability in fatigue life at a given stress amplitude but that this "scatter" usually decreases in the short-life or *low-cycle-fatigue* (LCF) regime as the cyclic stress is increased. Although test results indicate the nominal "fatigue life" for polished specimens, this time period is actually comprised of the formation and growth of small microscopic cracks that develop at local imperfections and/or regions of plastic slip in the test material. The fact that there is less "scatter" for smooth specimens tested at the larger cyclic stresses

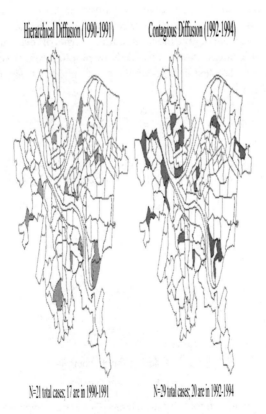

Hierarchical Diffusion (1990-1991) Contagious Diffusion (1992-1994)

N=21 total cases; 17 are in 1990-1991   N=29 total cases; 20 are in 1992-1994

Figure 9.5. (*a*) Schematic representation of *S–N* curve obtained from stress-life tests; (*b*) typical test specimen (broken).

is attributed to early cracking at those large loads. In contrast, cracks do not form as quickly at lower stress levels, and so variations in specimen quality (e.g., the presence of material flaws, surface scratches) play a significant role in the actual crack nucleation in the long-life or *high-cycle-fatigue* (HCF) regime and lead to more disparity in life.

Some materials (steels in particular) demonstrate an *endurance limit* S_e, defined as the largest stress amplitude that can be applied without having a fatigue failure. Specimens subjected to cyclic stresses below this value are "run-outs" that do not fail (i.e., have "infinite" life). Endurance limits for steels are typically on the order of half the ultimate stress. Although some materials (e.g., aluminum) do not show endurance limits and eventually develop cracks at very small cyclic stresses, it is common to define an *effective endurance limit as the stress amplitude that gives a "long" fatigue life* (e.g., 10^6 or 10^7 cycles). It must be emphasized, however, that while endurance limits are an important measure of fatigue resistance, this property is measured in small, polished specimens subjected to constant-amplitude loading and may not accurately reflect the fatigue behavior of large components that contain initial material and/or structural imperfections, residual stresses, and so on.

The original semilog format of the *S–N* curve may be put in another form that is useful for subsequent mathematical modeling. First, express the applied stress amplitude in terms of "true stress" $\Delta\sigma/2$ rather than "engineering stress" $\Delta S/2$ (recall Equation 2.22). Next, give the fatigue life in terms of reversals $2N_f$ rather than cycles N_f (one cycle equals two reversals). Now plot $\log(\Delta\sigma/2)$ versus $\log(2N_f)$ as shown in Figure 9.6. The test data above the endurance limit may often be modeled with a simple power law:

$$\frac{\Delta\sigma}{2} = \sigma'_f (2N_f)^b \tag{9.1}$$

Equation 9.1 is known as Basquin's rule. Here σ'_f is the fatigue strength coefficient (or Basquin's coefficient) and b is the fatigue strength exponent (or Basquin's

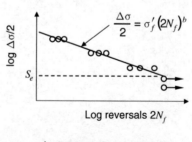

σ'_f = fatigue strength coefficient

b = fatigue strength exponent

Figure 9.6. Log-log plot of stress–life data in terms of true stress amplitude versus fatigue life measured in reversals showing definition of Basquin's rule.

exponent), which is typically in the range $-0.12 < b < -0.05$. Note the dimensions of these empirical constants are such that b is dimensionless and σ_f' has units of stress. Since the basic S–N curve (or Basquin's rule) is obtained from small, polished specimens subjected to completely reversed constant-amplitude loading, it is quite limited in scope. As described below, one must account for specimen size, stress gradients, surface condition, mean stress, notches, and variable-amplitude loading for many practical applications.

Several factors that result in differences between the fatigue behavior of small specimens and larger structural components are related to the "weakest link" concept. Simply summarized, larger volumes of material have a greater probability of containing a crack-nucleating imperfection, resulting in shorter fatigue lives for large specimens compared to smaller ones tested at the same cyclic stress. This phenomenon is particularly prevalent for HCF where the crack nucleation period is a larger portion of total fatigue life. Size effects are typically observed in specimen diameters ranging between 0.3 and 10 in (0.7–2.5 cm) [1].

A related problem that confused early fatigue testing was how to treat the different stress gradients resulting from axial and bending loads. An axial "push–pull" specimen, for example, has a uniform stress over the specimen cross section, whereas a bending specimen, such as the "rotating-beam" tests used for much of the early fatigue research, sees a tensile stress on one side of the specimen, a compressive stress on the opposite side, and zero stress at the neutral axis. Thus, a bend specimen with the same diameter as an axially loaded specimen and subjected to the same peak cyclic stress actually has a much smaller volume of highly stressed material and by the weakest link concept often has a longer fatigue life (particularly in the endurance limit regime). In order to avoid these confusing stress gradient problems, modern fatigue research usually emphasizes axial rather that bending loads. Of course, specimen alignment is a critical issue in this case to prevent unintentional bending stresses.

In the same vein, surface condition has a critical role on fatigue life, and any deviations from the smooth, polished surface employed for the original S–N specimens will, in general, have the effect of serving as local crack-nucleating sources that shorten fatigue life. Thus, differences in surface finish associated with various forms of grinding or machining, plating and coatings, damage due to "nicks and dings," or corrosion can all have a tremendous effect on fatigue life. It should be noted that there are many empirically based fatigue "correction" factors to account for the size effect, stress gradient, and surface finish issues described above [1–3].

The effect of mean stress is an important consideration for applications that do not involve the completely reversed cyclic loading usually employed for S–N testing. Here mean stresses may be associated with the applied load, as for example, the high-frequency vibratory stresses that are superimposed on the large centrifugal tensile stress seen by a turbine engine fan blade. Or, residual mean stresses may result from various manufacturing processes (e.g., shot peening, grinding, welding) or from prior inelastic loading.

This issue is shown schematically in Figure 9.7, where a series of S–N curves are presented for various values of mean stress. Here the $S_{mean} = 0$ result is the original S–N curve obtained for the completely reversed loading discussed previously.

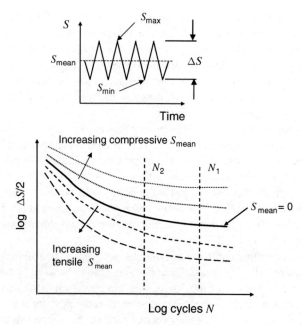

Figure 9.7. Family of S–N curves obtained for various mean stresses showing beneficial effect of compressive and detrimental effect of tensile mean stresses.

If additional constant-amplitude testing is performed with other mean loads, one finds that tensile mean stresses decrease life, whereas compressive means increase life for a given stress amplitude. Moreover, mean stress has a larger influence on HCF than on LCF, and the extent of the effect increases with the magnitude of the mean stress.

From a design standpoint, it is useful to demonstrate the influence of mean stress by constructing constant-life (Haigh) diagrams that relate combinations of stress amplitude and mean stress that give the same fatigue life. These results are obtained from the family of S–N curves by cross-plotting the data from constant lives (i.e., N_1 or N_2), as shown in Figure 9.8a (only the effect of detrimental tensile mean stresses are shown here).

Since fatigue testing is time consuming and expensive, considerable effort has focused on models to determine the influence of mean stress from the completely reversed loading results, with particular emphasis on the endurance limit behavior. Figure 9.8b, for example, shows several models to predict how mean stress alters the endurance limit. The rationale behind these models is based on the expected behavior for limiting cases of zero-mean-stress and zero-stress amplitude. Note that when the mean stress is zero, the allowable stress amplitude for infinite life is given by the completely reversed endurance limit S_e. As a tensile mean stress is applied, however, the allowable cyclic stress amplitude $\Delta S/2$ will decrease. When the applied stress amplitude approaches zero, one no longer has a fatigue test and might expect the curve to intercept at a static property such as the tensile yield stress S_{ys}, the ultimate

"Haigh" constant-life diagram

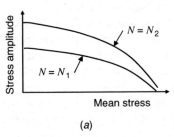

(a)

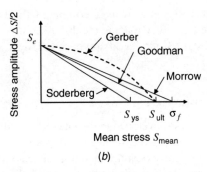

(b)

Figure 9.8. Constant-life diagrams: (a) combinations of stress amplitude and mean stress that give same fatigue lives; (b) stress amplitude versus mean stress for infinite-life (endurance limit) conditions.

stress S_{ult}, or the true fracture stress σ_f. These three intercepts are shown in Figure 9.8b, and the various lines shown there are given by the following expressions:

Soderberg Model

$$\frac{\Delta S/2}{S_e} + \frac{S_{mean}}{S_{ys}} = 1 \qquad (9.2)$$

Goodman Model

$$\frac{\Delta S/2}{S_e} + \frac{S_{mean}}{S_{ult}} = 1 \qquad (9.3)$$

Gerber Model

$$\frac{\Delta S/2}{S_e} + \left(\frac{S_{mean}}{S_{ult}}\right)^2 = 1 \qquad (9.4)$$

Morrow Model

$$\frac{\Delta S/2}{S_e} + \frac{S_{mean}}{\sigma_f} = 1 \qquad (9.5)$$

These plots are collectively known as Goodman diagrams and purport to predict the reduction in endurance limit associated with tensile mean stresses. Note that the only material properties required for these models are the completely reversed endurance limit and the monotonic stress–strain properties. It has been reported that the Soderberg equation is usually conservative, whereas the Goodman and Gerber models typically bound test data [1–3]. Although these mean stress models are for endurance limit behavior (i.e., infinite life), they can, in principle, be modified for finite-life regimes by replacing the endurance limit S_e in Equations 9.2–9.5 with the appropriate completely reversed cyclic stress amplitude that corresponds to the fatigue life of interest.

Other practical issues associated with determining fatigue lives for notched components or for variable-amplitude loading are discussed in Sections 9.2.4 and 9.2.5. First, however, consider the strain-based approach to fatigue summarized in the following subsection.

9.2.3 Strain–Life Approach

The strain–life approach was developed in the late 1950s and 1960s to handle shortcomings of the stress–life method associated with plastic deformation encountered during LCF. This method involves testing smooth specimens under constant-amplitude "strain control" as shown in Figure 9.9 and measuring the number of *reversals* to failure $2N_f$. As discussed in Section 9.2.1, the completely reversed strain limits are maintained by means of an extensometer mounted to the smooth test specimen that measures and *controls* the strain history by adjusting the applied stress peaks as needed. Although the strain amplitude $\Delta\varepsilon/2$ is kept constant, the corresponding stress amplitude $\Delta\sigma/2$ may change as the material undergoes inelastic deformation.

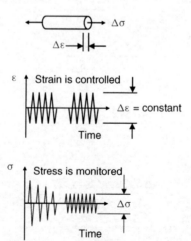

Figure 9.9. Strain-controlled fatigue test of smooth specimen showing change in applied stress needed to maintain fixed strain limits (cyclic softening).

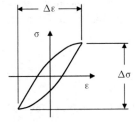

Figure 9.10. Hysteresis loop showing stable stress–strain response that develops in a strain-controlled fatigue test.

Indeed, one of the objectives of the strain-controlled experiment is to determine how the stress–strain response of the material is modified by cyclic loading.

In the example shown in Figure 9.9, the stress required to maintain the constant-strain peaks decreases initially as this material "cyclically softens" but eventually stabilizes at a constant amplitude. Other materials may require *increasing* stress amplitudes for the constant-strain cycles (i.e., the material "cyclically hardens") or the cyclic stress may not change at all. Stable stress–strain behavior is usually observed, however, by mid–fatigue life and results in the schematic stress–strain plot shown in Figure 9.10. Continued cycling between the completely reversed strain limits gives the same stress–strain trace, known as a *hysteresis loop,* for the majority of the remaining fatigue life. Thus, two pieces of information are obtained from the strain-controlled fatigue test: the stable stress amplitude and the fatigue life for the given applied strain amplitude. These data are now used to characterize the fatigue behavior of the material.

First, the stable hysteresis loops provide a *cyclic stress–strain* curve by plotting the steady-state stress amplitude $\Delta\sigma/2$ versus the completely reversed applied strain amplitude $\Delta\varepsilon/2$ for a series of tests, as shown in Figure 9.11. This cyclic stress–strain curve may be compared with the conventional monotonic (i.e., single-load) stress–strain curve to determine differences in material response for cyclic as compared to single-load applications (cyclic softening occurs for the case shown in Figure 9.11).

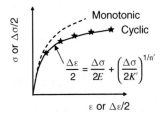

$$\frac{\Delta\varepsilon}{2} = \frac{\Delta\sigma}{2E} + \left(\frac{\Delta\sigma}{2K'}\right)^{1/n'}$$

E = elastic modulus
K' = cyclic strength coefficient
n' = strain-hardening exponent

Figure 9.11. Cyclic stress–strain curve obtained by plotting stable stress amplitude as function of applied strain amplitude for several strain-controlled fatigue tests. Comparison with conventional monotonic stress–strain curve shows cyclic softening in this case.

The cyclic stress–strain curve may be modeled for subsequent fatigue analyses by dividing the total applied strain amplitude $\Delta\varepsilon/2$ into elastic and plastic components as in the equation

$$\frac{\Delta\varepsilon}{2} = \frac{\Delta\varepsilon_{\text{elastic}}}{2} + \frac{\Delta\varepsilon_{\text{plastic}}}{2} \qquad (9.6)$$

For *uniaxial* loading, Hooke's law gives the *elastic* component of strain in Equation 9.7, where E is the elastic modulus:

$$\frac{\Delta\varepsilon_{\text{elastic}}}{2} = \frac{\Delta\sigma}{2E} \qquad (9.7)$$

Now, Equations 9.6 and 9.7 can be solved for the plastic strain amplitude, which is then plotted versus the stable stress amplitude $\Delta\sigma/2$, as shown in Figure 9.12. Fitting these data with a power law defines the empirical constants K' (cyclic strength coefficient) and n' (cyclic strength exponent) in the equation

$$\frac{\Delta\sigma}{2} = K' \left(\frac{\Delta\varepsilon_{\text{plastic}}}{2} \right)^{n'} \qquad (9.8)$$

Equations 9.6–9.8 are now combined to give the cyclic stress–strain curve:

$$\frac{\Delta\varepsilon}{2} = \frac{\Delta\sigma}{2E} + \left(\frac{\Delta\sigma}{2K'} \right)^{1/n'} \qquad (9.9)$$

While Equation 9.9 characterizes the material's stable cyclic stress–strain response, it does not contain fatigue life information. Fatigue life behavior is obtained by relating the total applied strain amplitude or the plastic strain amplitude to the lives of individual specimens tested to various strain amplitudes. First, following the pioneering work of Coffin and Manson [4, 5], plot the plastic strain amplitude $\Delta\varepsilon_{\text{plastic}}/2$

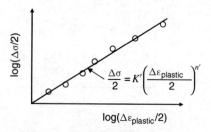

K' = cyclic strength coefficient
n' = cyclic strain-hardening exponent

Figure 9.12. Plot of stable cyclic stress versus applied plastic strain showing definition of cyclic strength coefficient and cyclic strain-hardening exponent.

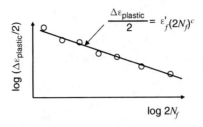

$$\frac{\Delta\varepsilon_{\text{plastic}}}{2} = \varepsilon'_f (2N_f)^c$$

ε'_f = fatigue ductility coefficient
c = fatigue ductility exponent

Figure 9.13. Plastic strain amplitude versus fatigue life in reversals.

for each specimen versus fatigue life (given in reversals $2N_f$) as shown schematically in Figure 9.13. These data may be represented with the following power law:

$$\Delta\varepsilon_{\text{plastic}}/2 = \varepsilon'_f (2N_f)^c \tag{9.10}$$

Known as the Coffin–Manson expression, Equation 9.10 defines the fatigue ductility coefficient ε'_f and the fatigue ductility exponent c. Here fatigue life is given as a function of plastic strain amplitude, which is the controlling parameter for LCF applications that are characterized by large plastic strains.

Also note that the elastic component of strain may be correlated with fatigue life by combining Equations 9.1 (Basquin's rule) and 9.7 (Hooke's law):

$$\frac{\Delta\varepsilon_{\text{elastic}}}{2} = \frac{\sigma'_f}{E} (2N_f)^b \tag{9.11}$$

Now, the total strain amplitude in Equation 9.6 can be expressed in terms of fatigue life by adding Equations 9.10 and 9.11:

$$\frac{\Delta\varepsilon}{2} = \frac{\sigma'_f}{E} (2N_f)^b + \varepsilon'_f (2N_f)^c \tag{9.12}$$

This summation is shown schematically in Figure 9.14 and is the key to the powerful strain–life approach to fatigue. Note, for example, that the total strain–life approach effectively combines the original S–N curve (as reflected by Basquin's rule) with the Coffin–Manson plastic strain–life approach. In the LCF regime, elastic strains are small and fatigue life is dominated by the plastic strain amplitude (Equation 9.10). In the HCF regime, the applied loads are much smaller and the plastic strains are negligible, so that fatigue behavior is controlled by the elastic strain term (Equation 9.11). Note that the fatigue life corresponding to the point where the elastic and plastic strain amplitudes are equal provides a convenient benchmark to separate HCF from LCF. This "transition life" $2N_t$ is readily obtained by setting Equations 9.10 and 9.11 equal to each other and gives the result

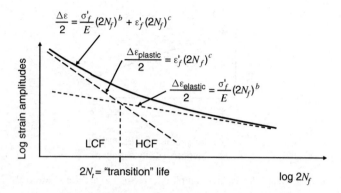

Figure 9.14. Elastic, plastic, and total strain amplitudes versus fatigue life in reversals.

$$2N_t = \left(\frac{\varepsilon'_f E}{\sigma'_f} \right)^{1/(b-c)}$$ (9.13)

Note that the strain–life approach has defined the following fatigue properties based on the results of completely reversed, constant-amplitude strain-controlled tests:

$$K' = \text{cyclic strength coefficient}$$

$$n' = \text{cyclic strength exponent}$$

$$\varepsilon'_f = \text{fatigue ductility coefficient}$$

$$c = \text{fatigue ductility exponent}$$

In addition, the stress–life tests gave the following Basquin's rule properties:

$$\sigma'_f = \text{fatigue strength coefficient}$$

$$b = \text{fatigue strength exponent}$$

These six fatigue properties are defined by Equations 9.1, 9.8, and 9.10. If these simple power laws are "perfect" fits to the test data, only four of the six constants are independent. Solving Equations 9.1 and 9.10 to eliminate fatigue life $(2N_f)$ and then comparing the result with Equation 9.8 give the following relations between these six fatigue properties:

$$n' = \frac{b}{c}$$ (9.14)

$$K' = \frac{\sigma'_f}{(\varepsilon'_f)^{n'}}$$ (9.15)

Since fatigue testing is expensive, many attempts have been made to relate these various fatigue properties to those obtained from the conventional monotonic tension test. Several of those estimates are given in References 1–3. A particularly ambitious effort has been obtained by Muralidharan and Manson, who proposed the following expression based on test data obtained from 47 materials, including steels, aluminum, and titanium alloys [6]:

$$\frac{\Delta\varepsilon}{2} = 0.623 \left(\frac{S_{ult}}{E}\right)^{0.832} (2N_f)^{-0.09} + 0.0196(\varepsilon_f)^{0.155} \left(\frac{S_{ult}}{E}\right)^{-0.53} (2N_f)^{-0.56}$$

(9.16)

Here S_{ult} is the ultimate tensile stress, E is the elastic modulus, and ε_f is the true fracture ductility of the material obtained from the standard monotonic tension test (recall Section 2.5).

Note that Equation 9.16 suggests that the fatigue strength exponent $b = -0.09$ and the fatigue ductility exponent $c = -0.56$ and are constant for all materials. Tests with individual materials indicate that these dimensionless quantities generally fall in the ranges $-0.14 < b < -0.06$ and $-0.7 < c < -0.5$. Here the lower bounds are appropriate for ductile materials, whereas the upper bounds better represent high-strength materials [1–3]. Equation 9.16, as well as the other estimates for the cyclic material properties reported in Reference 1–3, including Equations 9.14 and 9.15, should be used with caution, however, and only when actual fatigue test data are unavailable.

So far, discussion has been limited to completely reversed cycling without the presence of mean stresses. As, in the stress–life approach, however, mean stress also influences strain–life behavior, with compressive mean stresses being beneficial and tensile means detrimental. In addition, mean stresses are usually not as effective in the LCF regime, where they tend to cycle out at the large plastic strains. The influence of mean stress on strain–life behavior has been modeled by the three following modifications to the original strain–life equation 9.12:

Morrow [7]

$$\frac{\Delta\varepsilon}{2} = \frac{\sigma_f' - \sigma_{mean}}{E}(2N_f)^b + \varepsilon_f'(2N_f)^c$$

(9.17)

Manson–Halford [8]

$$\frac{\Delta\varepsilon}{2} = \frac{\sigma_f' - \sigma_{mean}}{E}(2N_f)^b + \varepsilon_f'\left(\frac{\sigma_f' - \sigma_{mean}}{\sigma_f'}\right)^{c/b}(2N_f)^c$$

(9.18)

Smith–Watson–Topper [9]

$$\sigma_{max}\frac{\Delta\varepsilon}{2} = \frac{(\sigma_f')^2}{E}(2N_f)^{2b} + \sigma_f'\varepsilon_f'(2N_f)^{b+c}$$

(9.19)

In Equations 9.17–9.19, σ_{mean} is the mean stress, and σ_{max} in Equation 9.19 is the peak stress, defined as the stress amplitude plus the mean stress (Equation 9.19 is only valid when $\sigma_{max} > 0$). Note that Equation 9.17 only adjusts the elastic component for strain, whereas the other two expressions modify both the elastic and plastic components of strain. These three equations are empirically based and should be compared with test data when possible to determine which model is most appropriate for the material and test conditions of interest.

9.2.4 Notch Fatigue

The stress–life and strain–life methods discussed in Sections 9.2.2 and 9.2.3 both employ *smooth* test coupons to characterize the fatigue behavior of the test material. Actual structures contain many holes, fillets, "nicks and dings," and other forms of stress concentrations that will, in general, hasten formation of fatigue damage at those critical locations. Thus, the ability to apply smooth specimen test data to more realistic structural configurations is an important issue in fatigue analysis. As might be expected, *notch fatigue* is a daunting problem, and although a brief summary of some key points are provided here, the reader is encouraged to consult References 1–3 for additional details on this complex topic.

As a first approach to the notch problem, compare results of *S–N* (or strain–life) tests with notched and smooth specimens as shown in Figure 9.15. Here results are presented for specimens made from the same material and subjected to the same remote (i.e., nominal) stress amplitude $\Delta S/2$. As shown, the notched *S–N* curve typically falls below the smooth specimen result, although the difference may diminish in the LCF regime for ductile materials where plastic deformation tends to minimize the effect of the notch. One may use these data to define the fatigue notch concentration factor K_f given by

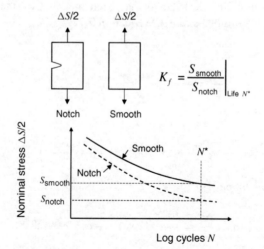

$$K_f = \frac{S_{smooth}}{S_{notch}}\bigg|_{\text{Life } N^*}$$

Figure 9.15. Comparison of *S–N* curves for smooth and notched specimens showing definition of fatigue notch concentration factor K_f.

$$K_f = \frac{S_{\text{smooth}}}{S_{\text{notch}}}\bigg|_{\text{Life } N^*} \qquad (9.20)$$

Here K_f is the ratio of the smooth and notched fatigue strengths corresponding to a particular fatigue life N^*, usually taken in the HCF regime (i.e., 10^6–10^7 cycles). The fatigue concentration factor will, in general, vary as $1 \leq K_f \leq K_t$, where K_t is the usual elastic stress concentration factor for the given notch. Note that when $K_f = K_t$, the notch is said to be fully effective in reducing the fatigue strength of the material, and when $K_f = 1$, the notch has no effect on fatigue life (a desirable result that is, unfortunately, often not the case). If K_f always equaled K_t, the notch problem would be simple to analyze by just reducing the fatigue strength by the elastic stress concentration factor for the geometry of interest. This approach is, however, overly conservative for cases where $K_f < K_t$, and, moreover, K_f is usually not constant for all fatigue lives.

The fatigue notch concentration factor depends on the material, notch size, and fatigue life of interest. In general, high-strength materials are more notch sensitive than ductile materials, and large notches are more detrimental than smaller ones (with the same K_t). This latter effect is associated with the "weakest link" concept discussed previously, where the stress gradient associated with a larger notch is distributed over a larger volume so that more material sees the large peak stresses. There are several empirical models given in References 1–3 that may be used to estimate K_f for various materials and notch geometries.

The fatigue notch factor has been called a "crutch with which one can limp from the unnotched specimen to notched component" [10].[†] While K_f is useful from an engineering standpoint in the HCF regime, where stresses are nominally elastic, it has much to be desired for LCF, where plastic deformation will be expected to occur at the notch root. If, however, one could determine the local cyclic strains at the notch, it is conceivable that the strain–life data from smooth specimens could be used to estimate notch fatigue life. The situation of interest is shown schematically in Figure 9.16, where points remote from and next to the notch are compared. Designate the cyclic stress and strain ranges at the notch root by $\Delta\sigma$ and $\Delta\varepsilon$, respectively, and the corresponding "remote" values outside of the influence of the notch by Δs and Δe. The ratios between the notch and remote (i.e., nominal) quantities define the following *concentration* factors:

$$K_\sigma = \frac{\sigma}{s} = \frac{\Delta\sigma}{\Delta s} = \text{notch stress concentration factor} \qquad (9.21)$$

$$K_\varepsilon = \frac{\varepsilon}{e} = \frac{\Delta\varepsilon}{\Delta e} = \text{notch strain concentration factor} \qquad (9.22)$$

Here, Equations 9.21 and 9.22 are defined for inelastic as well as elastic behavior and are distinguished from the usual *elastic* stress concentration factor K_t, which is equal

---

[†]Reference 10 is highly recommended for those interested in a history of fatigue, beginning with the first published account of fatigue in 1837.

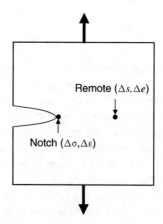

Figure 9.16. Comparison of nominal and notch stresses and strains.

to Equation 9.21 but is limited to elastic behavior. Note that for elastic conditions $K_\sigma = K_\varepsilon = K_t$, but when yielding occurs at the notch, K_σ will decrease and K_ε will increase. Indeed, finding K_σ or K_ε for conditions of cyclic yielding is the challenging stress analysis problem at the heart of the notch fatigue issue.

An engineering approach to notch fatigue analysis was proposed by Topper, Wetzel, and Morrow [11], who applied an earlier result by Neuber for static plasticity of notches in torsion to the fatigue problem. Neuber proposed that the product of the stress concentration and strain concentration factors remained constant after yielding and are related by

$$K_t = \sqrt{K_\sigma K_\varepsilon} \qquad (9.23)$$

Taking the cyclic definitions of the stress and strain notch concentration factors and replacing K_t with the fatigue concentration factor K_f, Equations 9.21–9.23 combine to give

$$(K_f)^2 \, \Delta s \, \Delta e = \Delta\sigma \, \Delta\varepsilon \qquad (9.24)$$

Note that Equation 9.24, known as Neuber's rule, relates the remote stress and strain ranges with the notch stress and strain ranges through the fatigue notch concentration factor. Furthermore, if one assumes that the remote behavior is elastic and uniaxial, then $e = s/E$ and Equation 9.24 becomes

$$\frac{(K_f)^2 \, \Delta s^2}{E} = \Delta\sigma \, \Delta\varepsilon \qquad (9.25)$$

Now Equation 9.25 (or 9.24) can be solved with the cyclic stress–strain curve (Equation 9.9) and the strain–life relation (Equation 9.12) to find the fatigue life of the notched member assuming that the nominal stress range Δs and the fatigue notch concentration factor K_f are known for the problem of interest. Note that other interpretations of Neuber's rule employ K_t rather than K_f in Equation 9.24, and there are other similar notch relationships proposed for the fatigue problem [2]. Although these approaches do provide relatively simple means to relate the remote and notch stresses and strains and, thus, estimate fatigue life, design analyses should be verified by fatigue testing whenever possible.

One final observation regarding notch fatigue deals with the fact that the stress–strain response at the notch root often undergoes *strain-controlled* cycling even though the remote loads are applied under *stress control*. Assume, for example, that the remote stress and strain ranges Δs and Δe in Figure 9.16 are elastic but that yielding occurs at the notch so that the corresponding notch stress and strain ranges $\Delta \sigma$ and $\Delta \varepsilon$ are inelastic. This means that if an elastic remote stress is applied to the member and removed, the nominal strain will return to zero when the stress returns to zero. At the notch root, however, yielding has occurred, so that the notch strains are not zero when the remote load is removed, providing a potential incompatibility in strain. Since the inelastic behavior is confined to a small region at the notch root in this case, the much larger volume of remote elastic strains actually constrains the stress–strain response in the notch area, so that it undergoes strain-controlled rather than stress-controlled cycling. This point explains why the stress–life approach to notch fatigue often breaks down in the LCF regime. It is also important to note that the local stress ratio R in the notch area can be different from the remotely applied R.

9.2.5 Variable-Amplitude Loading

Now consider the important issue of variable-amplitude fatigue life prediction by the stress–life or strain–life methods. Although this topic might best be approached by fatigue tests with the actual load history of interest, such experiments are cost prohibitive for all possible loadings, so there is great interest in predictive methods that employ the basic, completely reversed constant-amplitude data. Again, however, variable-amplitude fatigue analysis is a complex subject that can only be overviewed here, and the reader is referred to References 1–3 for additional details.

Recall that the effects of mean stress on constant-amplitude loading are described by Equations 9.2–9.5 for stress–life testing and by Equations 9.17–9.19 for the strain–life approach. Those equations provide the first step for extending the completely reversed fatigue test data to more complex loading. Now consider the more general case where the applied load consists of various *blocks* with different stress amplitudes and mean stresses (or strains) or, in the limit, completely random loading.

The simplest approach here is to assume that fatigue damage accumulates in a "linear" manner that is independent of load history. This assumption may be stated by Equation 9.26 and is known as linear cumulative damage or Miner's rule:

$$\sum \frac{n_i}{N_{if}} = 1 \qquad (9.26)$$

As shown in Figure 9.17, n_i is the number of *applied* cycles with the ith loading block that has a given stress range ΔS_i and mean stress and N_{if} is the fatigue life

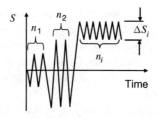

Miner's rule: $\Sigma(n_i/N_{if}) = 1$

n_i = number of applied cycles of stress range ΔS_i
N_{if} = fatigue life for ΔS_i cycling only

Figure 9.17. Definition of Miner's rule for variable-amplitude loading.

that would result if only the ith block of loading were applied to the specimen. Basically, Equation 9.26 states that n_i/N_{if} is the percentage of fatigue life exhausted for the ith loading block and that when the damage summation for all blocks equals 1 (i.e., 100%), the total fatigue life is expended. Note that N_{if} for the ith block incorporates the effect of both mean stress and stress range but does not include potential load history effects (i.e., residual stresses) that could be developed by prior loading blocks. Miner's rule can also be employed with the strain–life approach by simply expressing the fatigue lives in terms of reversals (i.e., $2n_i$ and $2N_{if}$) and using the strain–life relationships (Equations 9.17–9.19) to compute the percentage of damage accumulated during each loading block.

Although Miner's rule has seen wide use in industry, it has significant limitations that can lead to inaccurate results. Chief among these shortcomings is the failure to recognize that the *sequence* in which loads are applied can have a significant influence on life through development of residual stresses. Recalling the load history effects associated with fatigue crack growth discussed in Section 7.4, it is not surprising that similar issues also exist for the fatigue "nucleation" phenomena discussed in this chapter.

Crews [12], for example, reports the results of fatigue tests with 12×35-in (30×87.5-cm) 2024-T3 aluminum plates that contained 2-in- (5-cm-) diameter open holes. One reference specimen (Figure 9.18a) was subjected to a constant-amplitude remote stress that varied from 0 to 20 ksi (0–139 MPa) remote stress and had a fatigue life of 230,000 reversals. Two identical specimens were tested to the same stress history, except for a few initial reversals of ±40 ksi (275 MPa) stress, as shown in Figure 9.18. Although similar, these two "high–low block" loads differed in one critical detail: the sign of the last large peak before the load amplitude was decreased.

This subtle change in loading sequence had a large influence, however, on fatigue life. In Figure 9.18b, 20 reversals of ±40 ksi (±138 MPa) loading were applied before reducing to the 0–20 ksi (0–138 MPa) load *after* the *last large compressive* (-40 ksi $= -276$ MPa) peak. That specimen had a subsequent life of 126,000 reversals, less than the reference case without the large preloads. Figure 9.18c was identical to

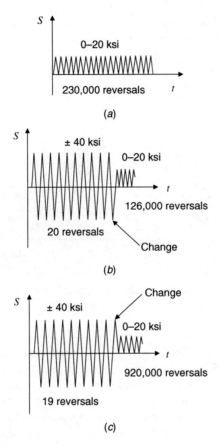

Figure 9.18. Stress–time histories showing effect of load sequence on fatigue life for notched aluminum specimens tested by Crews.

Figure 9.18*b* except that the high–low load transition was made *after* the last 40-ksi (138-MPa) *tensile* peak and only involved 19 reversals of the larger preload. Note that this specimen resisted 920,000 additional 0–20-ksi (0–138 MPa) reversals, over seven times as many as the sequence in Figure 9.18*b* and four times as long as the reference in Figure 9.18*a*. Thus, the large preloads actually *increased* the fatigue life in Figure 9.18*c* as compared to Figure 9.18*a* without the large loads. Clearly, load sequence effects play a significant influence on fatigue life that is not predicted by Miner's rule, which would give essentially the same total life for Figures 9.18*b, c.*

The high–low load block sequencing phenomenon results from mean stresses created by overloads, as shown schematically in Figure 9.19. Here two completely reversed-strain blocks are applied to a smooth test specimen, and the resulting stress history required to maintain the strain peaks is shown. Note that the first large strain block develops a stable hysteresis loop, and continued cycling about this loop results in the large, *completely reversed* stress history shown.

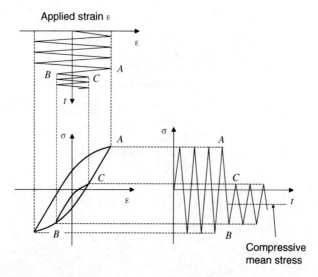

Figure 9.19. Strain–life sequence showing development of compressive mean stress under high–low strain loading sequence where the last large strain peak is tensile.

When, however, the strain amplitude is reduced after a tension peak (i.e., the *ABC* sequence shown in Figure 9.19), subsequent strain-controlled cycling follows the smaller hysteresis loop *BC* contained in the original large loop. Note that this sequence develops a *compressive mean stress* for subsequent completely reversed cycling at the smaller strain amplitude. This compressive mean stress will increase the fatigue life for the small-amplitude strain block. If, instead, the sequence for changing from high to low strain amplitudes is reversed, as shown by the *DEF* sequence in Figure 9.20, a tensile mean stress is created. Note that here the last large peak before changing amplitude was *compressive* and subsequent strain cycling follows hysteresis loop *FE*, embedded in the original large loop. In this case, the *tensile mean* stress will have a deleterious effect on remaining fatigue life.

Thus, as shown by the applied strain sequences of Figures 9.19 and 9.20, it is possible to develop tensile or compressive mean stresses during sequences of completely reversed strain cycling. These mean stresses may be estimated by the cyclic stress–strain curve (Equation 9.9) for *simple* block loadings and incorporated into Miner's rule by modifying the N_{if} term in the denominator of Equation 9.26 to account for the presence of a tensile or compressive mean stress. Details for these types of calculations are given in References 1–3.

One may ask how this description of mean stresses that develop under *strain-controlled* loading sequences explains the load sequence effect observed by Crews for his *stress-controlled* tests. The key point here is that although the loads shown in Figure 9.18 are applied *stress* sequences, they represent the *remote* behavior and not the stress–strain response that occurs next to the holes (i.e., notches) and led to fatigue failure in the Crews specimens. As discussed previously, inelastic yielding at

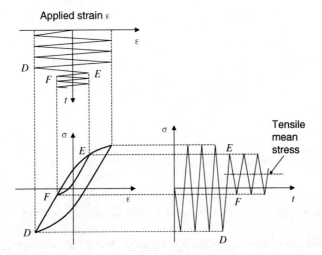

Figure 9.20. Strain–life sequence showing development of tensile mean stress under high–low strain-loading sequence where the last large strain peak is compressive.

the notch root can actually result in strain-controlled conditions at the notch root even though the elastic material surrounding the notch undergoes stress control. Thus, the loading sequence Figure 9.18*b* led to a local tensile mean stress at the notch root, which decreased fatigue life in the Crews tests, while the load sequence in Figure 9.18*c* led to a local compressive mean stress that increased life.

9.2.6   Cycle Counting

Although Figure 9.4 gives precise definitions for load reversals, ranges, peaks, and valleys, characterization of a *load cycle* is problematic. In the context of fatigue analysis, a load cycle is a sequence of two reversals that begin and end at the same value of stress (or strain). The cycle has a definite minimum and maximum load that can be used to compute the cyclic amplitude, mean stress, stress ratio R, and so on. In variable-amplitude loading, however, one often has the situation where subsequent valleys (or peaks) are not equal, so it is not obvious what constitutes a load cycle.

This situation is described schematically in Figure 9.21. Note that the original load–time history on the left starts at zero load (point 0), progresses to peak A, reverses to valley B, reverses again to peak C, and then returns to zero load at point D. Since valley B does not return to zero stress, a load cycle is not completed at this point. The various segments of the load–time history may be combined, however, to define two load cycles, as shown by count 1 or 2 on the right. Note that both of these cycle counts use all segments of the original load–time spectrum and result in two well-defined cycles, as recorded in Table 9.1. The cycles obtained by these two counting methods are not, however, the same, and they would cause different fatigue damage. As indicated in Table 9.1, method 2 defines one cycle with amplitude 1.5 and another with amplitude 0.5 (arbitrary units). The two cycles defined by method

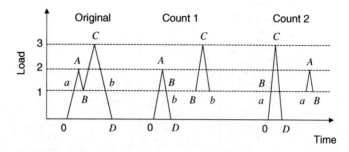

Figure 9.21. Two methods for resolving a variable-amplitude load history into complete cycles.

1 both have amplitude 1. Thus, method 1 omitted the largest stress amplitude and, since fatigue life is primarily dependent on the *cyclic* stress (or strain), would predict less fatigue damage than the cycles defined by method 2. Obviously, when analyzing the fatigue life caused by a load spectrum that consists of millions of peaks and valleys, it is essential that cycles be defined in a consistent and rigorous manner.

There have been many counting schemes developed over the years to analyze variable-amplitude load histories [1–3, 13]. The goal of these procedures is to reduce a complex sequence of peak and valley loads to an equivalent set of well-defined cycles (i.e., specified minimum and maximum loads) that may then be analyzed by an appropriate fatigue crack growth or crack nucleation (i.e., stress–life or strain–life) procedure that employs Miner's rule. Desirable features for cycle counting include the necessity to count the largest range in the sequence (i.e., "don't miss the forest because of the trees"), define cycles in a manner that maximizes the number of load ranges, and count all segments of the load–time sequence without counting any segment twice.

The rainflow and range-pair methods described in Reference 13 give similar results for most load histories and are the generally preferred cycle-counting methods. (Count 2 shown in Figure 9.21 is a range-pair count.) They are two-parameter methods that yield both the cyclic range and mean value for all cycles that are counted. The reader is referred to References 1–3 and 13 for detailed descriptions of these counting methods. (Reference 1, for example, provides a simple computer code for accomplishing the rainflow-counting algorithm.) A final point here is that employing a cycle-counting algorithm to define the stress (or strain) cycles provides a means

TABLE 9.1   Summary of Cycles Counted by Two Methods in Figure 9.21

Quantity	Count Method 1		Count Method 2	
	Cycle 1	Cycle 2	Cycle 1	Cycle 2
Maximum load	2	3	3	2
Minimum load	0	1	0	1
Range	2	2	3	1
Amplitude	1	1	1.5	0.5
$R = $ min/max	0	$\frac{1}{3}$	0	0.5
Mean load	1	2	1.5	1.5

to account for sequence effects when using Miner's rule (Equation 9.26) to analyze variable-amplitude fatigue lives. Cycle counting is also a key step for defining a variable amplitude load history, for fatigue crack growth analyses by the cycle-by-cycle approach discussed in Section 7.9.2. The AFGROW life prediction software described in the Appendix, for example, has cycle counting provisions (see Section A.5.3.1).

9.3 CORROSION

Corrosion is a common form of material degradation that can occur during extended periods of service and, if unchecked, can significantly impair structural integrity. Corrosion prevention and repair often lead to extremely high maintenance costs that may require structural retirement for economic rather than safety reasons. This economic situation is particularly true for older structures (i.e., "aging aircraft") constructed without the benefit of modern corrosion resistant materials and protection systems. Dealing with airframe corrosion is, for example, the most expensive maintenance item for the USAF [14].

Corrosion is a particularly vexing process that can occur independent of structural use (e.g., aircraft can corrode whether they are flown or not). Corrosion is difficult to predict, and the fact that its natural mechanisms often take long time periods to develop complicates laboratory evaluations of alloys and protection methods. Although accelerated testing schemes are available [15], one must be concerned with how accurately those tests mimic the natural deterioration that results from decades of service. Indeed, some organizations find it necessary to maintain outdoor facilities to monitor corrosion over many years of specimen exposure to "natural" environments.

Corrosion is an extremely complex failure mechanism that is strongly dependent on material, component geometry, operating environment (e.g., chemical, thermal, stress), and maintenance actions and can occur in many forms. Fontana [16], for example, has identified eight types of corrosion: uniform general attack, galvanic, crevice corrosion, pitting, intergranular corrosion, selective leaching, erosion corrosion, and stress corrosion. While Fontana's classification is based on visual characteristics, Craig and Pohlman [17] provide the following five alternate categories based on the mechanisms of corrosion attack:

- *General Corrosion*   Atmospheric; galvanic; stray-current; general biological; molten salt; liquid metals; high-temperature (oxidation, sulfidation, carburization, other forms).
- *Localized Corrosion*   Filiform; crevice; pitting; localized biological.
- *Metallurgically Influenced Corrosion*   Intergranular; dealloying.
- *Mechanically Assisted Degradation*   Erosion, fretting, cavitation, and water drop impingement; corrosion fatigue.
- *Environmentally Induced Cracking*   Stress corrosion cracking; hydrogen damage; liquid metal embrittlement; solid-metal-induced embrittlement.

Pitting, intergranular, exfoliation, stress corrosion cracking, corrosion fatigue, and uniform corrosion have been identified as particularly dangerous to aircraft structural integrity [18]. While a detailed discussion of all the many forms of corrosion is beyond the scope of this overview, some of the more common types are briefly summarized in the following subsection. (Study References 16 and 17 for more details of the many forms of environmental attack.)

9.3.1 Forms of Corrosion

General corrosion involves wide-area degradation that leads to overall thinning of the component and to a general increase in stress. Although localized attack can also occur by these various mechanisms, the main structural concern here is a general loss of load-bearing material that leads to increased stress and reduction in stiffness. Atmospheric corrosion results from exposure to air and its pollutants rather than by immersion in a liquid and is one of the most common forms of corrosion failure. Galvanic corrosion occurs when two or more dissimilar metals are in contact through an electrolyte that permits electron flow similar to that in a battery. Table 9.2 lists various materials with respect to their galvanic potential in seawater [16, 17]. Here the less noble material serves as the anode in the galvanic couple and is corroded away with respect to the more noble material in the joint (cathode). The further apart the two contacting materials are in the galvanic series, the greater the potential for this form of corrosion. Although the extent of general corrosion can be difficult to quantify, one study [19] has employed the following classifications for corrosion damage:

- Mild—corrosion depth < 0.001 in (25 μm)
- Moderate—0.001 in < depth < 0.01 in (25 μm < depth < 250 μm)
- Severe—corrosion depth > 0.01 in (250 μm)

Localized corrosion mechanisms involve high rates of chemical attack in small areas. These forms are harder to detect than general corrosion and can cause stress

TABLE 9.2   Galvanic Ordering of Some Materials in Seawater Arranged from Most to Least Noble

	← More noble (cathodic)	Less noble → (anodic—corrodes)
More noble (cathodic) ↑	Platinum, gold, graphite, titanium, silver	
	Hastelloy C, stainless steels (passive), Inconel (passive), nickel (passive)	
	Bronzes, copper, brasses, Inconel (active), nickel (active)	
	Tin, lead, stainless steels (active), cast iron, wrought iron, low-carbon steel	
	Aluminum alloys, cadmium, aluminum, galvanized steel, zinc	
Less ↓ noble (anodic corrodes)	Magnesium alloys, magnesium	

Note: The greater the separation of materials in the list, the more likely the less noble material will corrode when in contact with more noble material. See References 16 and 17.

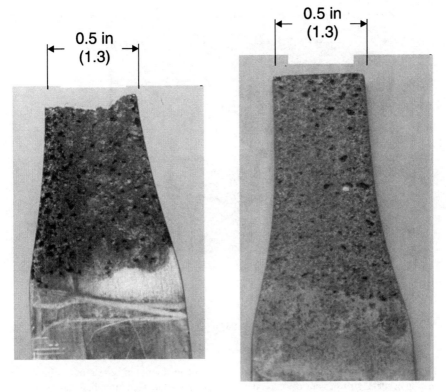

Figure 9.22. Example of pitting corrosion on two aluminum test specimens.

concentrations that accelerate development of fatigue cracks. Pitting, for example, is a form of localized corrosion that can lead to rapid development of many small "holes" on an otherwise undamaged surface. Although pit shapes may differ in their depth-to-width ratios, they can represent serious stress concentrations (see Figure 9.22) that may accelerate the development of widespread fatigue damage. Crevice corrosion is another form that occurs in tight gaps that allow the local environment to develop chemical concentrations that hasten corrosive attack. Filiform corrosion is a specialized form of crevice corrosion evidenced by small, randomly oriented "tunnels" beneath surface films that are filled with corrosion products. Corrosion "pillowing" is a consequence of corrosion products that build up in tight gaps between mechanically fastened lap joints. Here the corroded material, which may increase in volume by as much as 600%, causes bulging of the skin between the fasteners and, combined with thickness loss, can lead to significant "pillowing stresses" in the joint. The out-of-plane deformations caused by corrosion pillowing can aid in detection of this hidden form of corrosion.

Metallurgically influenced corrosion deals with features of the alloy composition and heat treatment. Grain boundaries frequently contain different concentrations of alloy constituents than the remaining matrix, for example, and are common sites for

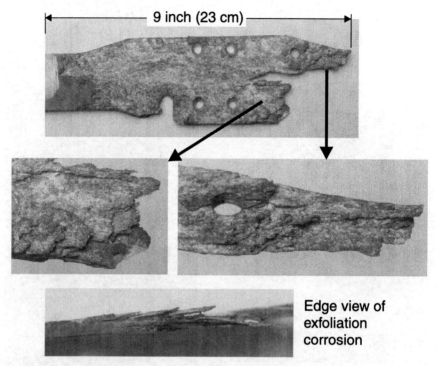

|◄——————— 9 inch (23 cm) ———————►|

Edge view of
exfoliation
corrosion

Figure 9.23. Several views of exfoliation corrosion of an aluminum tear strap showing delamination of metal strips as corrosion products wedge open elongated grain boundaries.

localized corrosion (e.g., a galvanic couple may develop between the grain boundaries and the contained matrix). Exfoliation is an exaggerated form of intergranular corrosion that occurs in aluminum alloys that have elongated grains oriented parallel to the surface. Corrosion products build up as the grain boundaries are eaten away and cause a wedging action that leads to delamination (exfoliation) of thin layers of metal from the component surface (see Figure 9.23). Exfoliation corrosion can be a particular problem in joints that contain fasteners made from dissimilar metals than the parent structure. Dealloying, or selective leaching, is another form of metallurgical induced corrosion that occurs when one alloy constituent is selectively removed, leaving an altered metallurgical structure. The resulting metal may develop porous areas with reduced strength, ductility, and hardness.

Mechanically assisted degradation involves chemical attack combined with wear or fatigue. Erosion, for example, occurs when material is removed by repeated impact with a fluid that contains many small solid or liquid particles. Erosion can be evidenced by a light "polishing" of the surface or by more aggressive localized attack that leads to rapid thinning and penetration of the component. Fretting is another failure mechanism that results when two surfaces are subjected to a clamping force combined with perpendicular vibratory stresses. Although the surfaces are in nominal contact, there are local areas of slip that lead to large stress gradients and subsequent crack

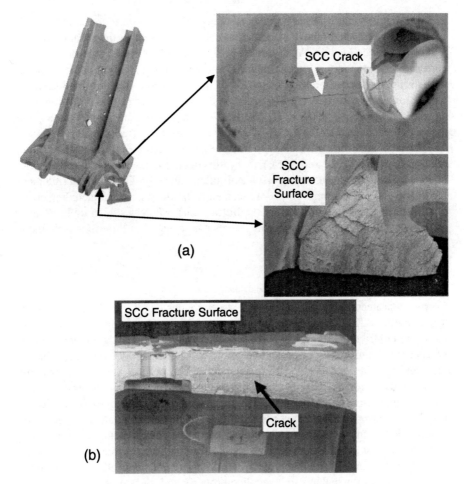

Figure 9.24. Examples of stress corrosion cracking fracture surfaces in 7075-T6 aluminum aircraft fin to fuselage attachment beam (top) and trunion fitting (below).

formation. Small wear fragments that are trapped between the contacting surfaces can accelerate the fretting process by abrasive action. If these debris particles also corrode, the resulting corrosion products (e.g., oxides) may become more aggressive abrasives, leading to a process known as fretting corrosion. When fretting-induced cracks grow to a size where they are driven by the remote cyclic loads, the process is known as fretting fatigue. Corrosion fatigue is another form of mechanically assisted degradation that involves a synergism of corrosion and fatigue damage mechanisms that is often more damaging than the two processes considered separately. Some aspects of corrosion fatigue were discussed previously in Section 7.8.

Environmentally induced cracking leads to explicit *crack* formation during environmental exposure. Stress corrosion cracking, for example, results from the interaction of *sustained tensile* stresses and environmental attack (see Figure 9.24).

This mechanism can be transgranular or intergranular in nature and is particularly susceptible to residual tensile stresses that could be present in the member regardless of its service loading. As discussed previously in Section 3.5, the rate of stress corrosion cracking in a particular material and environment can often be correlated with the applied stress intensity factor K. When $K < K_{ISCC}$ (the threshold stress intensity for stress corrosion cracking), cracks do not grow, and stress corrosion is not an issue. When $K_{ISCC} < K < K_c$ (the fracture toughness), cracks grow in a stable manner characterized by the $da/dt–K$ relationship for the material and environment. The period of stable growth leading to fracture can be calculated by integration of that relationship (recall Equation 3.25). Other forms of environmentally induced cracking include hydrogen damage, liquid metal embrittlement, and solid metal embrittlement. Hydrogen can cause several forms of mechanical trauma in metals, including cracking, blistering, and loss of ductility (embrittlement), whereas similar embrittlement can also result when normally ductile metals are coated with thin films of liquid or solid material.

9.3.2 Corrosion Protection

Corrosion protection is achieved through structural design [20] and by corrosion protection methods [21]. Insight into this important topic can be obtained by considering the many contributing causes to corrosion: basic component design, manufacturing procedures, deterioration of protective coatings, accidental contamination, the service environment, and inadequate maintenance. The original design, for example, must specify corrosion resistant materials and coatings for all components that are exposed to aggressive environments and avoid dissimilar-metal couples that could lead to galvanic corrosion. Tight crevices should be sealed, and adequate drainage should be provided to all areas that could collect corrosive environments. Manufacturing procedures must avoid introduction of tensile residual stresses that could accelerate stress corrosion and ensure that protective finishes are not damaged during assembly. It is also critical for sealants to be applied properly and for bonding processes to be carefully monitored to prevent open crevices that could attract corrosive fluids by capillary action.

Protective coatings are needed for many components and may be of two types: barrier or sacrificial. Barrier coatings, such as paints and corrosion inhibitors, are intended to block corrosive substances from reaching the parent metal but can be defeated by scratches or cracks that penetrate the coating surface. Sacrificial protection, on the other hand, is provided by applying thin layers of material that are anodic to the parent material, resulting in a galvanic couple that sees the coating corroded away while the more noble parent structure remains intact (recall Table 9.2). Alclad aluminum is one example of this form of protection where the aluminum alloy is plated with a thin layer of pure aluminum ($\sim$2.5–5% of total thickness). Galvanized steel (zinc coating) and cadmium plating of steel or titanium fasteners are other examples of this sacrificial corrosion protection.

Corrosion protection is also provided by carefully monitoring the service environment to avoid severe conditions when possible. Structures that operate near seacoasts or industrial sites plagued by acid rain are particularly susceptible, along with those that trap condensation or are in regular contact with corrosive cargo during normal service. The possibility for accidental contamination, such as spillage around lavatories and galleys (notorious corrosion-prone areas in aircraft) or chemical spills or fire residues, must also be prevented.

Finally, and perhaps most importantly, a rigorous maintenance system must be planned and executed throughout the life of the structure. Although corrosion processes often take years to cause physical damage, one must not be lulled into the temptation to avoid maintenance costs until after corrosion first appears. Indeed, resources initially invested on corrosion *prevention* (e.g., regular washing, painting, application of corrosion preventative compounds) are much more effective than those spent later on corrosion repair, for once corrosion begins, the maintenance costs needed for continued safe operation can increase dramatically. Although grinding out the affected areas and restoring protective coatings may repair minor areas of corrosion, material removal is usually limited to 10% of the original thickness and can lead to larger stresses that hasten other failure modes. Components with extensive areas of corrosion must be replaced or repaired by adding load-bearing members (e.g., patches or doublers).

9.4 TOTAL-FATIGUE-LIFE CALCULATIONS

Damage tolerance life assessments begin with the assumption that the structure contains preexistent cracks of a *specified size*, so that the fatigue life is determined on the basis of *crack growth* analyses. This chapter, however, has discussed damage that develops during service. In the event that one does not wish to implement the original crack assumption for this *service-induced* damage, a fatigue crack nucleation period could, in principle, be determined by the stress–life or strain–life formulations discussed in Section 9.2 and added to the crack growth period as shown schematically in Figure 9.25 and given by

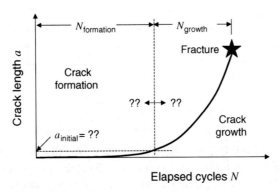

Figure 9.25. Total fatigue life separated into crack formation and crack growth periods showing difficulty in defining initial crack length that separates nucleation and growth lives.

$$N_{\text{total}} = N_{\text{nucleation}} + \int_{a_i}^{a_f} \frac{da}{F(\Delta K)} \qquad (9.27)$$

Here the total fatigue life N_{total} is simply the sum of crack nucleation and propagation lives. Recall that the limits of integration for the crack growth term are the initial and final (e.g., fracture) crack sizes a_i and a_f. Although conceptually straightforward, determination of the crack size a_i that divides the nucleation and propagation lives is problematic. As discussed previously, fatigue crack growth calculations are often extremely sensitive to the initial crack length, particularly when crack growth begins with small ΔK's that correspond to slow rates of cracking. Thus, selection of the initial crack size that defines the end of the nucleation and start of the crack periods will have a major influence on the total-life calculation.

A number of factors influence the choice of the nucleated crack size a_i. First, one might consider the crack length associated with the crack formation life measured for the stress–life of strain–life experiments used to establish Equations 9.1 or 9.12. Usually the "nucleation lives" associated with these tests are simply the period required to fracture the smooth test specimens, which are typically $\frac{1}{8}-\frac{1}{2}$ in (3–13 mm) in diameter. A more sophisticated crack formation criterion sometimes employed for strain-controlled experiments is based on a specified reduction in specimen compliance associated with crack formation (i.e., a sudden drop in the steady-state stress required to maintain the given strain range). Or, one could base the nucleated crack size assumption on *field-level* inspection limits for the *structure* in question. An empirical amalgamation of these and other considerations have led some engineers to define a "committee" (i.e., consensus of several criteria) crack length on the order of $\frac{1}{32}$ in (1 mm) as the size to begin crack growth calculations for total-life analyses. In any event, choice of the "nucleated" crack length should be given careful consideration when employing "total-life" analyses.

9.5 EQUIVALENT INITIAL CRACK SIZE

Another initial crack size that has proven useful for analysis of *originally uncracked* components is that of the *equivalent initial flaw size* (EIFS) concept shown schematically in Figure 9.26. Here fracture mechanics techniques are use to "back calculate" a fictitious initial crack size a_{eifs} that would give the total fatigue life on the basis of crack growth criteria:

$$N_{\text{total}} = \int_{a_{\text{eifs}}}^{a_f} \frac{da}{F(\Delta K)} \qquad (9.28)$$

Note that Equation 9.28 differs from Equation 9.27 in that the nucleation period is absorbed into the crack propagation period to define the equivalent initial crack size a_{eifs}. In order to compute a_{eisf} from Equation 9.28, one must know the total fatigue life N_{total} corresponding to a given final crack size a_f, the da/dN–ΔK relationship for the material of interest, the applied loads during the total-life period, and the stress

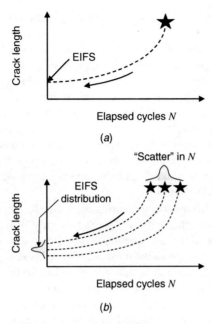

Figure 9.26. (*a*) Schematic representation of equivalent initial crack size calculated by assuming that total fatigue life of smooth or notched results from growth of fictitious preexistent crack. (*b*) Equivalent initial flaw size distribution to characterize scatter in fatigue life.

intensity factor solution for the *given configuration*. Recall that the stress intensity factor K is given by

$$K = \sigma\sqrt{\pi a}\beta(a) \tag{9.29}$$

Here $\beta(a)$ is the dimensionless factor that depends on crack size, shape, and structural configuration, σ is the applied stress, and a is the crack length. Thus, in order to compute a_{eifs} from Equation 9.28, one must *assume* a particular *crack geometry* to define the appropriate $\beta(a)$. This selection is usually based on the types of cracks that could be expected to develop in service and is often assumed to be a small surface crack, corner crack, or through-thickness edge crack.

It should be emphasized here that a_{eifs} is a *fictitious* crack that does not necessarily have physical significance. The EIFS concept was originally developed as a means to assess component quality in the context of LEFM. Early uses of this methodology, for example, were to evaluate various hole preparation processes on fatigue life [22]. Holes were drilled in flat specimens with, for example, various drill speeds and reaming procedures and then cycled to failure under constant-amplitude loading. The resultant fatigue lives were then interpreted as an indication of the "quality" of the particular manufacturing method (i.e., "long" lives represented "good"-quality manufacturing). Although these test results could have been used to define the fatigue notch concentration factor K_f discussed in Section 9.2.4, a fracture mechanics interpretation in terms of the potential for crack growth was desired, leading to the EIFS calculation.

Note that repeated fatigue tests with specimens prepared by the same process would be expected to give similar but not identical lives due to the well-known "scatter" inherent in the fatigue process. Thus, one would obtain not a single a_{eifs} from a group of repeated tests but rather several initial crack sizes (i.e., a *statistical distribution*) reflecting the variability in hole quality, as shown schematically in Figure 9.26*b*. The equivalent initial flaw size distribution is characterized by a mean crack size and coefficient of variation. Here, a good process would have both a small mean and a small coefficient of variation for the a_{eifs} distribution. Larger mean crack sizes and/or amounts of scatter could be an indication that the process parameters are not optimized or under tight control.

In addition to quantifying manufacturing methods, the equivalent initial flaw size concept has had several other applications—assessment of material quality, evaluation of teardown inspections, and quantification of corrosion damage. Variations in aluminum alloys associated with the presence of porosity or constituent particles, for example, have been examined in smooth specimen fatigue tests and used to compute EIFS distributions that help optimize materials processing and predict fatigue behavior for other specimen scenarios [23, 24]. Another EIFS application was demonstrated by Luzar and Hug [25], who conducted an intensive teardown inspection on two retired Boeing 707 aircraft wings to determine the types and sizes of cracks present in old airplanes. These final cracks were then grown back in time by assuming a typical service load history to arrive at equivalent initial flaw size distributions that would be representative of the new structure. These EIFS distributions were subsequently used as the starting point for a probabilistic analysis for the development of widespread fatigue damage [26].

The equivalent initial flaw size concept has had at least two applications related to the evaluation of corrosion damage. The breaking load test [27], for example, evaluates the resistance of alloys to stress corrosion by subjecting small smooth specimens to a fixed stress in the presence of an aggressive environment for a given period of time (several hours to weeks in length). The specimens are then loaded to failure, and fracture mechanics analyses are used to compute the size of a single crack that would cause fracture in an uncorroded specimen. This single "equivalent" crack is used to represent the total material degradation associated with the corroded specimen and effectively integrates thickness loss, pitting, exfoliation damage, and so on, into a single crack size metric. Again, repeated tests will give various final fracture loads and subsequent equivalent crack sizes due to the wide variation of corrosion damage incurred by similar specimens under the same nominal aggressive environment. The equivalent initial flaw size *distribution* determined in this manner is a measure of the resistance of the particular alloy to the given environment.

Another variation of the EIFS concept employed to determine the effects of corrosion on the subsequent fatigue life is described in References 28 and 29. Here smooth specimens were corroded to various degrees and subsequently cycled to failure. The EIFS distributions were then calculated from the total fatigue lives, and although there were fairly large variations in life, the EIFS distributions did correlate reasonably well with the length of corrosion exposure. It is proposed elsewhere [28, 29] that the effect of corrosion on remaining fatigue life could be assessed by two parameters: the EIFS, which is a measure of local corrosion (e.g., pitting), and a thickness loss resulting from general corrosion. Thickness loss leads to an increased stress, whereas the EIFS distribution serves as the starting point for fatigue crack growth calculations for the remaining life. (See Problems 17.17–17.20 at the end of Chapter 17 for more discussion of this proposed approach for evaluating corrosion damage.)

9.6 ACCIDENTAL AND DISCRETE-SOURCE DAMAGE

As a final source of service-induced damage, consider the possibility for accidents that result in structural cracking that must be arrested or contained before total loss of the structure can occur. Although it is impossible to guard against all sources of accidental (or malicious) damage that could happen in service, the engineer must provide protection against certain minimum levels that could reasonably be expected during routine operation. Bird strikes to an aircraft in flight or vehicle impact with a bridge structure could all, for example, be expected during the service life of those structures.

With respect to commercial transport aircraft, there are specific requirements in this regard. The possibility for discrete-source (i.e., accidental) damage requires explicit consideration in paragraph (e) of FAR 25.571, which states:

> The airplane must be capable of successfully completing a flight during which likely structural damage occurs as a result of (1) impact with a 4-pound bird . . . (2) Uncontained fan blade impact; (3) Uncontained engine failure; or (4) Uncontained high energy rotating machinery failure.

As discussed later in Chapter 16, this requirement leads to selection of damage tolerant materials and structural features that arrest (at least temporarily) such forms of discrete-source damage. The effectiveness of these damage containment features is subject to rigorous testing intended to simulate in-service accidents. The pressurized fuselages of commercial aircraft are, for example, punctured with harpoons to demonstrate that subsequent crack growth will not fail the entire fuselage structure. The ability of turbine engines to resist foreign object damage (FOD) in a controlled manner that does not propel shrapnel to other parts of the aircraft is another key damage tolerant feature that must be demonstrated by full-scale tests. In this case, birds (actual or simulated) are ingested into powered engines to demonstrate that all fractured components resulting from this high-speed impact are contained within the engine housing without causing other external damage to the aircraft.

9.7 CONCLUDING REMARKS

Damage tolerant designs provide margins of safety by selecting materials and sizing components with the premise that the new structure contains preexistent cracks.

While hopefully a conservative assumption for new construction, pristine members will, in fact, gradually deteriorate during service by fatigue and/or corrosion. Or the structure may encounter accidental damage that must be contained before growing to final fracture. Thus, there is need for life analyses that do not automatically invoke the initial crack size assumption but that also consider a crack formation period associated with cyclic loading of undamaged specimens.

Section 9.2 summarizes predictive techniques for "natural" fatigue crack formation in smooth or notched components. Both stress–life and the more general strain–life methods to calculate the time for fatigue crack initiation are summarized. These engineering approaches extend test results from completely reversed, constant-amplitude loading with smooth specimens to analysis of more general situations that involve notched members subjected to variable-amplitude loading. This crack formation period may then be added to the fatigue crack growth life determined by LEFM provided a reasonable estimate can be obtained for the crack size following the nucleation period. This crack size could be based on inspection limits or the "committee" crack discussed in Section 9.4.

The various forms of corrosion that can occur during service are discussed in Section 9.3 along with a brief summary of methods to resist corrosion. Corrosion results from many sources and is a particularly difficult problem to analyze and repair. Thus, every effort should be taken to prevent its occurrence through corrosion resistant design and corrosion preventative maintenance.

The equivalent initial crack size introduced in Section 9.5 provides a useful engineering approach for including fatigue crack formation and the effects of corrosion in the context of LEFM analyses. Although these fictitious cracks may not have physical basis, the EIFS distribution does provide a convenient computational tool to quantify initial component quality associated with manufacturing, material inhomogeneities, and/or corrosion. The EIFS distributions may then be used to estimate further damage to the component caused by subsequent cyclic loading.

PROBLEMS

9.1   Briefly discuss the term *initiation life* in the context of the *S–N* and strain–life approach to fatigue.

9.2   It has often been observed that endurance limits measured from small specimens (e.g., $\frac{1}{4}$ in diameter) subjected to rotating bending tests differ from endurance limits measured under other test conditions.

   (a) Define the term *endurance limit* and briefly discuss its significance.

   (b) Would the endurance limit measured under the following conditions be expected to increase, decrease, or remain the same (as compared to the $\frac{1}{4}$-in-diameter specimens tested in rotating bending)? Give a brief explanation for your answer.

   (i) Axial loading (push/pull) tests

   (ii) Two-inch-diameter specimens subjected to rotating bending

9.3 A particular steel has an ultimate stress of 100 ksi and a yield stress of 80 ksi.

  (a) Estimate the maximum completely reversed stress amplitude $\Delta S/2$ that can be applied to a smooth tension member that is to have an "infinite" fatigue life.

  (b) Estimate the maximum stress S_{max} per cycle that can be applied to another smooth tension member if the minimum stress per cycle $S_{min} = 0$ and there are to be no fatigue failures.

  (c) Repeat part b if $S_{min} = 20$ ksi.

9.4 Some *S–N* data obtained from MIL-Handbook 5 are given in Figure P9.4 for aluminum alloy 7075-T6. Note that the format for these smooth specimen results involves plotting the maximum stress per cycle S_{max} (rather than stress amplitude) versus log life measured in cycles.

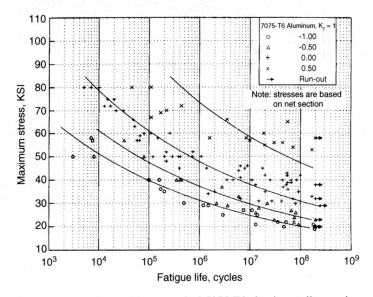

Figure P9.4. Best-fit *S–N* curves for unnotched 7075-T6 aluminum alloy, various product forms, and longitudinal direction [30].

  (a) Determine the conventional endurance limit (in terms of stress amplitude $\Delta S/2$) for stress ratios $R = -1.0, -0.5, 0.0, +0.5$.

  (b) Compare the $R = 0$ variation in fatigue life at cyclic stress levels of $S_{max} = 80$ ksi and $S_{max} = 40$ ksi.

  (c) Compare the $R = -1.0$ variation in fatigue life at cyclic stress levels of $S_{max} = 50$ ksi and $S_{max} = 25$ ksi.

  (d) Discuss a physical explanation for the observations made in parts b and c.

9.5 The *S–N* data from MIL-Handbook 5 are given in Figure P9.5 for 4340 steel (note format where S_{max} is plotted versus log life). This steel has a tensile ultimate stress of 200 ksi and a tensile yield stress of 190 ksi.

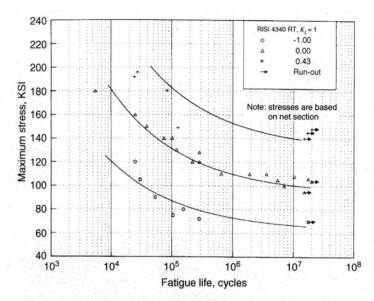

Figure P9.5. Best-fit *S–N* curves for unnotched AISI 4340 alloy steel bar and die forging; ultimate stress is 200 ksi; longitudinal direction [30].

(a) Use these data to estimate the fatigue strength coefficient σ'_f and the fatigue strength exponent b for this material. You may assume that the true stress and engineering stresses are the same for this problem.

(b) Use these data to prepare constant-life Haigh diagrams for this material (i.e., plot stress *amplitude* versus mean stress for fatigue lives of 10^5, 10^6, and 10^7 cycles).

(c) Prepare Soderberg, Goodman, and Gerber diagrams for this material to predict the relationship between stress amplitude and mean stress for endurance behavior. How well do these models predict the actual experimental results for this case?

(d) Repeat part c for a fatigue life of 10^5 cycles. How well do the models predict the results in this case?

9.6   Derive Equation 9.13 for the transition life $2N_t$.

9.7   Derive Equations 9.14 and 9.15 that relate the six strain–life fatigue properties. Why might these expressions not always agree with experimental data?

9.8   Note that Equation 9.16 estimates strain–life fatigue behavior in terms of *static* stress–strain properties.

(a) Use this expression to determine the six strain–life properties σ'_f, b, ε'_f, c, K', and n'. Discuss your results.

(b) Use this expression to determine the transition life $2N_t$.

9.9   It has been suggested that *strain-controlled* fatigue analyses should be used for analysis of notched specimens subjected to *stress-controlled* testing. Briefly discuss why strain–life analyses would be appropriate for such stress-controlled loading.

9.10 Results from strain-controlled fatigue tests ($R = -1.0$) with an experimental aluminum–lithium alloy are given in Table P9.10 [31]. The table gives the total applied strain *range* versus fatigue life in *cycles* along with the stable stress *range* developed in these strain-controlled tests. Monotonic properties obtained from standard stress–strain tests with this material include the following: elastic modulus $E = 11,530$ ksi, strength coefficient $K = 101.6$ ksi, and strain-hardening exponent $n = 0.094$.

TABLE P9.10 Results of Completely Reversed Strain-Controlled Fatigue Tests with Experimental Aluminum–Lithium Alloy

Applied Strain Range	Fatigue Life in Cycles	Stable Stress Range (ksi)
0.00381	2,804,131	45.8
0.00386	422,480	46.2
0.00542	184,447	63.0
0.00602	22,795	70.1
0.00703	4,980	82.8
0.00939	1,852	104.3
0.00944	1,320	104.0
0.00945	412	105.7
0.01211	271	119.9
0.01608	95	127.9
0.01698	41	126.9

Note: See Reference 31.

(a) Use these data to determine the fatigue strength coefficient σ'_f and the fatigue strength exponent b for the test material.

(b) Use these data to determine the fatigue ductility coefficient ε'_f and the fatigue ductility exponent c.

(c) Determine equations for the monotonic and the cyclic stress–strain curves and plot them on the same set of axes. Does the material cyclically soften, harden, or remain unchanged?

(d) Plot the total strain *amplitude* versus fatigue life in *reversals* and compare the test data with the numerical model that uses the results found in parts a and b. How well do these data agree with the model?

9.11 A threaded rod with 1.0 in *nominal* diameter is made from a material with the following properties: $\varepsilon'_f = 0.66$, $c = -0.65$, $\sigma'_f = 120$ ksi, $b = -0.089$, and $E = 30,000$ ksi. This rod fails in the *threaded* end after 10,000,000 reversals of an applied constant-amplitude *nominal* stress of ±5 ksi. Hourglass fatigue specimens are to be designed with the same threaded ends used for the rod as shown in Figure P9.11. It is desired that failure always occur in the reduced test section (diameter D) and not in the threaded ends that are used to grip the specimen to the test machine. Determine the maximum test section diameter D that can be used for a completely reversed LCF test specimen that is to last 1000 reversals. Failure is to always occur in the reduced section and not in the threaded ends. Do you expect this same specimen to be suitable for a HCF test which is to last 10,000,000 reversals? Why?

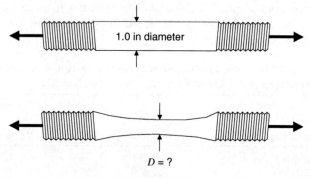

1.0 in diameter

$D = ?$

Figure P9.11

9.12 A student has been sentenced to a lengthy jail term for sleeping in class. He is assigned the jail cell formerly occupied by another prisoner who has recently escaped. This escaped prisoner was able to break one of the bars in his cell by bending it back and forth until a fatigue failure occurred. He bent the bar back and forth 1000 cycles a day. Unfortunately, it took 20 years for the fatigue failure to occur, and the broken bar has since been replaced with an identical new bar. The new prisoner would like to escape also, but he wants to be home in time for Christmas. He recalls from his fatigue class that notches hasten fatigue failures, so he decides to cut a notch in one of the cell bars. He consults his "Handbook of Useful Information for Prisoners" and discovers that the fatigue notch concentration factor for a notched jail bar steel is given by $K_f = 20x$. Here x is the notch depth measured in inches. He also notes that jail bar steel has the following properties: $b = -0.1$, $c = -0.5$, $\sigma'_f = 200$ ksi, $E = 30,000$ ksi, and $\varepsilon'_f = 0.5$.

(a) What is the minimum notch depth x if the cell bar is to be broken in 2 weeks? Assume that the new prisoner bends the bar back and forth in the same manner as his predecessor (e.g., some force, rate) and the new bars are made from the same material as the original. (Also assume the prisoner has the ability to notch the bars without being detected.)

(b) To pass the time while he is bending the bars back and forth, the prisoner has decided to compute the cyclic stress–strain curve for the jail bar steel. Determine an equation for the cyclic stress–strain curve for this material.

(c) Also determine the transition life $2N_t$ for this material.

9.13 Frustrated by their unsuccessful efforts to tear down the goal posts following several recent football victories, a group of students decide to *cut notches* in the posts to hasten the nucleation of fatigue cracks as they plan to bend the goal posts back and forth following their next victory. During a full-scale fatigue test of one of the notched posts, it is discovered that *10 students* can exert a nominal cyclic stress amplitude $\pm S$ that causes a crack to form in 10,000,000 reversals. Since the students do not expect the stadium guards to allow them to remain on the field for the time required to apply 10,000,000 reversals (and since the students are anxious to return to their studies following the game), it is decided to recruit more students in order apply a larger cyclic stress and, thus, hasten the fatigue failure of the goal posts. How many students are required to develop fatigue cracks in the goal posts in *100 reversals*? You may assume:

- The goal posts are made from a material with the following properties: elastic modulus $E = 10{,}000$ ksi, fatigue strength coefficient $\sigma'_f = 100$ ksi, fatigue strength exponent $b = -0.07$, fatigue ductility coefficient $\varepsilon'_f = 0.6$, and fatigue ductility exponent $c = -0.6$
- The completely reversed nominal stress amplitude is directly proportional to the number of students who rock the posts back and forth (i.e., 10 students cause $\pm S$, 20 students cause $\pm 2S$, etc.)
- Uniform notches are cut into the posts (e.g., same depth, radius of curvature)

9.14 A smooth specimen made from a material with $\sigma'_f = 220$ ksi and $b = -0.1$ is subjected to repeated blocks of axial loading. Each loading block consists of 200 reversals of ± 80 ksi, followed by 1000 reversals of 0–100 ksi stress, followed by 1000 reversals of -100–0 ksi stress. How many repeated blocks of this load can be applied before expecting a fatigue failure?

9.15 Fracture mechanics concepts have been used to estimate the traditional *S–N* curve originally developed for uncracked fatigue specimens. In this approach, the material is assumed to contain *an inherent initial crack of a given size* and the allowable stress amplitude $\Delta S/2$ is computed as a function of cyclic life N (see Figure P9.15). Use this "inherent initial crack size" approach to compute *S–N* curves for large specimens made from a material that is assumed to contain embedded circular (penny-shaped) cracks (see Figure 3.2*h*). Note that when the crack is circular in shape (i.e., $a = c$), the stress intensity factor

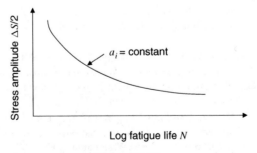

Figure P9.15

reduces to $K = (2\sigma/\pi)(\pi a)^{1/2}$. Assume that the initial circular crack radius $a_i = 0.05$ in, the cracks remain circular in shape as they grow, and the cyclic stress varies between 0 and the maximum value S. Repeat this calculation for initial crack radii of 0.02 in. Plot your results in the S–N curve format shown in Figure P9.15. Assume $da/dN = 5 \times 10^{-9}\Delta K^4$, $K_c = 50$ ksi-in$^{1/2}$, and $\Delta K_{th} = 4$ ksi-in$^{1/2}$. Discuss potential limitations of these calculations.

REFERENCES

1. J. A. Bannantine, J. J. Comer, and J. L. Handrock, *Fundamentals of Metal Fatigue Analysis,* Prentice-Hall, Englewood Cliffs, New Jersey, 1990.

2. R. I. Stephens, A. Fatemi, R. R. Stephens, and H. O. Fuchs, *Metal Fatigue in Engineering,* 2nd ed., Wiley, New York, 2000.

3. N. E. Dowling, *Mechanical Behavior of Materials,* 2nd ed., Prentice-Hall, Englewood Cliffs, New Jersey, 1999.

4. J. F. Tavernelli and L. F. Coffin, Jr., "Experimental Support for Generalized Equation Predicting Low Cycle Fatigue," *Transactions of the ASME, Journal of Basic Engineering,* Vol. 84, No. 4, p. 533, December 1963.

5. S. S. Manson, discussion of Reference 4, *Transactions of the ASME, Journal of Basic Engineering,* Vol. 84, No. 4, p. 537, December 1962.

6. U. Muralidharan and S. S. Manson, "Modified Universal Slopes Equation for Estimation of Fatigue Characteristics," *Transactions of the ASME, Journal of Engineering Materials and Technology,* Vol. 110, p. 55, 1988.

7. J. Morrow, "Fatigue Design Handbook," in *Advances in Engineering*, Vol. 4, Society of Automotive Engineers, Section 3.2, Warrendale, Pennsylvania, 1968, pp. 21–29.

8. S. S. Manson and G. R. Halford, "Practical Implementation of the Double Linear Damage Rule and Damage Curve Approach for Treating Cumulative Fatigue Damage," *International Journal of Fracture,* Vol. 17, No. 2, pp. 169–172, R35–R42, 1981.

9. K. N. Smith, P. Watson, and T. H. Topper, "A Stress-Strain Function for the Fatigue of Metals," *Journal of Materials,* Vol. 5, No. 4, pp. 767–778, 1970.

10. W. Schultz, "A History of Fatigue," *Engineering Fracture Mechanics,* Vol. 54, No. 2, pp. 263–300, 1996.

11. T. Topper, R. Wetzel, and J. Morrow. "Neuber's Rule Applied to Fatigue of Notched Specimens," *Journal of Materials*, Vol. 4, No. 1, pp. 200–209, March 1969.

12. J. H. Crews, Jr., "Crack Initiation at Stress Concentrations as Influenced by Prior Local Plasticity," *Achievement of High Fatigue Resistance in Metals and Alloys,* ASTM STP 467, American Society for Testing and Materials, West Conshohocken, Pennsylvania, 1970, p. 37.

13. "Standard Practice for Cycle Counting in Fatigue Analysis, ASTM Designation E-1049-85," in *Annual Book of ASTM Standards, Volume 03.01, Metals—Mechanical Testing: Elevated and Low-Temperature Tests; Metallography,* American Society for Testing and Materials, West Conshohocken, Pennsylvania, 2000.

14. *Aging of U.S. Air Force Aircraft,* Final Report, Committee on Aging of U. S. Air Force Aircraft, National Materials Advisory Board, National Research Council, Publication NMAB-488-2, National Academy Press, Washington, D.C., 1997.

15. D. O. Sprowls, "Corrosion Testing and Evaluation," in *ASM Handbook,* Vol. 13: *Corrosion,* ASM International, Materials Park, Ohio, 1987, pp. 193–317.

16. M. G. Fontana, *Corrosion Engineering,* 3rd ed., McGraw-Hill, New York, 1986.

17. B. Craig and S. L. Pohlman, "Forms of Corrosion," in *ASM Handbook,* Vol. 13: *Corrosion*, ASM International, Materials Park, Ohio, 1987, pp. 79–189.

18. G. K. Cole, G. Clark, and P. K. Sharp, *The Implications of Corrosion with Respect to Aircraft Structural Integrity,* Defence Science and Technology Organization Research Report DSTO-RR-0102, Aeronautical and Maritime Research Laboratory, Fishermans Bend, Victoria, Australia, March 1997.

19. J. W. Lincoln, "Corrosion and Fatigue: Safety Issue or Economic Issue," in *Fatigue in the Presence of Corrosion*, RTO-MP-18, AC/323 (AVT) TP/8, Corfu, Greece, October 7–9, 1998. NATO Research and Technology Organization, Neuilly-Sur-Seine, Cedex, France.

20. J. Q. Lackey, "Designing to Minimize Corrosion," in *ASM Handbook,* Vol. 13: *Corrosion*, ASM International, Materials Park, Ohio, 1987, pp. 321–374.

21. H. E. Townsand, T. W. Cape, and K. B. Tator, "Corrosion Protection Methods," in *ASM Handbook,* Vol. 13: *Corrosion*, ASM International, Materials Park, Ohio, 1987, pp. 377–505.

22. J. Rudd, *Applications of the Equivalent Initial Quality Method,* AFFDL-TM-77-58-FBE, Air Force Flight Dynamics Laboratory, Wright-Patterson AFB, OH, July 1977.

23. P. E. Magnusen, A. J. Hinkle, A. J. Rolf, R. L. Bucci, and D. A. Lukasak, "Methodology for the Assessment of Material Quality Effects on Airframe Fatigue Durability," in *Fatigue 90*, Vol. IV, Honolulu, Hawaii, July 15–20, 1990, Materials and Component Engineering Publications, Ltd., Birmingham, U.K., pp. 2239–2244.

24. A. J. Hinkle and A. F. Grandt, Jr., "Predicting the Influence of Initial Material Quality on Fatigue Life," in A. F. Blom (Ed.), *Durability and Structural Integrity of Airframes,* Vol. I, *Proceedings of 17th Symposium of the International Committee on Aeronautical Fatigue,"* Stockholm, Sweden, June 9–11, 1993, Engineering Materials Advisory Services Ltd, West Midlands, U.K., pp. 305–319.

25. J. Luzar and A. Hug, "Lower Wing Disassembly and Inspection Results of Two High Time USAF B707 Aircraft," paper presented at the 1996 USAF Structural Integrity Program Conference, San Antonio, Texas, December 3–5, 1996.

26. A. F. Grandt, Jr. and H. L. Wang, "Analysis of Widespread Fatigue Damage in Lap Joints," *SAE 1999 Transactions, Journal of Aerospace*, Section 1, Vol. 108, pp. 262–269, 1999.

27. R. J. Bucci, R. L. Brazill, D. O. Sprowls, B. M. Ponchel, and P. E. Bretz, "The Breaking Load Method: A New Approach for Assessing Resistance to Growth of Early Stage Stress Corrosion Cracks," American Society for Metals Paper 8517-044, presented at the International Conference and Exposition of Fatigue, Corrosion Cracking, Fracture Mechanics and Failure Analysis, Salt Lake City, Utah, December 2–6, 1985.

28. J. N. Scheuring and A. F. Grandt, Jr., "A Fracture Mechanics Based Approach to Quantifying Corrosion Damage," *SAE 1999 Transactions, Journal of Aerospace*, Section 1, Vol. 108, pp. 270–281, 1999.

29. J. N. Scheuring and A. F. Grandt, Jr., "Quantification of Corrosion Damage Utilizing a Fracture Mechanics Based Methodology," in *Proceedings of The Third Joint FAA/DoD/ NASA Conference on Aging Aircraft,* Albuquerque, New Mexico, September 20–23, 1999.

30. *Metallic Materials and Elements for Aerospace Vehicle Structures,* Military Handbook MIL-HDBK-5G, Department of Defense, Washington, D.C., 1994.

31. P. A. Blatt, "Evaluation of Fatigue Crack Initiation Behavior of an Experimental Ternary Aluminum-Lithium Alloy," M.S. Thesis, Purdue University, West Lafayette, Indiana, May 1990.

PART III

NONDESTRUCTIVE EVALUATION

The objective of the six chapters in Part III is to expand the discussion of nondestructive inspection methods for detecting manufacturing and service-induced damage introduced in Chapter 4. First, the six major NDE methods commonly employed in service are developed. Chapter 10 overviews visual and liquid penetrant inspection methods, Chapter 11 is devoted to radiography, Chapter 12 to ultrasonic techniques, Chapter 13 to eddy current methods, and Chapter 14 focuses on magnetic particle inspection. Chapter 15 then concludes Part III with a discussion of other specialized inspection methods.

CHAPTER 10

VISUAL AND LIQUID PENETRANT INSPECTIONS

10.1 OVERVIEW

This chapter discusses visual and liquid penetrant inspections—two relatively simple, yet powerful techniques for finding surface-breaking anomalies. Visual inspection is the most common damage detection procedure employed in practice. It has been reported that in some industries, for example, as many as 80% of cracks are found by visual means [1]. Clearly, visual NDE is the first line of defense for safety-related failures and is often used along with other more sophisticated inspection methods. Visual inspections become even more powerful when combined with aids that magnify or accentuate the indication of interest.

One particularly useful means for emphasizing surface damage entails soaking the component with a brightly colored liquid that penetrates into surface openings. The highlighted flaws are then detected visually. Although essentially an enhancement of visual inspection, the liquid penetrant method is usually considered as a distinct NDE method in its own right. Applicable to many types of materials, liquid penetrant inspection involves flooding the surface of the test object with a low-viscosity fluid. The part is allowed to soak for a period of time while the penetrant enters into cracks and other tight crevices by capillary action. The penetrant is then carefully cleaned from the test piece, followed by application of a developer to draw out penetrant that remains between crack faces. The developer provides both a "blooming" effect to enhance the penetrant remnants and a visually contrasting surface to view the crack penetrant indication. A variety of penetrant types are available, including fluorescent versions that are viewed under an ultraviolet or black light and those that can be applied in the field with small spray cans.

10.2 VISUAL INSPECTION

Spencer [1] defines visual inspection as "the process of examination and evaluation of systems and components by use of human sensory systems aided only by such mechanical enhancements to sensory input as magnifiers, dental picks, stethoscopes, and the like. The inspection process may be done using such behaviors as looking, listening, feeling, smelling, shaking, and twisting. It includes a cognitive component wherein observations are correlated with knowledge of structure and with descriptions and diagrams from service literature." Note that this definition includes other human senses along with sight and involves both a search and a decision process. Although inspection reliability is highly dependent on the capabilities of the observer, the human eye/brain is a very sensitive and discerning inspection system that can be trained to find and interpret subtle changes in structural components.

The visual inspection technique can be divided into two major groups: direct and indirect examinations. Direct visual inspections include the use of telescopes, borescopes, periscopes, optical projectors, comparators, real-time video, or other electronic image testing as well as dye penetrant or magnetic particle inspections. Indirect visual inspections include examining photographs, radiographs, metallographs, or videotapes taken for later interpretation.

Visual aids such as light sources, magnification devices, and sensors for remote viewing are often employed to aid detection. The intensity and wavelength of the light source as well as its direction and the background lighting are important in forming shadows that highlight anomalies. Magnification sources include small hand-held magnifying glasses (2–5×), toolmaker's microscopes (10–200×), or optical comparators (5–500×) that project the component silhouette onto a screen. Other image sensors such as photographs, videotapes, or computer-enhanced images are useful for remote viewing or for permanent records of the inspection.

10.2.1 Borescopes

Many visual inspections employ industrial telescopes known as borescopes to provide optical access to closed areas or locations that are difficult to reach without time-consuming disassembly [2]. As shown in Figure 10.1, borescopes come in a variety of sizes and configurations and have many practical uses. Turbine engines, for example, often have built-in borescope access ports to facilitate inspection of internal components, and the B-1 aircraft was designed to permit borescope inspection of much of the airframe. There are also many medical applications for borescopic examination of the human body.

Borescopes come in three general types: rigid, flexible, and charge-coupled devices (CCDs). Rigid borescopes are generally limited to a straight-line path and range in length from 0.5 to 100 ft (0.15–30.5 m). They provide a typical magnification of 3–4× (although 50× is possible) and often have an incandescent lamp at the viewing end or supply illumination through a fiberoptics light guide. Borescopes generally employ a series of lenses with diameters ranging from 0.15 to 2.75 in (0.38–7.0 cm). The typical field of view is 55° but can vary from 10° to 90°. It is possible to provide

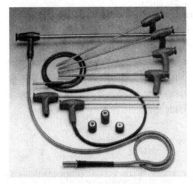

Examples of rigid borescopes   Flexible fiberscope with video display

Borescope inspection of jet engine

Figure 10.1. Examples of borescopes used to obtain visual access to enclosed quarters. (Courtesy of Olympus Industrial.)

right-angle or retrospective viewing along with direct or oblique forward views. Rigid borescopes also come in "mini" versions that are 4–7 in (10–18 cm) in length and 0.035–0.105 in (0.09–0.27 cm) in diameter.

Flexible fiberscopes are used for applications that do not have straight-line access between the viewer and test area. They are based on fiberoptics bundles that include light guides, objective lenses, interchangeable viewing heads, and remote controls for moving the tip. Diameters range from 0.055 to 0.5 in (0.14–1.3 cm), whereas individual optical fibers typically are 0.001 in (0.0025 cm) in diameter. Fiberscopes may be as long as 40 ft (12.2 m). Most provide adjustable focusing and a typical field of view of 40°–60°, although 10°–120° views are possible.

Charge-coupled device imaging provides video transmission to a remote monitor. The CCD resolution depends on aberrations in the device but is generally better than rigid or flexible fiberscopes. These sensors often have longer working lengths and can generate reticules on the screen to aid in point-by-point measurements. Since the electronic signal can be digitally enhanced, CCDs may be incorporated into automatic inspection systems. In addition, eye fatigue is often reduced given that the operator is examining the component on a video screen rather than through the eyepiece of a telescope.

Some examples of the use and limitations of visual inspections for aircraft applications are described in the following two case histories.

10.2.2   Japanese Airline Experience

Endoh et al. [3] summarize Japanese airline experience with visual inspections encountered during a three-year study (1988–1990) that involved standard field-level inspections with 3 commercial Japanese airlines and 10 aircraft types. A total of 1054 cracks were found visually in 159 individual aircraft during the three-year period. The detected crack lengths ranged from 0.02 to 14.5 in (0.05–36.8 cm). Since teardown inspections were not performed, however, the actual crack population was not determined, and it is not known what crack sizes were *missed* in these cases. Although it is not possible to obtain probability-of-detection results from these data, this study does provide an instructive benchmark on the merits of visual inspection for commercial aircraft.

The authors report that prior experience (knowledge of potential problems in the inspection area) was the biggest single factor associated with the ability to find cracks visually. Figure 10.2 compares the frequency distribution of all cracks detected when prior information was available with those found when no warnings were

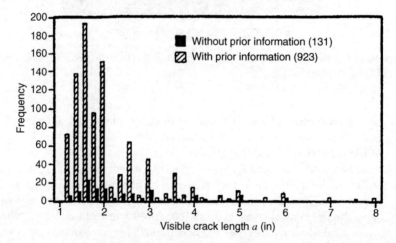

Figure 10.2. Frequency distribution of crack lengths found in commercial aircraft inspections showing effect of prior information of anticipated cracking in specific locations. Separate results for entire aircraft population, B-747 aircraft only, and all aircraft examined except for B-747 [3].

provided [3]. Note that prior experience involved 90% of all cracks located, including the smaller crack sizes. Moreover, the median detected crack size for the "prior experience" cracks was 0.67 in (1.70 cm), whereas the "no experience" cracks had a median length of 1 in (2.54 cm). Clearly, expectation that cracks could be present is an important human factor that influences the effectiveness of visual inspections.

The distance between the inspector and the test object was also significant, as more cracks were found when the inspector was within 20 in (50 cm) of the test piece. Surface cleanliness was important, and the authors also noted that more cracks were found at free edges than at holes.

Aircraft inspections are generally ranked from A to D level according to increasing inspection time and decreasing frequency. An A-level check, for example, involves only a few hours of time and is conducted quite often. A B-level inspection occurs approximately once a month and may require removal of access panels to examine various components. The C-level check follows approximately 5000 hours of service, and entails in-depth inspection and maintenance. The most extensive D-level check follows several years of service (approximately 20,000 flight hours) and might take several months to complete. Exterior paint is removed and the aircraft interior is completely stripped to provide detailed access for close inspection of major components.

Endoh et al. [3] found that the maintenance activity or the structural location had little or no influence on the inspection results. There was, for example, little difference between a "C-level check" or a higher inspection activity. Moreover, the results did not seem to depend on whether inspections were performed according to a service bulletin or resulted from a supplemental or additional inspection. Although most inspection locations were internal fuselage structure, when normalized with the number of occurrences per category, similar crack size distributions were found in fuselage, wing, empennage, pylon, and door locations.

10.2.3   Federal Aviation Administration (FAA) Benchmark Study

Spencer [1] describes the results of a 1995 effort by the FAA Aging Aircraft Validation Center to benchmark visual inspection capability. Although conducted in a quasi-laboratory environment, this study sought to quantify the capabilities of visual inspections for commercial aircraft. Inspections were conducted on a retired Boeing 737 as well as on several other large test panels designed to represent typical joints in aircraft structures. The B-737 had been retired after more than 38,000 hours of flight service and was in a stripped condition typical of a D check. The other manufactured test panels contained cracks that were well characterized with respect to their sizes, types, and locations.

Twelve commercial airline inspectors participated in the study. All were qualified as visual inspectors by their respective employer. While employed by one of four commercial airlines, they represented seven different maintenance facilities. Effort was taken to give each inspector the same equipment, tools, instructions, and so on, and to simulate the inspection environment that would be encountered in a typical airline maintenance depot.

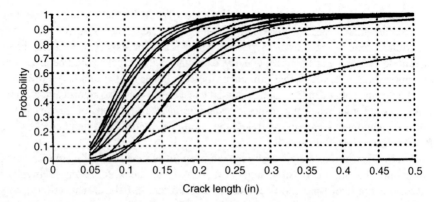

Figure 10.3. Mean probability of detection curves for 12 inspectors visually examining a large lap-splice panel [1].

The mean probability-of-detection (POD) curves obtained for the 12 inspectors as a function of crack length are given for one large lap-splice panel (Task 701) in Figure 10.3. Note that the mean 90% POD crack lengths for 11 of the 12 inspectors varied between 0.16 and 0.36 in (0.40 and 0.91 cm). The twelfth inspector performed very poorly on this specimen and resulted in a 90% POD crack size of 0.91 in (2.31 cm). Much more variation was found in the inspection of the actual B-737 aircraft, as shown in Figure 10.4, where the B-737 detection rates are shown by the X's as a function of crack size. (The Task 701 triangles represent the lap-splice panel results.) The 50% POD crack length for the B-737 inspections was 4.5 in (11.4 cm) in this case, as compared to 0.13 in (0.33 cm) for the lap-splice panels given above [1].

Spencer reports several observations and conclusions from this visual inspection benchmark. First, the inspection results can be very task dependent, and one must be cautious in applying results from one task to another. The inspector who performed

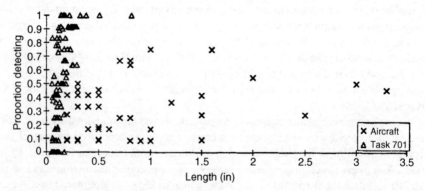

Figure 10.4. Comparison of visual detection rates for B-737 aircraft and a large test panel Task 701 [1].

the worst with the lap-splice panels, for example, was one of the better inspectors on the actual aircraft.

The search and decision processes also varied considerably among the 12 inspectors. Several stopped looking when they located a crack in a given area, stating that the entire component would be replaced or repaired. This approach led to many large cracks not being reported in the B-737 test aircraft inspections and was one cause for the significantly poorer POD results there. This practice was especially common when dealing with corrosion damage. While stopping inspection after a component has been tagged as "bad" may be an acceptable practice, there are potential concerns here. Does the inspector, for example, have sufficient knowledge that an entire component or assembly will be replaced or could an erroneous assumption be made in this regard? What if different inspectors are used before and after the repair? The main issue here is the need to ensure all damage in a given area is repaired, whether the inspector specifically highlights it or not.

Expectations of finding cracks in a given area were again a very important human factor. Prior knowledge of "hot spots" or guidance given by a task card led to much more diligence and success in those areas. Sometimes large cracks in nearby areas were missed, however, when too much focus (i.e., "tunnel vision") was given to a particular detail. The search component as compared to the decision process was determined to be of greater importance for these visual inspections. Experience was also judged to be a key factor.

In summary, the advantages of visual inspections are that they are relatively inexpensive and highly portable, provide immediate results, involve minimum training, and require no special safety precautions. Disadvantages of this procedure include the fact that only surface defects are located, the parts require cleaning, and there must be visual access for the inspector. Other problems include misinterpretation of scratches and sensitivity to the direction of illuminating light. Since a person's vision is affected by the color of light and human factors such as age, mood, attitude, and fatigue, calibration of visual inspections are difficult on a day-to-day basis. Thus, while visual inspections are a very important and widely used NDE method, their POD is low relative to the other more sophisticated NDE methods described in the following section and in subsequent chapters.

10.3 LIQUID PENETRANT INSPECTION

Liquid penetrant inspection is a relatively simple but effective technique for locating surface-breaking defects. Applicable to many types of materials, this NDE method involves flooding the surface of the test object with a low-viscosity fluid. After the part is allowed to soak for a period of time (dwell) so that the penetrant enters into tight crevices by capillary action, the penetrant is carefully cleaned from the surface of the test piece. A developer is then applied to draw out any penetrant that remains between crack faces. The developer provides both a contrasting background to view the penetrant indication and a "blooming" effect to enhance visibility of any penetrant remnants left in the crack. A variety of penetrant types are available,

TABLE 10.1   Crack Detection Limits for Dye Penetrants Inspections

Type of Inspection	Crack length in inches (mm)
Production parts—laboratory inspection	0.005 (0.127)
Production parts—production inspection	0.03 (0.76)
Structure—service inspection	0.05 (1.27)

Note: See Reference 4.

including fluorescent versions that are viewed under a black light and those that can be applied in the field with small spray cans.

The "oil and whiting" technique is an early version of the penetrant method developed in the late 1800s to inspect locomotive components [4]. Here the test objects were coated with lubricating oil and then painted with a mixture of white chalk and denatured alcohol. The alcohol evaporated, leaving the chalk coating. The presence of cracks was then indicated by oil that bled from cracks and discolored the white background. Colored dyes were added to the thinned oil during the 1930s, coining the term *dye penetrant.* Fluorescent penetrants were developed during the following decade to further enhance crack visibility. Penetrant inspections quickly became a complement to the magnetic particle inspections used at that time to locate surface cracks in ferromagnetic materials.

Although minimum crack size detection limits are not well defined, penetrant inspections are a powerful NDE tool. Crack detection limits, defined as *the minimum size or length of crack that can be overlooked,* are given in Table 10.1 [4]. This table only provides a rough indication of the crack sizes that can be found by penetrant inspections, however, and should not be taken as a rigorous evaluation of NDE capability.

A dye penetrant inspection consists of the steps outlined below and shown schematically in Figure 10.5. Note that these steps may be accomplished with penetrant materials that are available in small spray cans for portable field-level inspections or may be performed in larger facilities geared toward a production environment (see Figure 10.6).

Surface Cleaning   As shown in Figure 10.5, surface cleaning is the first (and key) step of a penetrant inspection. It is important to remove all coatings, dirt, grease, scale, and so on, from the surface of the test object, since successful inspection depends on the penetrant entering all discontinuities on the component surface. Penetrants are expensive, and because impurities could react with the penetrant, it is essential to keep the bath clean for reuse. Moreover, contamination could retain penetrant that leads to false indications or mask the discontinuity of interest. The surface must be thoroughly dried following cleaning to prevent the penetrant from being diluted by cleaning fluids. Mechanical cleaning should be used with caution due to the possibility of "peening" the component surface and closing crack faces. Shot

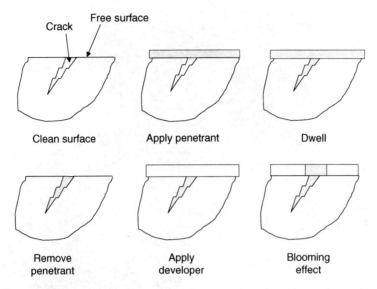

Figure 10.5. Schematic representation of the steps in a dye penetrant inspection.

cleaning is not permitted by most USAF facilities, for example, due to its "pinching" of crack surfaces.

Penetrant Application   Next, the penetrant is liberally applied to all surfaces of the test object by dipping, spraying, brushing, and so on. A dwell period then allows the low-viscosity penetrant to "soak" into surface flaws by capillary action. It is essential that there be an adequate period for the penetrant to enter all discontinuities. The dwell time depends on crack size, shape, type of penetrant, temperature, and surface inclination of test object. Dwell periods typically range from 3 minutes to 2 hours, depending on the particular application.

Penetrant Removal   The next step is to thoroughly cleanse the surface of all excess dye, leaving only penetrant that has soaked into surface defects. Although all background penetrant must be removed, it is essential to leave the penetrant that has entered crack surfaces. It is also important not to overwash at this stage. Overwashing can be a problem with shallow discontinuities or those that have wide openings. Penetrant removal is, perhaps, the most difficult and important step in the test procedure.

Developer   Application of a developer is the next step in this NDE process. The developer draws penetrant from the discontinuity to the surface and provides a contrasting background to view the penetrant indication. The penetrant also soaks the developer, resulting in a "blooming" effect that aids visual detection of the enlarged indication.

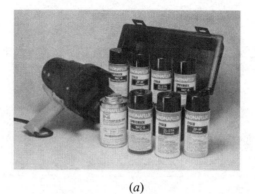

(*a*)

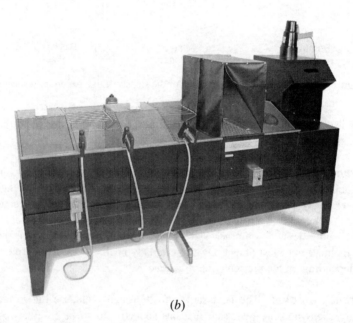

(*b*)

Figure 10.6. (*a*) Portable kit with spray cans of fluorescent penetrant materials and (*b*) commercial dye penetrant system for inspecting small components. Note series of work stages for cleaning specimen, applying and removing penetrant, application of developer and final viewing with a black light. (Photographs courtesy of Magnaflux, a division of ITW.)

Visual Examination   Visual examination of the component surface then locates discontinuities that are highlighted by the contrasting penetrant color (e.g., fluorescent penetrant excited by a black light or a red dye contrasted against a white developer). Visual aids such as magnifying glasses and scales can help locate and quantify the indications. The final step is to clean the surfaces once again to remove final traces of all penetrant materials and to protect the component against potential corrosion.

10.4   CLASSIFICATION OF PENETRANT METHODS

Penetrant inspection may be classified by the type of penetrant, removal method, and developer employed. Indeed, one of the strengths of this inspection method lies in the fact that there are several options in performing the various steps that can be adapted for particular applications [4–6]. While penetrant testing is deceptively simple, proper selection and application of penetrant, developer, cleaner, and so on, are essential for optimum results. Desirable properties for liquid penetrants include the following characteristics [4]:

- Ability to penetrate fine voids
- High wetting ability
- Surface tension lower than that of the component being tested so as to spread and enter crack faces
- Uniformity of properties
- Slow drying
- Noncorrosive
- Relatively low viscosity
- High flash point
- Easily removable
- High and stable brightness

10.4.1   Penetrant Types

There are two general types of penetrants: fluorescent (Type I) and nonfluorescent (Type II).

Type I. Fluorescent Penetrants   These penetrants fluoresce brightly under ultraviolet light when viewed in a darkened room. Fluorescent penetrants are usually green in color (due to greater eye sensitivity) and are available in several sensitivity levels (i.e., ability to give indications in dark areas):

- Level $\frac{1}{2}$—ultralow
- Level 1—low
- Level 2—medium
- Level 3—high
- Level 4—ultrahigh

Type II. Visible Penetrants (Nonfluorescent)   Nonfluorescent penetrants are usually red in color and are viewed under normal white light, avoiding the need for a black light or a darkened room. They are less sensitive than the fluorescent versions and are generally regarded to have the level 1 sensitivity of the fluorescent penetrants mentioned above.

10.4.2 Penetrant Removal Methods

As mentioned previously, it is very important to remove all penetrant from the component surface without affecting the penetrant that has seeped into discontinuities. Both Type I and Type II penetrants may be removed by the following four techniques.

Method A: Water-Washable Penetrant   These penetrants contain an emulsifier and may be removed directly by a water rinse following the dwell period. Although these are the simplest penetrants to remove, care must be taken to avoid overwashing.

Method B: Oil-Based (Lipophilic) Emulsifiers   This removal method employs an oil-based emulsifier to treat penetrants that are insoluble in water. The emulsifier, which is usually applied after a prerinse, makes the penetrant water soluble so that it may be washed off with water. Determining the proper emulsification time is a key step in this process.

Method C: Solvent-Removable Penetrant   Here the penetrant is removed with the same solvent used for the initial precleaning step. The solvent is usually sprayed or dipped and can be labor intensive. Method C is convenient for field use, however, and can be one of the most sensitive penetrant removal techniques. This method is only used for inspection of local areas or for on-site work.

Method D: Water-Based (Hydrophilic) Emulsifier   This penetrant removal technique is similar to method B in that a separate emulsifier is used to make the penetrant soluble in water. Water-based emulsifiers are employed here, however, and although they are slower acting than oil-based emulsifiers, it is easier to control the cleaning action. A prerinse step is required before applying the emulsifier, and water is again used for the final wash.

10.4.3 Developers

Developers increase the visibility of the penetrant indication by giving a uniform diffuse or matte background contrast for the test object surface and by providing a "blotting" action that draws penetrant from crack surfaces to yield an expanded indication. Developers also increase the brightness of fluorescent indications. Developers come in the following forms [5].

Form A: Dry Powder   Here the test piece is dried thoroughly after the penetrant has been removed, and a dry, white powder is dusted over the component. Although powder developers are widely used with fluorescent penetrants, they do not give suitable contrast for use with nonfluorescent dyes and are considered the least sensitive of the developers. Both the powder and test piece must be thoroughly dry to permit proper flow and dusting of the developer. Care must be taken with these powders to prevent inhalation or drying of the operator's skin.

Form B: Water Soluble Form B developers are used for both fluorescent and visible penetrants, although they are not recommended for water-washable penetrants. They are supplied as a dry powder concentrate that is mixed with water and have several advantages for penetrant inspections. The prepared bath does not require further agitation, and since they are applied prior to drying the component, development time is decreased. The dried developer is water soluble and easy to remove with a final water rinse.

Form C: Water Suspendible Form C developers are also used with both fluorescent and visible penetrants. They are supplied as a dry concentrate that is dispersed in water in a carefully controlled suspension. They are applied before drying the test piece following penetrant removal. The developing time is decreased because the dryer heat brings out the penetrant, which acts with the developer during the drying stage.

Form D: Nonaqueous Solvent Suspendible Form D developers are used with both fluorescent and visible penetrants. They produce a white coating on the test piece for good matte contrast. They are supplied in a ready-to-use form and are the most sensitive developer for fluorescent penetrants. The solvent action aids absorption and adsorption mechanisms, permitting contact with penetrant that is entrapped in tight cracks.

10.5 PENETRANT APPLICATIONS

By employing different combinations of penetrants, penetrant removal methods, and developers, one may tailor penetrant NDE for a particular application. Although many permutations are possible, the six major variations summarized below are most popular and are listed in order of decreasing sensitivity and decreasing cost [4]:

1. *Postemulsifiable Fluorescent Penetrant (Type I, Method B or D)* This combination is the most reliable and sensitive approach for locating tight cracks and is ideal for production work where large numbers of similar objects are to be examined. The approach is also well suited for wide, shallow discontinuities and for applications that require different sensitivity levels. The water-based emulsifier (method D) is preferred to the oil-based version (method B) in that it is more reliable and sensitive. The extra step of applying an emulsifier increases costs, however, and a black light and darkened area are required for detecting the indications.

2. *Solvent-Removable Fluorescent Penetrant (Type I, Method C)* This approach is recommended for spot inspections where it is inconvenient to use emulsifiers or water for removal of the penetrant. Although highly portable, solvent removal of the penetrant is time consuming for large areas. The method is, however, commonly employed for inspection of large aircraft in the field.

3. *Water-Washable Fluorescent Penetrant (Type I, Method A)*   The fastest of all the fluorescent penetrant procedures, the penetrant is removed here directly by water without employing postemulsifying agents or solvents. The approach is effective on large surface areas but does not have the sensitivity of the first two techniques described above.

4. *Postemulsifiable Visible Penetrant (Type II, Method B or D)*   Suitable for large surface areas, this combination is the most sensitive of the visible penetrant techniques. The extra emulsification step does, however, add time and cost.

5. *Solvent-Removable Visible Penetrant (Type II, Method C)*   The main advantage of this approach is its portability for field use, although solvent penetrant removal is labor sensitive. The method is usually limited to spot applications.

6. *Water-Washable Visible Penetrant (Type II, Method A)*   This is the fastest and simplest of all penetrant methods since the penetrant is simply washed off the test piece with water. The visible penetrant does not require darkened areas or power for a black light. While suited for large surface areas, it is, however, the least sensitive of all the penetrant NDE methods.

10.6 SAFETY CONSIDERATIONS

Changes in environmental protection standards have placed increased restrictions on many of the chemicals once used with penetrant inspections. Certain materials employed in previous fluorescent penetrants were irritants or even carcinogenic but have been replaced in modern fluorescent dyes. Although these restrictions have increased inspection costs in some cases, the NDE operator is no longer exposed to potentially hazardous cleaning compounds and solvents. As indicated below, however, care should still be exercised in working with penetrant NDE materials [4].

Gloves, aprons, masks, and/or filters should be worn when working with many of the chemicals. While not major health hazards, the toxicity of dyes can, for example, cause skin irritation and problems with allergies. Powder developers may dry skin, air passages, and the lungs. Water-washable penetrants and emulsifying agents may also irritate skin. Although black light is not harmful, the ultraviolet light from an arc lamp will burn skin or eyes if the filters are missing or broken. Moreover, many of the volatile solvents used for cleaning are highly flammable, and the cleaning area should be well ventilated to avoid accumulation of hazardous fumes.

The chemical compatibility of the test piece with various penetrant materials must also be considered to avoid component damage during the inspection. Although penetrants have been designed for testing specific materials (e.g., aluminum, steel, titanium), one must guard against potential corrosive action with the test object. Chlorides, for example, can be harmful to high-strength/high-temperature materials (e.g., nickel-based alloys, austenitic steels, titanium, aluminum). One must also anticipate the potential for adverse reactions with chemical residues left on the test piece. Inspection of liquid oxygen tanks, for example, requires selection of penetrants and cleaning materials that will not cause explosive reactions with the liquid oxygen residues.

10.7   SUMMARY OF LIQUID PENETRANT NONDESTRUCTIVE EVALUATION

The advantages and disadvantages of liquid penetrant NDE are briefly summarized below.

10.7.1   Advantages of Liquid Penetrant Inspection

1. Penetrant inspections are portable and well suited for fieldwork.
2. The technique is relatively inexpensive and requires minimal skills, although processing time can add to costs.
3. Penetrants are sensitive to very small surface discontinuities.
4. The method can be applied to many nonporous materials and to a wide range of test object sizes and irregular shapes.
5. Flaw orientation does not usually pose a problem.
6. Large objects can be spot checked with local application of penetrants or covered completely for full inspection.
7. There is a wide range of penetrant materials for various applications.
8. The flaw indication appears directly on the test piece rather than on another image or piece of instrumentation.
9. Penetrant inspection systems can be designed for high-volume production.

10.7.2   Disadvantages of Liquid Penetrant Inspection

1. Only *surface* anomalies that actually absorb penetrant are detected.
2. One must have access to the surface of the test object, and rough or porous surfaces cause problems.
3. Compressive residual stresses at the component surface (due to, e.g., shot peening and fatigue crack closure) can hamper penetration.
4. Significant surface preparation is required for good results (e.g., removal of finishes, sealants, contaminants).
5. Stricter environmental safety standards regarding the solvents used to clean specimen surfaces have significantly increased the costs associated with this technique.
6. Meticulous cleanliness is required to prevent contamination of penetrants.
7. The chemical compatibility of structural and penetrant materials must be considered (e.g., some polymers and rubbers may be damaged by various penetrant chemicals).
8. Penetrant NDE is a fairly slow method, requiring surface preparation, dwell period, clean-up, and so on.
9. Sharp corners and complex shapes can give false indications.

10. Proper lighting and ventilation are required.

11. Direct visual examination of surfaces is required to locate indications, introducing the opportunity for human fallibility. Indeed, a strong human element is involved in many inspection steps, leading to occasional problems with repeatability and the need for more automated systems.

12. Direct reflection of ultraviolet light from the component surface can cause the eye to fluoresce and hide penetrant indications.

10.8   CONCLUDING REMARKS

The human eye is the most common sensor for locating structural damage, and visual inspection is the most widely used of the major NDE methods. There are a variety of visual aids that increase the sensitivity and reliability of this approach—light sources, mirrors, magnifying devices, borescopes, and so on. Liquid penetrant inspection is an extension of visual detection that employs brightly colored dyes to highlight surface anomalies. There are many variations of penetrant inspection—fluorescent/nonfluorescent dyes, penetrant removal methods, different developers, and so on. Thus, liquid penetrant inspection has developed into a separate, stand-alone NDE method.

Both visual and liquid penetrant inspections are limited to detection of surface-breaking damage and require clean component surfaces to be successful. The inspector must have visual access to the test object, but both methods are highly portable and relatively easy to conduct. In both cases the technician directly examines the test piece for flaw indications rather than a reconstructed image or electrical signal. Both methods are applicable to a wide variety of structural materials and have few of the limitations inherent in some of the more sophisticated inspection techniques discussed in the following chapters. Thus, visual and liquid penetrant NDEs see wide applications in many different types of industries. Additional information about these two methods may be found in References 2, 4–8.

PROBLEMS

10.1   Why are visual inspections sometimes referred to as the first line of defense provided by NDE?

10.2   What are some aids that are frequently employed with visual inspections?

10.3   What other human senses besides sight contribute to the success of a visual inspection?

10.4   Use the Internet to locate the home page for a vendor that supplies borescopes for visual inspection purposes and record this company's address for future reference. Examine the borescope products and applications described on the web site and prepare a short report that describes the various capabilities and uses for the borescopes sold by this company.

10.5   Give some typical applications for borescopes.

10.6   Briefly describe the three general types of borescopes.

10.7   Discuss some human factors associated with visual inspection.

10.8   Give some examples of visual cues that might help an inspector find flaws by visual NDE.

10.9   Give examples of flaw characteristics that would make a particular type of flaw easier to detect visually. Give some example characteristics that would complicate visual inspection of a particular type of flaw.

10.10   Use the Internet to locate the home page for a vendor that supplies liquid penetrant inspection materials and record that company's address for future reference. Examine the NDE products and applications described on the web site and prepare a short report that describes the various capabilities and uses for the liquid penetrant items sold by this company.

10.11   Briefly discuss the principles and operation of the liquid penetrant inspection method.

10.12   Briefly summarize the classification procedure used to characterize liquid penetrants and their applications.

10.13   What role does the developer play in a penetrant inspection? Briefly discuss the various methods for applying developers.

10.14   Define the dwell period for penetrant NDE and discuss its significance.

10.15   What restrictions, if any, are associated with the component material that can be successfully examined by liquid penetrants?

10.16   What are some potential sources for false indications by the liquid penetrant method (Type I errors)? What are potential sources for Type II penetrant inspection errors that could lead to missing cracks in the test piece?

REFERENCES

1. F. W. Spencer, *Visual Inspection Research Project Report on Benchmark Inspections,* Technical Report DOT/FAA/AR-96/65, U.S. Department of Transportation, Federal Aviation Administration, Washington, D.C., October 1996.
2. "Visual Inspection," in *ASM Handbook,* Volume 17, ASM International, 1989, Materials Park, Ohio, pp. 3–11.
3. S. Endoh, H. Tomita, H. Asada, and T. Sotozaki, "Practical Evaluation of Crack Detection Capability for Visual Inspection in Japan," in *Durability and Structural Integrity of Airframes,* Proceedings of the 17th Symposium of the International Committee on Aeronautical Fatigue, Vol. I, Engineering Materials Advisory Services, Ltd, West Midlands, U.K., Stockholm, Sweden, June 9–11, 1993, pp. 259–280.
4. D. E. Bray and D. McBride, *Nondestructive Testing Techniques,* Wiley, New York, 1992, Chapter 19.

5. J. S. Borucki, "Liquid Penetrant Inspection," in *ASM Handbook,* Vol. 17, *Nondestructive Inspection and Quality Control,* ASM International, Materials Park, Ohio, 1989, pp 71–88.

6. R. Halmshaw, *Non-Destructive Testing,* 2nd ed., Edward Arnold, London, 1991, Chapter 7.

7. M. W. Allgaier and S. Ness, *Visual and Optical Testing—Nondestructive Testing Handbook*, 2nd ed., Vol. 8,c American Society for Nondestructive Testing, Columbus, Ohio, 1993.

8. N. Tracy and P. O. Moore, *Liquid Penetrant Testing—Nondestructive Testing Handbook,* 3rd ed., Vol. 2, American Society for Nondestructive Testing, Columbus, Ohio, 1999.

CHAPTER 11

RADIOGRAPHIC INSPECTIONS

11.1 OVERVIEW

This chapter deals with radiography, one of the oldest NDE methods for locating internal and external anomalies. The technique exploits the characteristic of many materials to differentially absorb and pass high-energy radiation. Variations in component density lead to different absorption rates as the radiation goes through the test object and creates an image of the internal structure on the recording plane. The source of radiation may be X-rays that are generated in a vacuum tube, gamma rays emitted from radiographic isotopes, or neutrons. Radiography is especially suited for locating nonplanar defects such as voids or porosity but can also identify planar flaws when they are located parallel to the test beam. Health hazards posed by human exposure to X-rays and gamma rays require compliance to strict safety procedures and significantly add to inspection costs and limit application of this NDE technique.

11.2 BACKGROUND

The electromagnetic spectrum is given in Figure 11.1 in terms of the wavelength λ, frequency f, and energy E of radiation. These electromagnetic quantities are related by the Equations

$$c = \lambda f \tag{11.1}$$

$$E = hf = \frac{hc}{\lambda} \tag{11.2}$$

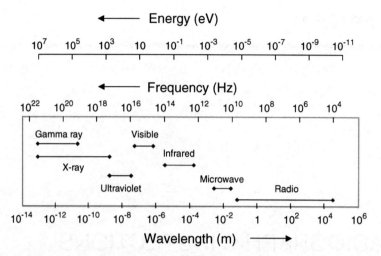

Figure 11.1. Electromagnetic spectrum in terms of wavelength, frequency, and photon energy, showing the extremely short wavelengths of X-rays and gamma rays.

where λ = wavelength of radiation (m)
f = frequency of radiation (Hz)
E = photon energy (J)
c = speed of light, 3×10^8 m/s
h = Planck's constant = 6.624×10^{-34} J-s

As discovered by Conrad Roentgen in 1895 [1, 2], X-rays penetrate many solid materials, allowing them to "see inside" solid objects. This penetration ability is due to the fact that X-rays and gamma rays lie at the short-wavelength (10^{-9}–10^{-14}-m) and high-energy end of the electromagnetic spectrum. A schematic drawing of an X-ray inspection is given in Figure 11.2. As indicated, radiation passes through the test object to the recording plane (e.g., film plate) where its intensity is measured. The radiation intensity depends on the distance traveled through the object and the

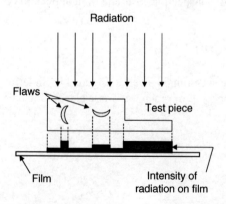

Figure 11.2. Schematic representation of X-ray image formed on film plane.

attenuation characteristics of the test material. Since inhomogeneities (e.g., voids, casting defects, cracks) absorb different amounts of radiation than the parent material, they result in different intensity levels at the image plane. Note that the maximum changes in intensity will occur when the major void dimensions are parallel to the X-ray beam.

Utilizing X-rays for NDE involves four major steps: (1) exposure of the test object to radiation, (2) processing image data from a sensing device placed on the opposite side of the test piece, (3) interpretation of the image, and (4) reporting of results in terms that are easily understood. The radiographic inspection process requires a radiation source, test piece, and sensing material. As discussed in Section 11.4, the radiation source could be X-rays (produced in a vacuum tube), gamma rays (from the decay of radioisotopes), or subatomic particles formed during nuclear decay (e.g., alpha and beta particles, neutrons). The recording medium could be film, a fluorescent screen, or some other electronic means. General X-ray techniques can be classified in the following three ways:

Film or Paper Radiography Here the image is recorded on film and developed for later viewing.

Real-Time Radiography In this case the image is immediately displayed on a viewing screen or monitor, providing the opportunity to manipulate the test piece during examination to ensure proper orientation with respect to potential flaws.

Computed Tomography (CT) This process involves generation of cross-sectional views that are easier to interpret than planar projections of radiation intensity and provides information regarding anomaly location, size, and shape.

11.3 DIFFERENTIAL ABSORPTION/SCATTERING

The key to radiographic inspection lies in the fact that radiation is differentially absorbed (or scattered) as it passes through the test object. Thus, the intensity of radiation that reaches the image plane varies and depends on the attenuation characteristics of the material along the line of passage. This variable absorption by the inspected object leads to the radiographic image of interest for NDE purposes.

The intensity of radiation traveling in a vacuum decreases with the inverse square of the distance from the source. In addition, as given by Equation 11.3, the radiation intensity decreases further as it passes through an object and is scattered or absorbed:

$$I_x = I_0 e^{-\mu x} \qquad (11.3)$$

where I_0 = initial radiation intensity at surface of object
I_x = intensity at distance x below surface
μ = coefficient of linear absorption (in reciprocal length)

Linear absorption coefficients (μ) for lead, iron, and aluminum are shown in Figure 11.3 [3] as a function of the applied X-ray energy. Note that the coefficient of absorption depends on both the radiation energy and the test material, with lead

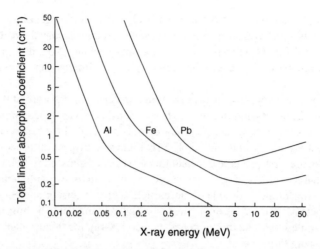

Figure 11.3. Total linear absorption coefficients as function of X-ray energy for lead, iron, and aluminum. (From [3], © 1991 R. Halmshaw, reproduced by permission.)

absorbing more radiation than either iron or aluminum. Equation 11.3 may be used to determine the thickness of material that will decrease the radiation of a given energy by a specified amount. Solving Equation 11.3 for the distance of travel x gives

$$x = -\frac{\ln(I_x/I_0)}{\mu} \tag{11.4}$$

Suppose, for example, that one is interested in finding the distance of travel in a given material that will decrease the X-ray intensity by 50%. Now

$$\ln\left(\frac{0.5}{1}\right) = -0.693$$

and the half-value thickness (HVT) is given as

$$x = \frac{0.693}{\mu}$$

Thus, from Figure 11.3, one can determine that a 0.5-MeV X-ray will lose half of its energy after traveling approximately $0.693/1.7 = 0.4$ cm (0.16 in) through lead. By comparison, the corresponding HVT of aluminum is $0.693/0.23 = 3$ cm (1.2 in). The tenth-value thickness TVT (i.e., the distance required to lose 90% of the X-ray energy) is readily obtained from Equation 11.4 as $x = 2.303/\mu = $ TVT. Again, from Figure 11.3, the TVT for 0.5-MeV X-rays passing through lead and aluminum are approximately 1.4 and 10 cm (0.56 and 4 in), respectively.

Although absorption and scattering of radiation energy are very complex phenomena, the four general mechanisms described below are important to radiography. The particular operating mechanism(s) for a given situation depends on the energy E of the X-ray (or gamma-ray) photon.

Rayleigh Scattering   This mechanism, also known as coherent scattering, occurs at low radiation energies ($E < 0.1$ MeV) but has a minor role in radiography. In Rayleigh scattering, the X-ray photon is deflected but does not change energy. The photon scattering is mostly in the forward direction.

Photoelectric Effect   This absorption mechanism predominates for low-energy X-rays ($E < 0.5$ MeV). Here the incipient photon strikes an inner shell electron of the object atom, resulting in ejection of that electron. *All* of the photon energy is transferred to the kinetic energy of the electron, and, thus, the incident X-ray is completely absorbed by the test object atom.

Compton Scattering   This phenomenon, also known as incoherent scattering, occurs at intermediate energy levels (0.1 MeV $< E < 3.0$ MeV) and results in the applied radiation photons being deflected and losing a *portion* of their energy. Compton scattering occurs when the incident photon strikes an outer shell electron in the object atom. These electrons are scattered, with some (but not all) of the photon energy being transferred to the scattered electron. Although the original incident photon loses energy, it continues to travel in a deflected direction. This "secondary" radiation may reach the image plane in an indirect manner, however, and cause "background" noise or it may collide with another atom, resulting in multiple scattering in the test object.

Pair Production   This scattering mechanism only occurs at X-ray energy levels greater than 1.02 MeV. In this case, the incident photon collides with the atom nucleus, resulting in the ejection of an electron and a positron. The positron is short lived, however, and soon collides with another electron. This subsequent collision of the positron and electron results in emission of two new photons. Both new photons have the same energy $E = 0.51$ MeV but travel in opposite directions. Thus, this process again results in radiation with reduced energy and different directions of travel. The new photons may, of course, again scatter by one of the other mechanisms described above.

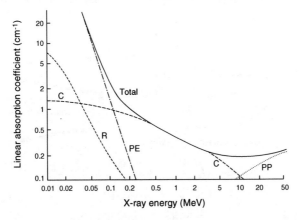

Figure 11.4. Components of linear absorption coefficient for iron as a function of X-ray energy: C, Compton scatter; R, Rayleigh scatter; PE, photoelectric scatter; PP, pair production. (From [3], © 1991 R. Halmshaw, reproduced by permission.)

A comparison of the relative contribution of these various scattering mechanisms to the linear absorption coefficient μ for iron is given in Figure 11.4 [3]. Note that there is a particular energy level associated with minimum attenuation (μ), and, thus, an optimum radiation energy for maximum depth of penetration.

11.4 SOURCES OF RADIATION

The goal of this section is to describe the sources of radiation used for nondestructive inspection. As indicated below, one may employ X-rays generated in a vacuum tube, gamma rays emitted from radioisotopes, or neutrons.

11.4.1 X-Rays

X-rays are produced in a vacuum tube when high-energy electrons strike a high-density target [3–5] embedded in a copper anode, as shown schematically in Figure 11.5. The target is usually tungsten, but gold and platinum can be used if the target is cooled properly. The filament current controls the filament temperature and the quantity of electrons emitted and determines how much time is needed for the exposure. The tube voltage (i.e., the anode/cathode potential) controls the energy of the electrons that strike the target and the energy of the subsequent X-ray beam.

Electrons striking the target atoms collide with their nuclei, leading to rapid deceleration of the electron and resulting in emissions of quanta of X-rays. Some electrons may, however, first collide with various shell electrons around the atom and lose energy before striking the nucleus. These collisions result in the "specific" and "broad-spectrum" types of X-rays shown in Figure 11.6. *Specific wavelength* (or characteristic) X-rays are associated with shell electrons that "jump" from one orbit to another to fill voids created by ejected electrons. These are the L and K characteristic X-rays of the target material that are used for X-ray diffraction and spectrometry measurements but are of minor importance for industrial NDE. *Broad-spectrum*

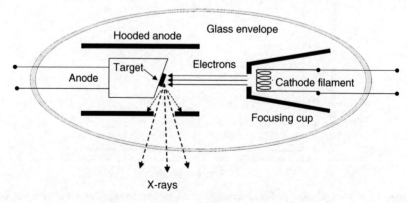

Figure 11.5. Schematic drawing of X-ray tube.

wavelength X-rays are formed with many wavelengths and energy levels since the electrons that eventually strike the nucleus have different energies. These are called "white radiation" or "breaking radiation" and are the X-rays used for NDE.

The number of X-rays formed depends on the quantity of electrons that strike the target and is controlled by the tube current. Although many wavelengths are emitted from the tube, X-rays are frequently characterized in terms of their minimum wavelength λ_{min} (or maximum energy level). Since the tube voltage specifies the electron energy striking the target and controls the X-ray wavelengths, X-ray output is often defined with respect to the voltage associated with λ_{min} of the X-ray spectrum (i.e., 200-keV or 20-MeV X-rays). Recalling Planck's law from Equation 11.2, λ_{min} describes the highest energy X-ray ($\lambda_{min} = hc/E$) and, thus, specifies the maximum penetrating power of the X-rays emitted by the tube.

There are several target considerations associated with operation of X-ray tubes: target heating, the duty cycle, and the source width. Target heating is significant in that only 1–10% of the electron energy that impacts the target is converted to X-rays, whereas the rest is given off in heat. Thus, the target must be cooled in some manner (e.g., air, water, oil), and heat removal is a serious consideration in tube design. The duty cycle (DC) specifies the percentage of time that the X-ray tube can be energized due to heat buildup. Inspection of field welds might require a duty cycle of 30%, whereas laboratory uses might need a DC of 60–70%. Fluoroscopic applications require a DC of 100%.

X-ray tubes may also be classified with respect to the focal spot size (i.e., source width), and in general, sharper images are obtained with smaller spot sizes. Conventional X-ray tubes have focal spot size diameters of approximately 0.08–0.2 in (2–5 mm). There are also minifocus tubes with focal spot sizes of 0.008–0.03 in (0.2–0.8 mm) and microfocus tubes with adjustable spot sizes of 0.0002–0.002 in (0.005–0.05 mm). Since these microfocus tubes are hard to cool, they must operate at lower voltages and yield less penetration. In addition, the targets must be replaced often.

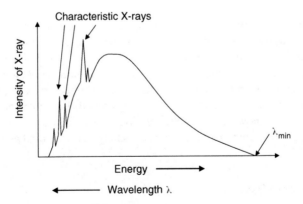

Figure 11.6. Distribution of X-ray intensity as a function of wavelength and voltage showing specific (characteristic) and broad-spectrum wavelengths.

Microfocus tubes are used to inspect integrated circuits and other miniature electronic components and can be placed inside very small areas (e.g., tubing and aircraft structures).

High-Energy X-rays Conventional X-ray tubes are limited to approximately 450 kV, and special designs are required to go beyond that output power.

Tuned Transformers These resonance transformer devices were developed in the 1940s for 1- and 2-MeV X-rays. They used an air-core transformer in a gas-insulated tank with tuned windings and a sectionalized X-ray tube. They are no longer made because of problems in constructing the very large gas tubes.

Van de Graaf Generators These devices work on electrostatic principles and have been employed to develop 1–8-MeV X-rays. Their main advantages lies in their small focal spot size [0.008 in (0.2 mm)], but they are used more for irradiation work than for NDE purposes.

Betatrons These are electron accelerators that can produce X-rays in the 1–30-MeV range. They can have a small spot size but are quite large and cumbersome and have generally been replaced by *linear electron accelerators* (linacs). Linacs are the most useful devices for producing high-energy X-rays ($1 < E < 25$ MeV) and can provide very high outputs of radiation with a small focal spot [< 0.08 in (2 mm)].

Selection of Energy Levels

Conventional Voltage (≤ 1 MeV) Most conventional voltage X-ray work is performed at 40–400 kV, although NDE of some polymer matrix composites may employ energies as low as 25 kV. At these low energy levels, one has the advantages of economy, easier shielding, and a rate of absorption that is more differentiated by changes in the test object material. The disadvantages here, however, are limited penetrating power and longer exposure times required by the lower energy levels.

High-Energy Radiography ($E > 1$ MeV) The advantages of high-energy radiation for NDE purposes are greater penetration power and faster film exposures. In addition, a sharper image may be obtained since most of the scattered radiation is in the forward direction, reducing the amount of scatter and indirect radiation reaching the image plane. High-energy radiation is better for NDE of complex shapes, including those with mixed thicknesses, and for examining areas that are not in contact with the recording film. It is, however, more difficult to get good contrast on the image plane at high-energy levels, and there are increased safety concerns (and associated costs) in this case.

11.4.2 Gamma Rays

Gamma rays have many of the same characteristics as X-rays but are generated during the decay of radioisotopes. Natural radioactive sources are almost never used

TABLE 11.1   Summary of Radioactive Elements Used for Gamma-Ray Inspections

Radioactive Element	Half-Life	Photon Energy (MeV)	Applications
Ytterbium (Yb) 169	31 days	0.063, 0.308	Thin sections—used in Europe
Thulium (Tm) 170	127 days	0.084, 0.54	Light alloys, 0.5 in. (1.3-cm-) thick steel
Iridium (Ir) 192	74 days	Several rays between 0.2 and 1.06	2.5-in- (6.4-cm-) thick steel, most common source used in North America (accounts for 95% of applications)
Cesium (Cs) 137	33 years	0.66	3.5-in- (9-cm-) thick steel
Cobalt (Co) 60	5.3 years	1.77, 1.33	9-in- (23-cm-) thick steel, accounts for approximately 4% of North American applications

for radiography, however, and are banned in some countries. Artificial radioisotope sources obtained by fission or irradiation in a nuclear reactor are usually employed for NDE purposes and are summarized in the Table 11.1. Note that iridium 192, with a half-life of 74 days, accounts for most North American NDE applications.

Gamma source requirements for NDE purposes include a reasonable half-life to avoid frequent replacement or recharging of the isotope, high specific activity (high output from a small source), and reasonable cost. Since gamma rays cannot be turned off, they require special handling. There are specific regulations that govern their shipment, and gamma sources must be shielded when not in use. As shown in Figure 11.7, there are a number of storage devices (isotope cameras) used to keep the isotope covered until needed to expose the radiograph [3]. Note that the physical size of the isotope corresponds to the focal spot size in an X-ray tube.

11.4.3   Neutron Radiography

Neutrons are subatomic particles that are characterized by a relatively large mass and neutral charge. They were discovered by Chadwick in 1932, and Kallman made the first neutron radiographs in 1935 [5]. Neutrons can be obtained from nuclear reactors, particle accelerators, and certain radioactive isotopes and can have a wide range of energy levels (see Table 11.2). Thermal neutrons are most commonly used for NDE purposes. They are easy to detect and have "favorable" attenuation characteristics (i.e., are absorbed by some materials that are transparent to X-rays).

Neutron radiography complements X-ray and gamma-ray inspections but is not a substitute for them. Although the processes are similar in many respects, the attenuation characteristics for the two types of energy do differ. Recall that the NDE process depends on the differential absorption of radiation by the inspected object and potential inclusions, voids, cracks, and so on. If radiation is not absorbed differently

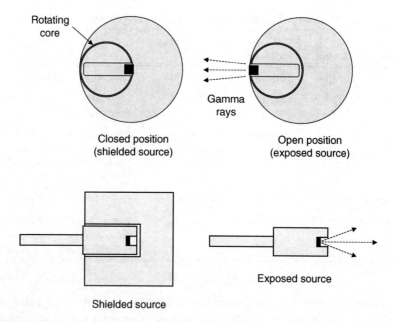

Figure 11.7. Typical isotope containment devices or isotope cameras. (After Reference 3.)

through the test piece by these anomalies, no useful image is formed. Some materials have poor X-ray attenuation characteristics (i.e., they are completely transparent to X-rays), however, and do not exhibit the partial absorption needed to reflect density changes on the radiograph.

Recall that X-rays and gamma rays interact with electrons orbiting *around* the atom, and their attenuation increases with density and atomic number, as shown in Figure 11.8 [5]. Neutrons, on the other hand, interact with the atomic *nucleus*, and their attenuation increases with material density and the total neutron cross section. Thus, as shown in Figure 11.8, neutron attenuation has a random relationship with atomic number. This means that some low-density materials (e.g., hydrogen, lithium,

TABLE 11.2 Classification of Neutrons by Energy

Neutron Classification	Energy Level
Cold	< 0.01 eV
Thermal	0.01–0.3 eV
Slow	0–10 keV
Epithermal	0.3–10 keV
Resonance	$1–10^2$ eV
Fast	10 keV–20 MeV
Relativistic	> 20 MeV

Note: See Reference 5.

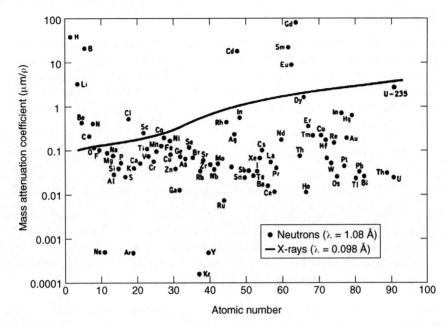

Figure 11.8. X-ray and neutron mass attenuation coefficients versus atomic number. (From [5], copyright © 1992 by John Wiley & Sons, Inc. This material is used by permission of John Wiley & Sons, Inc.)

and boron), which easily pass X-rays, strongly attenuate neutrons. Conversely, neutrons easily penetrate some heavy elements (e.g., lead, uranium, bismuth) that readily absorb X-rays. These different attenuation characteristics allow neutrons to inspect materials that are not readily examined by X-rays (and vice versa).

Neutrons cannot be detected with X-ray film or paper, so scintillators that, for example, emit light when exposed to thermal neutrons or track etch detectors are necessary to produce images. Applications of neutron inspections include rubber O-rings, metal-jacketed explosives (ordnance devices), and corrosion products in metal assemblies or adhesive-bonded honeycomb/aluminum assemblies. Residual core material in investment cast turbine blades is frequently detected when gadolinium oxide (Gd_2O_2) is used to dope the core material (1–3% by weight) in order to aid neutron inspection.

11.5   IMAGE FORMATION

The radiation that reaches the recording plane to form the image depends on its initial energy, the absorption characteristics of the test material, and the exposure level. This section reviews techniques to detect the intensity of radiation that has passed through the test piece. Attention then focuses on procedures for creating an image that can be used to assess the quality of the object in question.

11.5.1 Film

Film is the most common technique to record X-ray or gamma-ray intensity after it passes through the test object. Silver halide crystals in the film emulsion are excited by the radiation and, upon development, become darker with increased radiation exposure and the number of crystals exposed. The photograph density and exposure control the image quality. Note that only about 1% of the incident beam contributes to image formation.

Density   Photographic density D is defined as

$$D = \log_{10}\left(\frac{I_0}{I_t}\right) \tag{11.5}$$

where I_0 = intensity of light incident on the film negative
   I_t = intensity of light transmitted through the negative
If, for example, 1% of the incident light is transmitted through the negative, Equation 11.5 gives a photographic density $D = \log(100/1) = 2$.

Exposure   The exposure E is the dose of radiation striking the film and is defined by

$$E = \text{intensity} \times \text{exposure time} \tag{11.6}$$

Exposure Curve   Since photographic density does not, in general, vary linearly with exposure, exposure curves may be obtained for various films and development

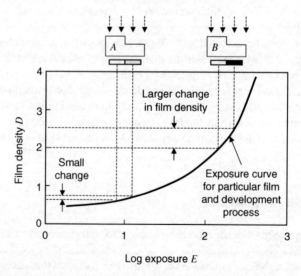

Figure 11.9. Characteristic density-versus-exposure curve for particular film and development process. Note that X-rays of the test piece for different exposures A and B lead to different film density contrast of the thickness step.

processes, as shown in Figure 11.9 [3, 6]. Here different portions of the film are subjected to assorted exposures using a step wedge. The film is then developed and used to obtain a plot of density D versus log exposure for that particular film/processing combination. The resulting curve may be employed to obtain the correct exposure and density to give the most discriminating image for a particular application. Since the relation between exposure and density is nonlinear, a desired change in film density can be selected for a particular exposure gradient. Note, for example, that exposure B in Figure 11.9 leads to a greater change in film density and thus more contrast between the changes in test object thickness than exposure level A. Thus, one can process radiographs to emphasize or deemphasize particular thickness changes in the component, depending on the desired degree of emphasis.

11.5.2   Fluorescent Screens

Fluorescent screens may also be employed to obtain a real-time radiographic image. Here, the fluorescent screen converts X-ray energy to visible light, which produces the image, although it is often faint and limited in fine detail. Fluorescent imaging is usually limited to low-density inspections (e.g., letters, plastics) but is also used on most titanium radiographs. It is possible to record the fluorescent screen with a TV camera, however, and to enhance the image by digital processing.

11.5.3   Linear Detector Arrays

Linear detector arrays are grids of small photodiodes covered with fluorescent material. Digital signals obtained from the individual diodes can be processed and used to form images. These devices are fairly complex and expensive but are becoming more popular with advances in signal processing.

11.6   IMAGE QUALITY

This section discusses issues associated with the quality of the radiographic image. Topics include control of scattered radiation, use of intensifying screens, geometric factors, film unsharpness, and image quality indicators.

11.6.1   Scattering Control

Radiant energy that is scattered as it passes through the test object may degrade the desired image. This indirect radiation includes, for example, *internal scatter* resulting from X-ray passage through the test piece, *side scatter* from X-rays striking walls, support brackets, and *back scatter* from X-rays that pass through the image plane and are reflected back from surrounding objects (e.g., floor, table). Since the scattered radiation reaches the recording plane by indirect paths, it is not image forming but does contribute to the background noise level.

In order to characterize this undesired radiation, it is useful to define a *buildup factor* BF as

$$BF = \frac{I_d + I_s}{I_d} = 1 + \frac{I_s}{I_d} \tag{11.7}$$

Here I_d and I_s are, respectively, the intensities of the direct and scattered radiation reaching the image plane. The buildup factor can vary between 2 and 20 and is smaller with high-energy radiation. Since the scattered radiation has lower energy than the primary image-forming X-rays that pass directly through the object, it may often be filtered with appropriate shields (e.g., lead apertures) placed around the X-ray source, the test object, and the image-recording device. Minimizing beam width also prevents stray radiation from reaching the imaging plane.

11.6.2 Intensifying Screens

It is also possible to enhance the X-ray image with "controlled" scatter energy obtained from lead or fluorescent intensifying screens placed next to the recording medium. Lead intensifying screens, for example, are thin lead foils [0.0008–0.008 in (0.02–0.2 mm) thick] placed on top and bottom of the film. Some radiation passing through these screens is scattered and adds to (or "boosts") the direct energy striking the film. Since these screens are right next to the film, however, the scattered energy does not have the opportunity to stray from the path of the direct, image-forming radiation. This "enhancement" of the image by the intensifying screen reduces the required film exposure times. Fluorescent intensifying screens (salt screens) are thin fluorescent layers on the film. Visible light formed by fluorescence caused by the X-rays again "intensifies" the film image. This process does, however, cause a loss in image quality and is not widely used in industrial radiography.

11.6.3 Geometric Factors

Figure 11.10 shows how image size is controlled by the relative placement of the radiation source, test object, and image plane. Simple geometric considerations, for example, indicate that the image height W_i is scaled by the factor $(a + b)/a$, where a is the source-to-object distance and b is the object-to-image plane distance. Note that increasing the object–image distance b magnifies the size of the test piece on the radiograph.

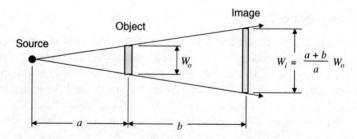

Figure 11.10. Schematic drawing showing how image size depends on relative placement of the radiation source, test object, and image plane.

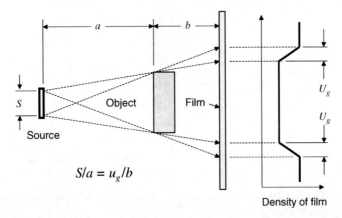

Figure 11.11. Geometric unsharpness U_g caused by finite source width S.

Other geometric issues also influence image sharpness. Note in Figure 11.11 that radiation from a source with finite spot size S passes through sharp corners on the test object at various angles and arrives at different locations on the image plane. These assorted paths prevent formation of a sharply defined thickness change on the image plane. The width u_g of this intensity gradient is defined as the *geometric unsharpness* shown in Figure 11.11 and is given by

$$u_g = \frac{Sb}{a} \qquad (11.8)$$

Geometric unsharpness is minimized by employing a radiation source with a small spot size S, minimizing the object-to-film plane (or detector) distance b, and by increasing the source-to-object distance a. Note, however, that the desire for a minimum object-to-plane distance b here is offset by the image magnification preference for a large b described in the preceding paragraph.

11.6.4   Film Unsharpness

The film unsharpness u_f shown in Figure 11.12 results when X-rays passing through the film excite electrons in the photographic emulsion on the film. The local excitation of the emulsion occurs in all directions and causes loss of sharpness between light and dark areas on the negative. This process is inherent to the film and increases with the energy of radiation. Film unsharpness can be calibrated as a function of radiant energy, and the resulting relationship can be used to select the appropriate X-ray energy for a particular application.

The total unsharpness of the X-ray is a combination of the geometric unsharpness u_g and film unsharpness u_f. Although these two factors combine in a complex way, the following two cases are often defined to optimize various set-up parameters for the X-ray:

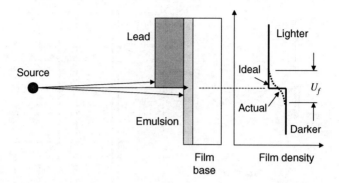

Figure 11.12. Film unsharpness caused by excitation of film emulsion by X-rays.

High-Sensitive Technique ($u_g = u_f$)   In this case, one selects the source width S, source–object distance a, object–plane distance b, film (and development process), and X-ray energy level so that the geometric unsharpness u_g equals the film unsharpness u_f.

Low-Sensitive Technique ($u_g = 2u_f$)   In this case, the various radiograph parameters are set so that the geometric uncertainty is twice that of the film.

11.6.5   Image Quality Indicators

Image quality indicators (IQIs) are small objects placed on top of the test piece to assess the quality of the X-ray image. Also called penetrameters (or pennies), these reference objects are made from the same material as the test member. They come in a variety of shapes, including wires of various diameters, step/hole configurations, or uniform thicknesses with various hole sizes. The IQIs are much thinner than the test object (usually 1–2% of the test object thickness) and are X-rayed along with the inspection piece. The ability of the resulting radiograph to discern known geometric features of the IQI forms the basis for quantifying the final image quality.

11.6.6   Sample Radiographs

The typical radiographs shown in Figure 11.13 demonstrate a few of the many capabilities of radiography. Note, for example, the liquid level is readily apparent in the X-ray of the aerosol canister (Figure 11.13a) along with the water that has been absorbed through an improper repair to the aluminum skin/honeycomb panel (Figure 11.13b). Porosity is clearly present in the weld shown in Figure 11.13c along with the ends of the IQI placed on top of the weld (i.e., the DIN62FE and IOISO16 labels). Although the small individual wires connecting the ends of the IQI may not be readily apparent in this photograph, they are visible on the original X-ray and are used to assess the quality of the image. Cracks are also visible in two of the eight cages that form the four-wheel-drive rear-wheel bearing shown in Figure 11.13d.

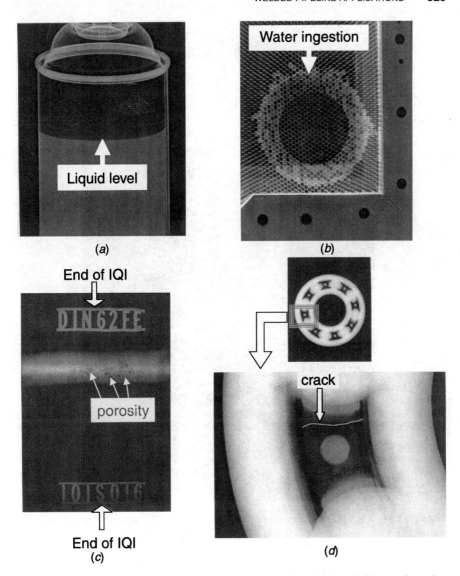

Figure 11.13. Sample radiographs: (*a*) aerosol canister showing fluid level; (*b*) water ingestion in an aluminum skin/honeycomb core panel; (*c*) porosity in a weld; (*d*) rear-wheel bearing of a four-wheel-drive vehicle showing broken cage (crack is shown in white for clarity).

11.7 WELDED PIPELINE APPLICATIONS

The fact that radiography can readily locate internal flaws in metal components and can be employed under adverse field conditions with portable equipment makes it particularly well suited for the inspection of welded joints in large pipelines (see

Figure 11.14. Radiographic inspection of welded pipeline. (Photograph courtesy of AGFA-Gevaert N. V.)

Figure 11.14). As described by Halmshaw [3], there are four basic methods to examine pipe welds, depending on the relative placement of the radiographic source, the film, and the pipe in question. Examining Figure 11.15a, the first method involves placing the source inside the pipe and wrapping the film around the outside. If the diameter is large enough for a good image to be obtained with the source at the center of the pipe, it is possible to obtain a complete 360° scan of the butt weld with a single panoramic exposure. If, however, the source–film distance a is too small for proper focus (recall the geometric unsharpness u_g given by Equation 11.8), it may be necessary to move the source to one side. In this case, multiple exposures will be required as only a segment of the weld circumference can be examined in a single set-up. (Also note that for minimum geometric unsharpness the film should be placed against the pipe and the source width should be as small as possible—minimizing S and b in Equation 11.8.)

Figure 11.15b shows an alternate set-up where the film in placed inside the pipe and the source is on the outside. Again, the source-to-film distance must be large enough to minimize geometric unsharpness, and the source/film must be repositioned several times to obtain complete circumferential coverage of the weld. Figures 11.15c,d show two arrangements where both the film and source are on the outside, avoiding the necessity for internal access to the pipe. In both cases, the radiation penetrates two walls of the pipe at the same time. In Figure 11.15c the source is placed close to one wall so that geometric unsharpness causes it to be out of focus and only the back wall is imaged on the film, although one still sees "ghost" images from the front wall. In Figure 11.15d, however, the source is far enough away from the front wall that both it and the back side are in focus at the same time, resulting

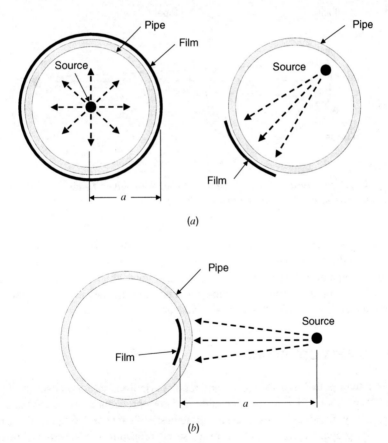

(a)

(b)

Figure 11.15. Various configurations for radiography of pipe: (a) film outside with source inside (two versions); (b) film inside with source outside; (*continued*)

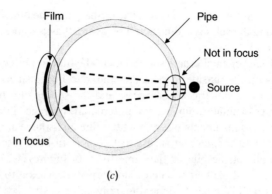

(c)

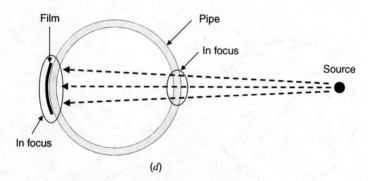

(d)

Figure 11.15. (*continued*) (*c*) source and film outside with only one wall in focus; (*d*) source and film outside with double image of two walls.

in superimposed images of both walls on the film. Finally, it should be noted that for all of the multiple-segment cases shown in Figure 11.15, when the X-rays intersect the curved pipe at different angles, the "effective" pipe thickness varies, depending on the placement of the source and pipe diameter, limiting the circumferential length of film that can be examined in a single exposure.

11.8 SAFETY CONCERNS

Working with radiation is hazardous and requires rigorous safety precautions to protect NDE personnel. Strict licensing and regulatory procedures are needed to ensure compliance with safety issues, often resulting in significant costs to satisfy these regulations. By proper adherence to appropriate safety procedures, it is, however, possible to conduct radiographic examinations in a safe manner.

In order to determine allowable radiation exposures, one must first quantify the amount of radiation received by an individual [6]. The rad (radiation absorbed dose), for example, is defined as the absorption of 100 ergs of energy per gram of radiated material and applies to any form of penetrating radiation. The rbe (relative biological effectiveness) ranks the danger of various types of radiation to human safety and is defined as follows:

$$
\text{rbe} = \begin{cases} 1 & \text{for X-rays, gamma rays, and beta particles} \\ 5 & \text{for thermal neutrons} \\ 10 & \text{for fast neutrons} \\ 20 & \text{for alpha particles} \end{cases}
$$

Finally, the rem (Roentgen equivalent man) unit is the product of the radiation absorbed dose and the relative biological effectiveness (i.e., rem = rad × rbe). The International System (SI) equivalent for the rem is the sievert, where 1 Sv = 100 rem.

Medical considerations specify a maximum acceptable accumulated radiation dosage that increases with age (i.e., banking effect). In general, no penetrating radiation exposure is allowed before age 18. After 18, one can safely accumulate an additional 5 rem/year (0.05 Sv/year), although one should not accumulate any more than 12 rem in any given year (0.12 Sv).

Reference 6 suggests that there are two main aspects of radiation safety: protecting personnel and monitoring radiation dosage. Radiation protection involves keeping exposure levels as small as possible by limiting the intensity and time of radiation. Recalling that radiation intensity decreases as the inverse square of the distance from the source, it is also important to limit access to radiation areas. Restriction of access might consist of signs, roped-off areas, or more stringent controls. In addition, it is often necessary to shield work areas with lead or concrete. Recall from Section 11.3, for example, that the TVT for lead was 0.56 in (1.4 cm) for 0.5-MeV X-rays. For 0.1 MeV, however, the TVT drops to 0.16 in (0.04) cm, indicating how effective lead is for shielding low-level radiation.

The amount of radiation allowed to escape from restricted areas must also be controlled, with the following maximum doses as typical [6]:

- 2 mrem total dose maximum in 1 hour
- 100 mrem total dose maximum in 7 consecutive days
- 500 mrem total dose maximum for calendar year

Radiation monitoring of the test facility and NDE personnel is also a required aspect of the safety program. The potential for X-ray leakage from the facility must be periodically examined by monitoring each radiation source under maximum exposure operating conditions. In addition, personnel within restricted areas are required to wear film badges and/or pocket dosimeters that can be read daily. The radiation doses must then be recorded and calibrated at regular intervals (e.g., monthly).

11.9 COMPUTED TOMOGRAPHY

Computed tomography (CT) is a special imaging method that provides a cross-sectional view of internal features in a solid object [2, 7, 8]. Also known as computer-assisted tomography (CAT) scan, this technique can be used with a variety of energy beams, including ultrasonic, electrons, protons, alpha particles, lasers, and microwaves. Its use with radiography is, however, the most important NDE application. The Austrian mathematician Johann Radon first described the mathematical principle of CT in 1917. The first practical applications did not occur, however, until the 1960s and 1970s, when Allan Cormack (South Africa) and Godfrey Hounsfield (England) developed CT for use in the medical field.

Conventional radiography compresses structural information from a three-dimensional object into a two-dimensional view. Although this procedure allows a large volume of material to be examined, only limited information is obtained. One cannot, for example, specify flaw shapes or locations, and interpretation of the radiographic image can be difficult. A tomogram, on the other hand, provides a complete three-dimensional view of internal features by obtaining a series of cross-sectional images of the test object. A schematic view of the CT process is given in Figure 11.16 [7]. Note that in this case the energy beam, detector, and surface being examined all lie

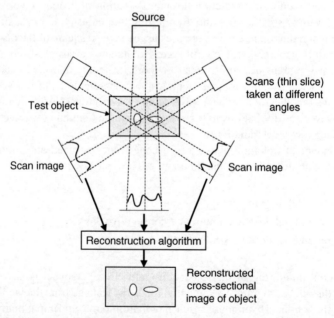

Figure 11.16. Schematic representation of CT process showing how scan information obtained along a flat plane from many scan angles is used to reconstruct cross-sectional image of test piece.

in same plane, whereas with conventional X-ray, the beam is perpendicular to the examination plane.

A CT scanning system consists of a radiation source, detector, precision manipulator, and reconstruction algorithm. The radiation source is collimated to form a thin fan beam that can be scanned in a specified plane. Radiation attenuation along this plane is then measured by a detector, which is often comprised of a linear array of individual sensors. A precision manipulator moves the object and/or source/detector system to record scans from many different angles. Finally, the reconstruction algorithm generates a two-dimensional image of the examination plane from the set of one-dimensional radiation measurements obtained for various angles. These calculations generally require a computer system, and advances in CT methods have paralleled developments in computing power since the 1970s. Reconstruction algorithms include *transform* techniques based on Radon's original theorem and *iterative* techniques that use algebraic methods employing initial guesses. These latter algorithms are rarely used, however, except for situations where only limited data are available (as with weak gamma sources that would take too long to generate detailed scans).

Computed tomography is a very versatile technology that is not restricted to the shape or composition of the object being inspected. Advantages of CT systems include their ability to examine complex objects in great detail and the capacity to obtain three-dimensional views that characterize the size, shape, and location of internal features. (See the examples shown in Figure 11.17 [8].) Computed tomography images are easy to interpret and provide a permanent record of the inspection. Limitations of CT include requirements for a stable radiation source, precise manipulators, a sensitive and linear X-ray detector system, and a high-quality image display. In addition, one often requires 360° access to the test object, and the extensive computations required to reconstruct the image can be expensive and/or time consuming, although this latter disadvantage has been minimized with recent advances in inexpensive computing power. Thus, complete three-dimensional analysis applications are usually limited to low-volume/high-value test pieces.

Some NDE applications of CT are as follows:

- Early applications were in the medical field and provided high-quality images of internal features of the human body.
- The petroleum industry was one of first to use medical-type scanners to analyze core drilling samples. Other early applications include inspection of large rocket motors and small precision castings for aircraft engines.
- Computed tomography continues to be used for complex precision castings and forgings such as turbine blades and vanes.
- Since density changes as small as 1% can be detected, CT is also used to inspect ceramics and powder metallurgy products and for examining composite materials.
- The CT technique is quite useful for examining assembled structures such as helicopter rotor blades, detonators, and electronic components.

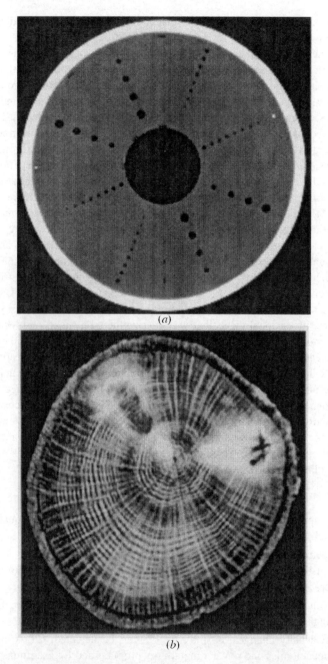

Figure 11.17. Examples of CT results. (*a*) This 120-mm-diameter simulated rocket motor shows voids and inclusions. The scan was taken at 320 kV. (*b*) This CT image of a log taken at 160 kV shows knots and cracks as well as the ring structure. (*continued*)

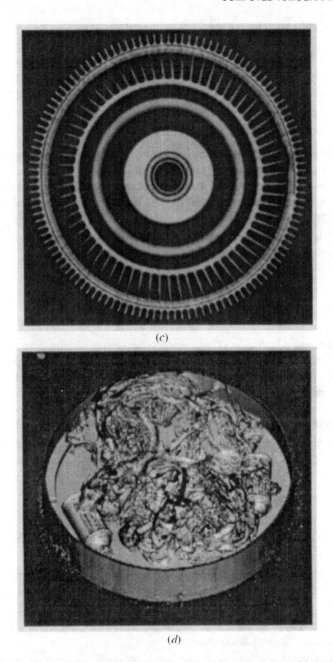

(c)

(d)

Figure 11.17. (*continued*) (*c*) This 2-MeV CT image of a cruise missile turbine engine shows erosion of the wall at the right. (*d*) This three-dimensional image of a section of a nuclear waste drum was made at 2 MeV and shows clothing and two spray cans. (Tomographs courtesy of Bir, Inc.)

11.10   CONCLUDING REMARKS

This chapter has overviewed radiographic NDE, and further details may be found in References 3–7 and 9. It is appropriate to conclude here with a brief summary of the advantages and disadvantages of radiographic inspections.

11.10.1   Advantages of Radiography

1. Radiography can detect internal and external damage such as material discontinuities, fabrication/assembly errors, or service-induced damage. X-ray inspection is especially good at locating nonplanar anomalies such as porosity and voids, although locating planar defects is also possible provided the plane is parallel to the X-ray direction.
2. Radiography is applicable to a wide variety of solid materials and can examine complex shapes and a wide range in component thickness.
3. A permanent record of the inspection is obtained.
4. Minimal part preparation is required.
5. Real-time viewing is possible.
6. Three-dimensional imaging is possible with CT technology.
7. Additional advantages of gamma-ray inspection include portability for site work (no external power requirements), lower costs than comparable X-ray units, and the availability of small devices that can be placed in locations accessible to X-ray units (e.g., inside pipes, engines). In addition, gamma-ray sources can have a variety of source sizes and some isotopes have good penetrating power.
8. Neutron radiography provides an alternate source of radiant energy that often complements the materials that can be inspected by X-rays or gamma rays.

11.10.2   Disadvantages of Radiography

1. There are significant safety issues due to the health hazards of radiation that add to the costs of inspection and limit application of radiography.
2. Two-sided access to the test piece is generally required.
3. Detection is very sensitive to flaw orientation in that the anomaly must be parallel to the direction of radiation, complicating detection of laminar defects and fatigue cracks.
4. Radiography requires a high degree of skill and experience, particularly in assessing marginal discontinuities.
5. Radiography is relatively expensive, particularly for thick components, and cannot be used with radioactive structures.
6. Determining the location of the anomaly requires special techniques.
7. There is usually some delay in obtaining the test results, even with mechanized processing.

8. Additional disadvantages of *gamma-ray* sources include the inability to turn off the source or to vary the energy level of radiation and the need to replenish the source periodically. In addition, gamma rays are often a greater health risk and require strict safety regulations regarding transport and use.

PROBLEMS

11.1 Use the Internet to locate the home page for a vendor that supplies radiographic supplies for NDE purposes and record this company's address for future reference. Examine the products and applications described on the web site and prepare a short report that describes the various capabilities and uses for some of the radiography products sold by this company.

11.2 Discuss the general classifications of X-ray radiography.

11.3 Briefly describe the four mechanisms of X-ray absorption that are important in industrial radiology and indicate their relative contribution to the total absorption/scattering process.

11.4 Determine the tenth-value thickness and the half-value thickness for an iron plate subjected to 0.5-MeV X-rays.

11.5 What factor(s) is frequently used to characterize the strength of the X-rays employed for a particular application?

11.6 What factors limit the maximum X-ray energy that can be obtained from a given X-ray tube?

11.7 Discuss the factors that would determine the choice of X-ray power for a given inspection situation.

11.8 What are gamma rays? How does their use compare with X-rays for NDE purposes?

11.9 Briefly compare and contrast neutron and conventional X-ray radiography.

11.10 Briefly discuss the factors that influence image quality of radiographic photographs. What techniques can be used to improve image quality?

11.11 What technique is frequently used to measure the quality of an X-ray photograph?

11.12 Define the "buildup factor" in radiographic inspection and briefly discuss what can be done to improve the situation it characterizes.

11.13 What is an exposure curve and how is it used for radiographic NDE?

11.14 It has been decided to shield an X-ray inspection system with sheets of a radiation-absorbing material. Determine the minimum thickness of material needed so that only 1% of the radiation will penetrate through the shielding. Briefly define all terms and/or material properties given in your answer.

11.15 Briefly compare and contrast "geometric" and "film" unsharpness in the context of radiographic inspections. How do these factors influence the quality of a radiographic inspection?

11.16 The film unsharpness for a particular film/process and X-ray output has been determined to be 0.01 in. The X-ray source has a 0.1-in spot size and the test object is a 1.5-in-thick plate. If the X-ray film is placed against the *back* of the test piece, determine the minimum distance that the X-ray source should be placed from the *front* of the test object for a "high-sensitive" inspection. What would be the minimum source distance in this case for a "low-sensitive" radiograph?

11.17 List and briefly explain the general principles that form the basis for radiographic safety procedures.

11.18 What is computed tomography? How does it differ from conventional radiography?

REFERENCES

1. H. Berger, "100 Years of X-Rays—Industrial Use of NDT and the Role of ASNT," *Materials Evaluation,* pp 1253–1260, November 1995.

2. T. A. Heppenheimer, "Medical Imaging: The Inside Story," *American Heritage of Invention and Technology,* Vol. 15, No. 3, pp. 54–63, 2000.

3. R. Halmshaw, in "Radiological Methods," in *Non-Destructive Testing,* 2nd ed., Edward Arnold, London, 1991, Chapter 3.

4. *Nondestructive Testing, Radiographic Testing,* Classroom Training Handbook CT-6-6, 2nd ed., General Dynamics Convair Division, 1983.

5. D. E. Bray and D. McBride (Eds.), *Nondestructive Testing Techniques,* Wiley, New York, 1992.

6. "Radiographic Inspection," in *ASM Handbook,* Vol. 17: *Nondestructive Evaluation and Quality Control,* ASM International, Materials Park, Ohio, 1989, pp. 295–357.

7. M. J. Dennis, Industrial Computed Tomography," in *ASM Handbook,* Vol. 17: *Nondestructive Evaluation and Quality Control,* ASM International, Materials Park, Ohio, 1989, pp. 358–386.

8. C. R. Smith, Vice President marketing, BIR Inc., Lincolnshire, Illinois, personal communication, January 1994.

9. R. H. Bossi, F. A. Iddings, and G. C. Wheeler, *Radiographic Testing—Nondestructive Testing Handbook,* 3rd ed., Vol. 4, American Society for Nondestructive Testing, Columbus, Ohio, 2002.

CHAPTER 12

ULTRASONIC INSPECTION

12.1 OVERVIEW

Ultrasonic inspections measure the travel of high-frequency sound waves that are introduced into the test object at various surface locations. The propagation of these electronically controlled sound pulses through the member is detected at specific points where the acoustic energy is converted back to an electronic signal. Interpretation of the reflection and refraction of these sound waves at particular boundaries leads to information regarding discontinuities in the test object. Since sound waves travel in many materials, ultrasonic inspection is one of the most widely used NDE tools for detecting both internal and surface-breaking flaws.

Typical NDE ultrasound frequencies range between 200 kHz and 25 MHz and are inaudible to the human ear, which is limited to frequencies between 20 and 20,000 Hz. Sound waves do not, in general, represent significant health hazards, unlike the X-rays and gamma rays considered in Chapter 11. Moreover, since they are generated by and converted back to electronic pulses, there are significant opportunities for signal processing of the test data.

A schematic view of a typical pulse-echo ultrasonic inspection is shown in Figure 12.1. Although this figure is discussed in more detail later, observe now that the nominal travel and reflection of the sound wave is altered by the presence of a discontinuity. Both the existence and location of the flaw (relative to the time axis) can be determined. Also note that maximum detection is obtained when the flaw is oriented perpendicular to the sound beam, which is opposite to the X-ray inspection method considered in Chapter 11.

This chapter begins with a brief discussion of the types of sound waves used for nondestructive evaluation and then considers the propagation and interaction of

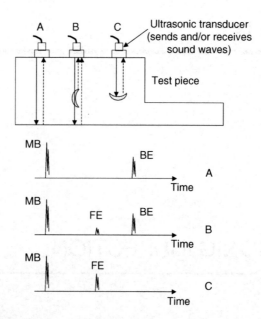

Figure 12.1. Schematic representation of an ultrasonic inspection (pulse-echo technique) showing main bang (MB), back-echo (BE), and flaw echo (FE) reflections determined at the measurement point as a function of time.

these waves with various interfaces. Methods to interrogate structures with ultrasonic waves are then described, along with a brief discussion of procedures to generate and sense ultrasound. The chapter concludes with a summary of the advantages and disadvantages of ultrasonic NDE.

12.2 PROPERTIES OF SOUND WAVES

Sound waves are mechanical (not electromagnetic) vibrations that propagate in most materials. Their velocity depends on the density and elastic properties of the parent material, and they may be reflected, focused, or refracted. Nondestructive evaluation ultrasonic waves are usually generated and detected by piezoelectric transducers that are coupled to the test object [1–4]. Piezoelectricity, discovered in 1880 by Currie, involves converting electrical signals to mechanical vibrations and vice versa. The sound waves are generated (sent) when an electrical signal is transformed to mechanical vibrations in a transducer. They are then measured (received) when mechanical vibrations in the transducer are converted back to electrical signals. This process and other methods for generating and recording sound waves are discussed in more detail later.

The velocity (V) of sound travel depends on the wavelength (λ) and the frequency (f) of vibration and is given by

$$V = \lambda f \qquad (12.1)$$

The typical wavelength of NDE ultrasound is on the order of 1 mm. [Recall for comparison purposes that λ for visible light is approximately 10^{-3} mm (4×10^{-5} in) and varies between 10^{-6} and 10^{-9} mm (4×10^{-8}–4×10^{-11} in) for X-rays.] As discussed below, sound waves can propagate through an object by several different modes.

Longitudinal Waves Longitudinal sound waves, also called compression waves, propagate so that particle vibrations are parallel to the direction of wave travel, as shown in Figure 12.2. Longitudinal waves are the most commonly used waveform for NDE and occur when the ultrasound beam enters nearly perpendicular to the surface of the test object. They are easily generated and detected and can travel in solids, liquids, or gases. The compression wave speed V_c in a large body is given by

$$V_c = \left(\frac{E(1 - v)}{\rho(1 + v)(1 - 2v)} \right)^{1/2} \qquad (12.2)$$

where V_c = velocity of compression wave
  E = elastic modulus of the material
  v = Poisson's ratio of material
  ρ = density of material
  Equation 12.2 assumes isotropic material behavior and that the wave travels in an infinite elastic body. The compression wave speed in a thin rod whose diameter is small with respect to the wavelength is given by

$$V_c = \left(\frac{E}{\rho} \right)^{1/2} \qquad (12.3)$$

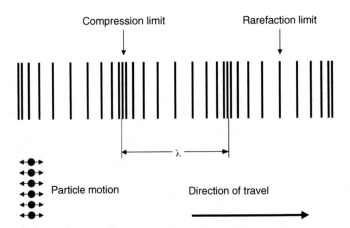

Figure 12.2. Schematic representation of a longitudinal (or compression) wave.

Ultrasonic shear wave

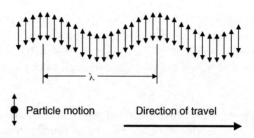

Particle motion   Direction of travel

Figure 12.3. Schematic representation of a transverse (or shear) wave.

Shear Waves Shear waves, also known as transverse waves, propagate so that particle motion is perpendicular to the wave motion, as shown in Figure 12.3. Shear waves are generally limited to solids but can occur in highly viscous liquids. They are formed when the sound beam enters the object surface at a moderate angle. They can be polarized, and their speed V_s is given by

$$V_s = \left(\frac{E}{2\rho(1 + v)} \right)^{1/2} = \sqrt{\frac{G}{\rho}} \qquad (12.4)$$

Here G is the shear modulus, and the other terms are the same material properties as before. Note that the shear wave speed is approximately one-half of the compression wave speed V_c given by Equation 12.3. Typical values of the compression and shear wave speeds in various materials are given in Table 12.1 [3]. (Data for other materials are given in References 1 and 2.) Since shear waves have a slower speed and shorter wavelength for a given frequency, they are more sensitive to small anomalies than

TABLE 12.1   Shear, Compression, and Surface Wave Speeds for Various Materials

Material	Density (g/cm³)	Compression Wave Speed		Shear Wave Speed		Surface Wave Speed	
		in/s	m/s	in/s	m/s	in/s	m/s
Aluminum	2.7	248,00	6,300	122,000	3,100	114,000	2,900
Copper	8.9	185,00	4,700	89,000	2,260	76,000	1,930
Magnesium	1.75	226,00	5,750	122,000	3,090	113,000	2,870
Mild steel	7.85	232,000	5,900	127,000	3,230	118,000	3,000
Titanium	4.5	240,000	6,100	123,000	3,120	110,000	2,790
Plexiglass	1.18	105,000	2,670	44,100	1,120	44,500	1,130
Concrete (28 days)	2.4	177,000	4,500				
Water (liquid)	1.0	58,700	1,490				
Water (ice)	0.9	157,000	3,980	78,300	1,990		
Air	0.001	13,400	340				

Note: See References 1–3.

compression waves. They are, however, more susceptible to scatter, resulting in less penetration.

Surface Waves   Surface, or Rayleigh, waves propagate so that particle motion follows an elliptical path, having *both* longitudinal *and* shear (transverse) motions, as shown in Figure 12.4. Surface waves occur when the sound beam enters a solid at a shallow angle, and their velocity is approximately 90% of the shear wave velocity V_s. Rayleigh waves travel near the surface of the object, and their energy decreases rapidly below the surface, ceasing to exist after a depth of two wavelengths (2λ). Although Rayleigh surface waves will travel around curves and surface contours, they are reflected by corners. They are also sensitive to changes in hardness, platings, surface cracks and to dirt and grease on the test piece.

Lamb Waves   Lamb waves, also known as plate waves or guided waves, are combinations of compression and shear waves that only occur in thin sheets (i.e., thickness of a few wavelengths). Their character depends on the material density and elastic properties, component geometry, thickness, and frequency. Lamb waves are dispersive and exist in several modes, such as the symmetrical (dilatational) or asymmetrical (bending) first modes shown in Figure 12.5. They may propagate at various velocities in a given material. Lamb waves are particularly useful for detecting longitudinal separations in metal or composite laminates.

Standing Waves   Standing waves result from the interference of other waves that have the same frequency but different origins and directions of travel. Under certain conditions, it is possible for particles to remain at rest following the interference of these propagating waves and form stationary nodes. Since these standing waves can confuse issues associated with propagation of continuous waves, most NDE techniques employ short bursts (or pulses) of sound for the interrogation procedure.

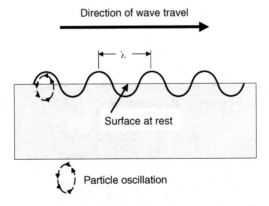

Figure 12.4.   Schematic representation of Rayleigh surface wave formed by oscillatory particle motion that has both longitudinal and transverse components. Ultrasonic wave travels at the surface of a solid.

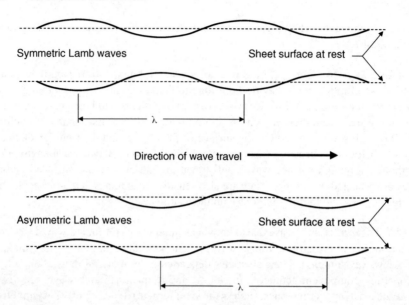

Figure 12.5. Symmetrical (dilatational) and asymmetrical (bending) Lamb waves propagating in thin sheets.

12.3 BOUNDARY INTERACTIONS

Ultrasonic NDE is based on the interpretation of test signals received from reflections and refractions of sound waves at various boundaries and discontinuous surfaces. This section describes issues associated with these wave–boundary interactions.

12.3.1 Acoustic Impedance

The specific acoustic impedance Z is given by

$$Z = \rho V \tag{12.5}$$

Here ρ is the density of the material and V is the wave velocity. When a sound wave strikes a plane boundary between two media, the percent energy transmitted E_t and energy reflected E_r are given by the equations

$$E_t = \frac{4Z_1 Z_2}{(Z_1 + Z_2)^2} \times 100 \tag{12.6}$$

$$E_r = \left(\frac{Z_1 - Z_2}{Z_1 + Z_2}\right)^2 \times 100 \tag{12.7}$$

Here Z_1 and Z_2 are the acoustic impedances for materials 1 and 2 given by Equation 12.5. These expressions are valid for *normal incidence* of both compression and transverse waves at a boundary. Since transverse (shear) waves cannot be sustained in a liquid or gas, however, effectively 100% energy reflection is obtained at solid–liquid or solid–gas interfaces. The data in Table 12.1 can be used to calculate acoustic impedances for various material interfaces, and Equations 12.6 and 12.7 may be used to compute the energy transmission and/or reflection at various material interfaces typically encountered in NDE.

Steel–Water Interface   The energy transmitted (E_t) and reflected (E_r) at a steel–water interface, for example, are given below:

$$Z_1 = \rho_{\text{water}} V_{c,\text{water}} = (1.0)(1490) = 1490$$

$$Z_2 = \rho_{\text{steel}} V_{c,\text{steel}} = (7.85)(5900) = 46{,}315$$

$$E_t = \frac{4(46{,}315)(1490)}{(46{,}315 + 1490)^2} \times 100 = 12.1\%$$

$$E_r = \left(\frac{46{,}315 - 1490}{46{,}315 + 1490}\right)^2 \times 100 = 87.9\%$$

Other cases of interest are recorded in Table 12.2. Note that complete reflection occurs at a metal–air interface and complete transmission results when the two interfaces are the same material. Water–metal interfaces are of particular interest since they are encountered in immersion ultrasonic NDE. Observe that under ideal conditions only 12.1% of the sound energy is transmitted at a water–steel boundary and only 29.6% at a water–aluminum interface. Plexiglass–metal interfaces are encountered in contact ultrasonics when the transducer is mounted to a plastic probe that then contacts the metal test object (see Figure 12.6a). Although ideally more energy is transferred in this case than with direct water contact (23.9% for steel and 52.7% for aluminum), these levels of sound transfer are rarely achieved in practice due to rough surfaces. Thus, a couplant (frequently oil) is usually required for contact ultrasonic testing. (Note that the use of a couplant prevents direct application of Equation 12.6.)

TABLE 12.2   Ultrasonic Energy Transmitted and Reflected at Various Material Interfaces (Normal Incidence)

Interface	Transmitted Energy (%)	Reflected Energy (%)
Steel–water	12.1	87.9
Aluminum–water	29.6	70.4
Steel–plexiglass	23.9	76.1
Aluminum–plexiglass	52.7	47.3
Steel–air	0	100
Steel–steel	100	0

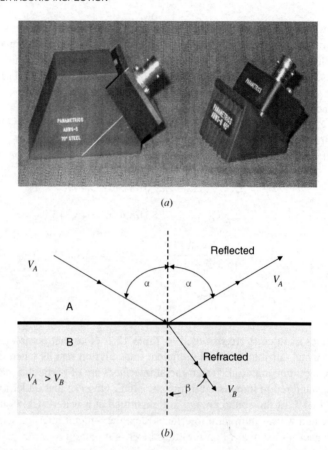

Figure 12.6. (*a*) Photograph of ultrasonic transducers mounted on plastic wedges that control angle of incidence to test piece. (*b*) Incident sound wave in material A results in reflected and refracted waves of same mode in materials A and B.

12.3.2 Oblique Incidence

The last section considered transmission and reflection of sound *energy* at an interface. This section looks more closely at the sound waves that result from incidence with a material boundary. As indicated in Figure 12.6, both reflected and refracted (i.e., transmitted) waves may result from oblique incidence of a sound wave at a material interface (Figure 12.6*b* follows the nomenclature of Reference 3). Here the ultrasonic transducer is mounted to a plastic wedge that is configured to transmit sound through the plastic at a predetermined angle of incidence to the test piece.

Note that when an incident sound wave traveling in material A with velocity V_A strikes the interface between materials A and B, one obtains both a *reflected* wave in material A and a *refracted* wave in material B. Both of these waves have the same mode (i.e., compression or shear) as the original incident wave. The reflected wave has the same velocity V_A and the same reflection angle α as the incident angle, whereas the refracted wave in material B has the appropriate velocity for material B (V_B) and refraction angle β. The relation between the reflected and refracted velocities V_A and V_B and the incident and refracted angles α and β are given by Equation 12.8 (Snell's law). Note that Equation 12.8 holds for both incident compression and shear waves:

$$\frac{\sin \alpha}{\sin \beta} = \frac{V_A}{V_B} \tag{12.8}$$

12.3.3   Mode Conversion

The reflected and refracted waves discussed above may also *change modes* (i.e., switch from compression to shear or vice versa) depending on the original velocity, material, and angle of incidence. Consider the following cases.

Incident Compression Wave   As shown in Figure 12.7, assume that a *compression* wave strikes the interface between materials A and B with an incident angle *i*. (It has been assumed in Figure 12.7 that $V_A > V_B$.) Note that the incident compression wave in material A can result in a combination of *four* possibilities:

- reflected compression wave with velocity V_{CA} in material A,
- reflected shear wave with speed V_{SA} in material A,
- refracted compression wave with speed V_{CB} in material B.
- refracted shear wave with velocity V_{SB} in material B.

Following the nomenclature of Reference 3, Snell's law leads to the following relations between these ultrasonic waves:

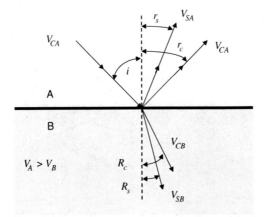

Figure 12.7. Incident compression wave in material A results in compression and shear waves in materials A and B.

$$\frac{\sin i}{V_{CA}} = \frac{\sin r_s}{V_{SA}} = \frac{\sin r_c}{V_{CA}} = \frac{\sin R_s}{V_{SB}} = \frac{\sin R_c}{V_{CB}} \qquad (12.9)$$

where i = angle of incidence

r_c, r_s = angles of reflection for compression and shear waves in material A

R_c, R_s = corresponding angles of refraction in material B

V_{SA}, V_{CA} = velocities of reflected shear and compression waves in material A

V_{SB}, V_{CB} = refracted shear and compression wave speeds in material B

Incident Shear Wave As shown in Figure 12.8, a similar situation occurs when the original incident wave in material A is a *shear* wave. (In this case, it has been assumed that $V_B > V_A$.) Again, Snell's law gives the following relations between the reflected and refracted wave velocities and angles:

$$\frac{\sin i}{V_{SA}} = \frac{\sin r_c}{V_{CA}} = \frac{\sin R_c}{V_{CB}} = \frac{\sin R_c}{V_{CB}} \qquad (12.10)$$

12.3.4 Critical Angles

Depending on the incident wave speed, angle, and properties of materials A and B, some of the waves refracted into material B may be suppressed. This situation leads to the concept of a *critical angle* that may be used to specify existence (or absence) of certain refracted waves. As shown in Figures 12.9 and 12.10, these critical angles depend on the particular material combination. The water–metal and plexiglass–metal interfaces encountered in immersion or contact NDE are of particular interest. Critical angles provide the means to intentionally introduce (or eliminate) a particular wave mode for a given NDE application.

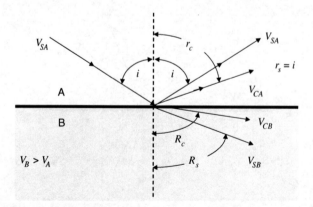

Figure 12.8. Incident shear wave in material A results in compression and shear waves in materials A and B (sketch assumes $V_B > V_A$).

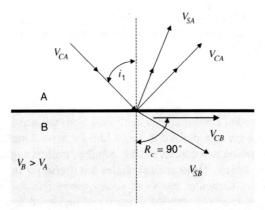

Figure 12.9. Incident compression wave at first critical angle i_1 in material A results in transmission of compressive wave along interface and shear wave across interface (assumes $V_B > V_A$).

First Critical Angle Note in Figure 12.9 that for a certain incident angle i_1 the refraction angle R_c for the compression wave in material B will be 90°. In this case, only the shear wave penetrates into the depth of material B, and the compression wave travels along the A–B interface. From Snell's law, $\sin(R_c) = 1$, and the first critical angle i_1 is given by the equation (assume $V_B > V_A$)

$$\sin i_1 = \frac{V_{CA}}{V_{CB}} \tag{12.11}$$

Second Critical Angle A similar situation occurs when the incident angle i_2 causes the refracted shear wave V_{SB} to propagate along the A–B interface (i.e., $R_s =$

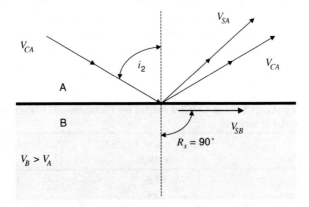

Figure 12.10. Incident compression wave at second critical angle i_2 results in transmission of shear wave along interface in material B (assumes $V_B > V_A$).

90°), as shown in Figure 12.10. For incident angles $i > i_2$, only surface waves (Rayleigh waves) can exist in material B. Snell's law gives the following value for the second critical angle i_2:

$$\sin i_2 = \frac{V_{CA}}{V_{SB}} \tag{12.12}$$

The first and second critical angles computed from the wave speed information given in Table 12.1 are summarized in Table 12.3 for several material combinations of interest to immersion and contact testing. Similar "critical angles" can be defined for incident shear waves. These critical angles are useful for designing transducer couplers to transmit particular modes of sound waves into the test object. Compression waves, for example, applied at angles of incidence that exceed i_1 cannot transmit compression waves into the test piece. Similarly, shear waves cannot be transmitted into the test piece by compression waves applied at larger incident angles than the second critical angle. (The 45° shear wave inspection is the most common NDE mode.)

The energy of an incident wave is partitioned between the various reflected and refracted waves in a complex manner that depends on the angle of incidence and materials involved. Assuming no coupling losses, the acoustic energy is conserved. For normal incidence energy transfer is given by the impedance calculations described previously in Equations 12.6 and 12.7.

The amplitudes of the shear and compression waves transmitted from plexiglass into aluminum by a compression wave are given as a function of angle of incidence in Figure 12.11 [3]. Note that a compression wave cannot be transmitted for incident angles greater than 25.1° and that a transmitted shear wave does not exist for incident angles greater than 59.5°. For incident angles above 59.5°, all of the applied compression wave is reflected at the plexiglass–aluminum interface, and no sound waves enter the aluminum. Also recall from Table 12.2 that for perpendicular incidence $(i = 0)$ Equations 12.6 and 12.7 indicate that 52.7% of the compression wave energy is transmitted and 47.3% is reflected at the plexiglass–aluminum boundary. These values correspond to the $i = 0$ intercepts shown in Figure 12.11. (It should be noted, however, that the couplant interfaces play a key role in the actual energy transmitted in practice.)

TABLE 12.3   First and Second Critical Angles for Compression Waves Incident to Various Material Interfaces

Material Interface	First Critical Angle, α_1 (deg)	Second Critical Angle, α_2 (deg)
Water–steel	14.6	27.5
Water–aluminum	13.7	28.7
Plexiglass–steel	26.9	55.8
Plexiglass–aluminum	25.1	59.5

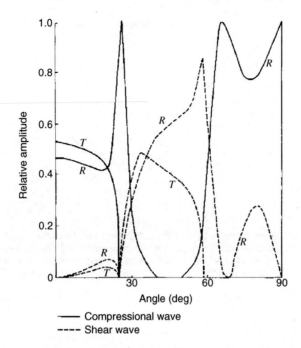

Figure 12.11. Relative amplitude of shear and compression waves transmitted into aluminum from a compression wave originating in plexiglass as a function of angle of incidence. Solid and dashed lines indicate compression and shear waves respectively, and T and R refer to transmitted and reflected beams. (From [3] © 1991 R. Halmshaw, reproduced by permission.)

12.3.5 Double Interface

Sound waves passing through two interfaces undergo transmission and reflection, and possible mode conversion, at both interfaces [3]. Note, for example, in Figure 12.12 that sound waves from material A are transmitted into material C through a liquid couplant B. At the first A–B interface, it is possible for both shear and compression waves to be reflected back to material A, whereas only compression waves can refract into the liquid couplant B. The compression waves continuing through liquid B will be reflected and refracted at the B–C interface. Some sound energy is transmitted directly to material C, with both shear and compression modes possible. Some energy is also reflected back to the A–B interface in the form of a compression wave and may return to material A in the form of shear and/or compression waves and/or be reflected back to the B–C interface.

Since transmission and/or reflection can occur again at the B–C juncture, the total energy transmitted into material C results from multiple reflections and refractions. Also note that waves going through the B–C interface travel different path lengths (due to the multiple reflections in material B). Thus, summation of the total energy transmitted by these direct and indirect routes must also consider the *phase*

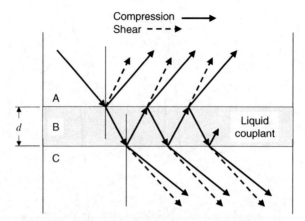

Compression wave from material A transmitted through liquid couplant
B into material C undergoes mode conversion and reflection in A–B
and B–C boundaries resulting in shear and compression waves
transmitted into material C and reflected into material A

Figure 12.12. Ultrasonic wave at an interface between two materials (A and C) with a liquid coupling layer (B) between them. (After Reference 3.)

of the individual waves, with in-phase waves adding and out-of-phase waves cancel-ing. The difference in path length depends on the thickness d of the coupling ma-terial B. When d is an integral multiple of $\lambda/2$, the direct and indirect waves are in phase and *maximum* energy transmission occurs. When d is an odd multiple of $\lambda/4$, the direct/indirect waves are out of phase and *minimum* energy transmission occurs [3].

These transmission/reflection issues are of particular concern when designing couplants between ultrasonic transducers and layered test pieces. For the best sound transmission, the couplant thickness d should be an even multiple of $\lambda/2$. (But, since this case occurs for a small portion of a broadband pulse, it is not usually considered during inspection of other than layered structures.) Recalling that the wavelength $\lambda = V/f$, the couplant thickness d depends on the inverse of the frequency f and the wave speed V (which depends on the material).

Note that if media B is an *air-filled crack*, it is possible for sound energy to propagate through the crack (instead of being reflected by crack surfaces and being detected). This situation can occur when the crack face separation is very small (i.e., a tight crack), and although not a major problem in practice, "tight" cracks can transmit rather than reflect sound energy [3]. (Recall from the discussion of fatigue crack closure in Sections 7.5 and 7.11 that fatigue crack surfaces are often tightly closed by residual deformations associated with crack tip plasticity. Thus, ultrasonic detection of fatigue cracks is enhanced if a preload is applied to the test piece in order to separate the crack faces.)

12.3.6 Corner Reflections

When sound waves impinge on a "corner," the reflections and refractions at the two corner surfaces can lead to complex mode conversions and considerable loss of energy. Consider, for example, the reflection/refraction of the incident compression wave shown in Figure 12.13 as it approaches right-angle surfaces 1 and 2. Note that reflection and refraction can occur at each surface, along with mode conversion. At surface 1, both a compression wave and a shear wave can result. There can also be compression and shear waves transmitted across the interface. (These refracted waves are ignored in the subsequent discussion.) At surface 2, the reflected compression wave can again reflect as shear and compression waves. The compression wave will emerge from the corner with the same direction as the original incident beam, but the shear wave will be reflected at a different angle. There can also be compression and/or shear waves transmitted across surface 2 (which are ignored in the subsequent discussion).

The shear wave formed by mode conversion at surface 1 can also reflect as a shear or compression wave at surface 2. This reflected compression wave will return at a different angle than the other compression wave reflected by the incident compression wave. The reflected shear wave will also have a different direction than the other sound waves. Again, the incident shear wave can also result in waves that are transmitted across the boundary. Thus, the original incident compression wave can return from the corner in the form of compression waves with the original angle of incidence (traveling in the opposite direction), compression waves traveling in a different direction, and shear waves traveling in two different directions. Similar scenarios can also be developed for an incident shear wave. Also note that all of these considerations are vastly more complex for nonisotropic materials such as composites.

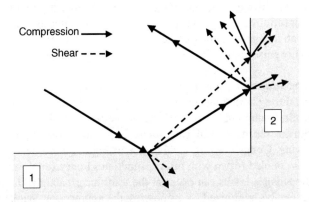

Figure 12.13. Compression wave approaching corner results in two reflected compression waves and one reflected shear wave. Shear and compression waves are also transmitted across boundaries 1 and 2, resulting in loss of ultrasonic energy.

The intensities of corner reflections depend significantly on the angles of incidence, with "poor" reflections encountered for 30° and 60° angles [3]. Thus, corner reflections have significant practical consequences for ultrasonic testing. They represent opportunities for "bad" reflections and large losses of sound energy. By similar reasoning, flaw orientation can also strongly influence the signal returned from the ultrasonic test pulse.

12.4   ULTRASONIC ATTENUATION

Ultrasonic energy is attenuated as it passes through the test object, and as a result, the intensity measured by the receiver is often significantly less than that initially transmitted. This loss in acoustic energy depends on the test material, frequency, coupling, component, and/or flaw geometry. Attenuation of ultrasonic beams can occur by the following means.

Beam Spreading   Since the beam spreads as it travels from the source, it is distributed over a larger area and loses local intensity. The intensity lost due to beam spreading is proportional to the inverse of the distance from the source squared.

Absorption   Absorption of a sound wave occurs as its mechanical energy is converted to heat. This heat loss is permanent and cannot be used for NDE. This heat loss is often anisotropic with respect to beam travel and, for most materials, increases directly with frequency.

Scattering   Scattering of ultrasonic energy occurs at grain boundaries, small cracks, and other material inhomogeneities. When the material grain size (g.s.) is less than 1% of the ultrasonic wavelength λ, scattering losses are negligible. If the grain size exceeds 10% of λ, scatter losses are on the order of grain size cubed and can account for serious energy losses [1]. Energy lost to scattering could, however, be interpreted as an indication of unacceptable grain sizes and be reason for component rejection. Highly anisotropic alloys such as Ti-6-4 are particularly bad in this regard.

Acoustic Impedance   Acoustic impedance losses occur between the transducer and test object during both transmission and reception of the ultrasonic signal. In the immersion testing technique described later in Section 12.5, water is used to couple the object with the transducer, and water–metal impedance losses occur both on sending and receiving. Contact testing involves enclosing the transducer in a plexiglass housing that is coupled (often with oil, or sometimes honey for shear waves) to the test object. Impedance losses can occur at the transducer/plexiglass/coupling/object interfaces but can be minimized by employing the appropriate couplant for the test material (i.e., by impedance matching).

Recall in Section 12.3.1 that the ideal energy transfer for a steel–water interface was 12.1% whereas that across a steel–plexiglass boundary was 23.9%. Thus, considerable energy is lost in both immersion and contact testing. Rough surfaces represent a multitude of angles of incidence and can cause even more sound loss in contact testing. Other impedance losses can also occur when the sound wave interacts with free boundaries or material interfaces in the test object. Impedance losses that occur at discontinuities may, however, indicate the presence of a flaw.

Attenuation Intensity   Overall attenuation of sound wave intensity may be expressed by

$$I = I_0 e^{-\alpha x} \tag{12.13}$$

where I_0 = initial acoustic intensity at a given point
I = intensity at distance x from that point
α = attenuation coefficient

The attenuation coefficient α is often expressed in decibels per unit length. The decibel is a logarithm-based unit (decibel = 0.1 bel). Two acoustic pressures (or intensities) differ by n bels if

$$\frac{I_1}{I_2} = 10^n \quad \text{or} \quad n = \log_{10}\left(\frac{I_1}{I_2}\right) \quad \text{bels} \tag{12.14}$$

Acoustic power is proportional to the square of the amplitude A. Thus,

$$n = 20 \log_{10}\left(\frac{A_1}{A_2}\right)$$

where A_1 and A_2 are the acoustic signal amplitudes. Typical attenuation coefficients [1,4] may range as follows:

$\alpha < 10$ dB/m for hot rolled pearlitic steels,

$\alpha \sim 100$ dB/m for certain stainless steels and aluminum alloys, and

$\alpha > 300$ dB/m for polymers.

Acoustic attenuation can pose a problem for ultrasonic inspection of materials such as polymers, composites, wood, and concrete. This attenuation is more pronounced at higher ultrasonic frequencies and is more significant for shear waves than for compression waves.

12.5   ULTRASONIC INSPECTION TECHNIQUES

Ultrasonic NDE is an extremely versatile method that allows a variety of application techniques for a given inspection problem. One may, for example, employ compression, shear, surface, and/or plate waves for the interrogation signal. Moreover, as discussed below, the ultrasonic signal can be transmitted, received, interpreted, and displayed in a variety of ways.

12.5.1   Immersion and Contact Methods

Ultrasonic inspections may be broadly classified as immersion or contact methods. Contact testing involves holding the transducer directly against the test object with the assistance of a thin-film couplant, whereas immersion testing employs a fluid (often water) to couple the ultrasound transducer with the test object. Advantages of immersion testing include speed of inspection, ability to control and direct sound beams, and adaptability to automated testing. As described below, there are three general immersion techniques: conventional immersion, water column, and wheel transducer.

Conventional Immersion Testing   Here both the transducer and the test object are immersed in a fluid as shown in Figure 12.14. Both straight-beam (compression wave) and angle-beam techniques that develop shear waves in the test piece can be employed for immersion testing. (The critical angles discussed in Section 12.3.4 define the angles of incidence for obtaining the desired wave type in the component.) Since the sound will travel approximately four times faster in the metal test piece than in water (recall Table 12.1), the water path distance between the transducer and the test object must be chosen to avoid spurious signals. The time the sound waves travel in water (i.e., the water path distance), for example, should exceed the scan depth time in the test object. One rule of thumb sets the water path equal to one-fourth of the part thickness plus $\frac{1}{4}$ in [5].

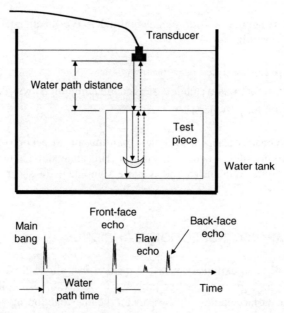

Figure 12.14. Immersion ultrasonic NDE setup showing water path distance between specimen and transducer.

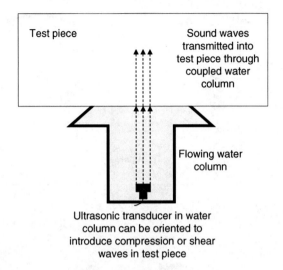

Figure 12.15. Water column or bubbler NDE procedure where ultrasonic transducer is coupled to the test specimen by a column of flowing water.

Water Column Testing   Also known as a squirter or bubbler, a water column is a variation of the immersion technique for applications where the entire member cannot be submerged. As shown in Figure 12.15, sound is projected from the transducer to the object through a column of flowing water, and it is not necessary to immerse the entire test object. The compression sound beam in the water column can again be applied perpendicular or at an angle to the surface to introduce compression and/or shear waves in the test piece.

Wheel Transducer   The wheel transducer technique uses a water-filled tire to transmit sound to and from the test object, as shown in Figure 12.16. The transducer is contained (immersed) in the water-filled tire that is then placed against the test piece. The rubber tire and/or transducer can be positioned at an angle to the member in order to transmit either compression or shear waves. The object and/or the tire may then be moved to scan the surface of the test piece.

Contact Testing   In contact testing, the transducer is placed directly against the test object with the assistance of a thin-film couplant. This contact arrangement allows the ultrasonic apparatus to be smaller and more portable than that employed for immersion testing. The ultrasonic probe may be hand held or the scanning can be automated. Contact angle-beam techniques apply the sound wave at a predetermined angle to introduce compression, shear, or surface waves in the test piece. The incident angle may, for example, be selected to eliminate compression waves, so that only shear or surface waves travel in the component. Surface (Rayleigh) waves result from large angles of incidence (incident angles greater than the second critical angle). These Rayleigh waves follow surface contours of the object and may be used to detect small surface flaws.

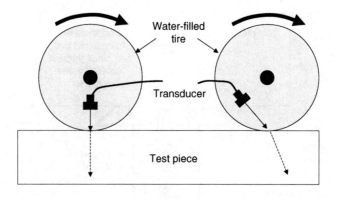

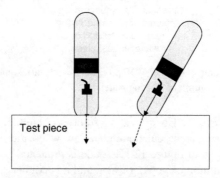

Angle of wheel and/or transducer can
be adjusted to introduce longitudinal
and/or shear waves into test piece

Figure 12.16. Wheel transducer NDE arrangement where ultrasonic transducer is contained in a water-filled tire that is then rolled across the test specimen. The wheel can be inclined at various angles to introduce longitudinal and/or shear waves into the test specimen (5).

12.5.2 Pulse Echo

Pulse-echo inspections employ a pulsed ultrasonic beam that is reflected back to the original (or adjacent) transducer by another surface in the test object, as shown in Figure 12.1. Both the initial ultrasonic signal ("main bang," MB) and the "back echo" (BE) from the reflection surface are displayed on an oscilloscope as a function of elapsed time, as shown schematically in Figure 12.1. Flaws in the test object also reflect the incident beam, resulting in a "flaw echo" (FE) that falls between the main bang and back echo on the oscilloscope trace. The time axis position of the FE pulse can be related to the physical flaw location in the test object. Note in Figure 12.1 (case C) that a large discontinuity can cause complete reflection of the pulse and no back-face echo BE is received.

Pulse-echo testing and its variants are, perhaps, the most commonly employed ultrasonic methods since only single-side access to the test object is required. Either single or double transducers may be employed for pulse-echo testing. In the first case, a single transducer sends and receives the ultrasonic signal. In the second instance, special dual-element probes employ separate transducers mounted together to send and receive. Since the individual transducers may be optimized for transmission or reception functions, dual-element probes allow increased directivity and resolution, particularly for near-surface flaws. If the flaw has an unfavorable orientation that prevents its echo, it may still be possible to detect flaw presence by a reduction in the amplitude of the main back echo. Pulse-echo techniques may be applied with compression, shear, surface, or Lamb waves and may be used with either immersion or contact testing.

12.5.3 Through Transmission

Through-transmission (also called pitch–catch) testing introduces the ultrasonic signal at one point and receives it at the opposite side of the test object, as shown in Figure 12.17*a*. In case B, for example, the flaw blocks a portion of the sound waves, reducing the strength of the received signal, whereas in case C, all of the sound is prevented from reaching the receiving transducer. Note that through transmission only measures signal attenuation. Thus, the presence of an anomaly causes a reduction in the received signal, but its location is not indicated. Two separate transducers are needed for sending and receiving, although they may again be optimized for those separate functions.

Since the ultrasonic signal only makes one pass through the test object, through-transmission testing is particularly suited for highly attenuative materials that absorb sound waves such as composites or concrete. Figure 12.17*b*, for example, shows a through-transmission setup to measure the compression wave speed in concrete specimens. The measured wave speed may then be used to determine the elastic modulus as the concrete cures (recall Equation 12.2).

Through transmission is also suited for detecting near-surface anomalies that do not provide a good reflective surface (e.g., inclusions, porosity, large grain boundaries). Through-transmission techniques also shield the receiver from front-surface reflections and the effects of beam spread. Disadvantages of through-transmission testing include transducer alignment problems and the requirement for coupling both the transmitting and receiving transducers to the test piece as they are scanned in tandem.

12.5.4 Contact Angle-Beam Techniques

Cracks that are normal to the component surface are difficult to detect with sound beams that are also applied perpendicular to the test object (recall Figure 12.1). If the cracks penetrate a free surface, they may be located with Rayleigh surface waves oriented to travel perpendicular to the crack plane (see Figure 12.18*a*). Subsurface cracks may also be detected by various angle-beam techniques designed to result in an appropriate reflection from the crack surface. As indicated in Figure 12.18*b*, sound waves may be "skipped" off various surfaces to reach the crack plane at

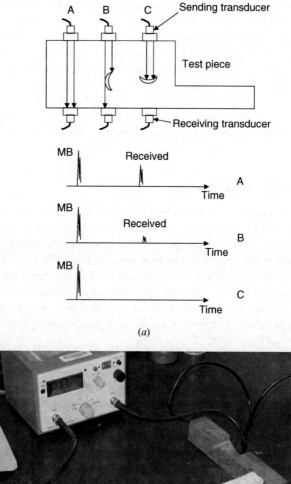

(a)

(b)

Figure 12.17. (a) Schematic representation of through-transmission (pitch–catch) inspection. Separate transducers are employed to send and receive the ultrasonic signal on opposite sides of the test object. (b) Ultrasonic pitch–catch arrangement used to determine elastic modulus in concrete test specimen by measuring compression wave speed.

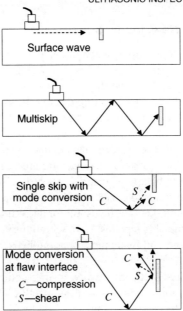

Figure 12.18. Examples of various angle-beam techniques to detect normal cracks: (*a*) surface wave interrogation signal; (*b*) multiskip interrogation; (*c*) single skip with mode conversion at back surface; (*d*) mode conversion at flaw surface.

an optimum angle for reflection. Note, however, that placement of the receiving transducer becomes more complicated when the test signal travels along an indirect path to and from the potential anomaly. The interrogation signal may also be designed to undergo mode conversion at free boundaries and/or the crack plane, as shown in Figures 12.18*c,d*. Thus, interpretation of the received signal can be especially difficult in complex bodies where the flaw reflection must be differentiated from other signals received from various free surfaces and component interfaces. (See Reference 3 for more details of the many various contact testing strategies.)

12.5.5 Critical-Angle Techniques

Critical-angle techniques exploit the unique reflection and refraction that occur at special angles of incidence (recall Section 12.3.4). Recall from Figure 12.11 that significant changes in acoustic wave pressure occur with small changes in the incident angle or test frequency near these angles. By calibrating the critical angle and frequency that cause total reflection in reference objects, it is possible to detect subtle changes due to material property variations, grain orientation, attenuation, and coldwork. The longitudinal wave speed is also sensitive to stress and can be used for residual stress measurements. Critically refracted shear, longitudinal, and higher order Rayleigh waves are also useful for near-surface inspections. But, this procedure is difficult to accomplish in practice and is subject to many errors.

12.5.6 Continuous-Wave Inspections

Continuous-wave inspections employ a continuously excited, single-frequency ultrasonic wave for the interrogation signal. This approach is limited, however, by problems with standing waves and the inability to determine flaw size or depth. Although continuous waves are not adept for locating discrete flaws, they are useful for measuring thickness, detecting disbonds in composite panels, and measuring material properties.

Specialized continuous-wave inspection techniques include resonance, eddysonic, and impedance testing. In resonance testing, the natural frequency of an object is measured and calibrated with various material properties. Eddysonic testing employs induced electrical eddy currents to generate and receive sound waves in metal test objects. This noncontact technique is quite useful for detecting bond flaws in composite materials that have thin metal faceplates. Here vibrations in the test object depend on the thickness of the object, and inadequate bonding between material layers results in anomalous vibrations.

Although contact is required, impedance testing complements eddysonic tests since it can be used on nonmetallic as well as metallic objects. Here, contact between the vibrating transducer and the ultrasonic conductor form a resonating system. Changes in the resonant frequency are then used to indicate the presence of anomalies. Bond defects in layered materials, for example, may be detected by changes in the acoustic impedance of the resonating system.

12.5.7 Time-of-Flight Diffraction

Another ultrasonic technique that has gained recent popularity, particularly for inspection of welds, is the time-of-flight diffraction (TOFD) method shown schematically in Figure 12.19. This approach differs in that *diffracted*, rather than reflected, sound waves are used to locate and size the defect in question [3, 6–8]. As shown in Figure 12.19, a wide-beam compression wave is introduced at an angle to the test piece, and four potential signals are received at the measurement point. The first signal to arrive is a mode-converted shear wave that travels along the component surface, and

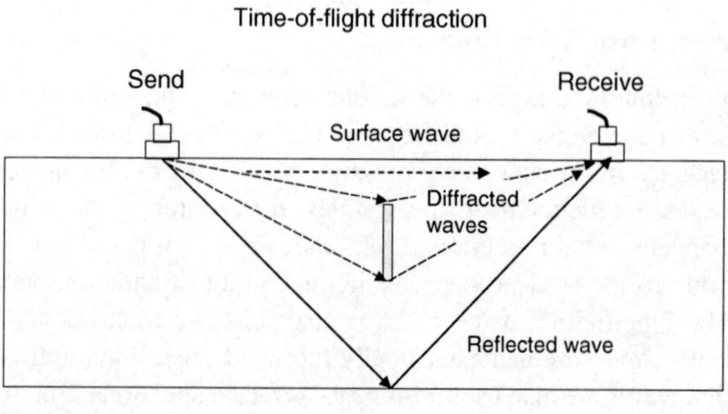

Figure 12.19. Schematic representation of TOFD ultrasonic inspection method.

the last is a reflected compression wave from the back surface. What distinguishes this method, however, is that if there is a crack present, two additional diffracted waves are also received. These latter signals originate from the ends of the defect and are very weak in comparison to the surface wave and back echo. They can be difficult to detect and may be obscured in coarse-grained materials. Measurement of their time of flight does, however, enable both detection of the presence of the flaw and determination of its size. The TOFD method does not depend on crack orientation, but when examining welds, one must have access to both sides of the weld. The method is currently in the process of being developed into codes and standards.

12.6   ULTRASOUND GENERATION AND DETECTION

This section summarizes various methods for generating ultrasonic waves for NDE purposes. Piezoelectric transducers, electromagnetic transducers, and lasers may all be employed to introduce ultrasound into the test piece.

12.6.1   Piezoelectric Transducers

Piezoelectric transducers are the most common source of NDE ultrasound. Piezoelectric materials have the ability to convert electrical signals to mechanical deformations and vice versa. In the send mode, an oscillating electric signal causes high-frequency vibrations in the transducer that are transmitted as sound waves into the test object. For reception purposes, sound vibrations coupled from the object to the transducer are converted back to electric signals that may then be recorded and analyzed. As reviewed below, there are several materials that exhibit the piezoelectric effect in a manner suitable for ultrasonic transducers [1, 3].

Piezoelectric Materials   *Quartz crystals* are natural materials that were once used for all ultrasonic transducers. Their piezoelectric properties are highly anisotropic and depend on the cutting orientation. The *x-cut* crystal vibrates primarily in the direction perpendicular to the cut and produces compressive waves when coupled to the test object. The *y-cut* crystal oscillates in the transverse direction, so that shear waves could be generated without the need for mode conversion. (However, NDE shear waves are usually introduced into the test piece with angled compression waves that cause mode conversion in the test object.)

Advantages of quartz crystals for transducer applications include good electrical and thermal stability, insolubility in most liquids, high mechanical strength and uniformity, and good resistance to wear and aging. Disadvantages of quartz crystals include relatively low electromechanical conversion efficiency, mode conversion interference, and the requirement for large voltages for low-frequency oscillation.

Lithium sulfate is another monocrystalline material used for ultrasonic transducers. The advantages of this material are excellent receiving characteristics (making them the most efficient receivers), intermediate conversion efficiency, negligible mode interaction, and resistance to aging. The disadvantages of lithium sulfate crystals lie in their fragility, a maximum service temperature limitation of approximately 165°F (74°C), and the fact that they are soluble in water and must be protected from moisture.

Polarized ceramics are sintered synthetic materials with high electromechanical efficiency and make the most efficient sound generators. This class of materials includes barium titanate, lead zirconate titanate, and lead metaniobate. Barium titanate ($BaTiO_3$) is mechanically rugged, but its properties vary with aging. It is often used for ultrasonic cleaning equipment. Lead zirconate titanate (PZT) is the best transmitter of this group and is usually preferred over $BaTiO_3$ but has limited resolution since its receiving characteristics are not as good. It is mechanically rugged and has good resistance to temperature and aging. Lead metaniobate (PMN) has good tolerance to temperature [it can be used up to 930°F (500°C)]. It also has low mechanical damping but a high dielectric constant that results in a high-capacitance transducer. Further comparison of the relative merits of various ultrasonic transducer materials are given in Reference 1.

Types of Transducers There are several potential configurations for piezoelectric transducers [5]. Transducers can, for example, be configured as angle beam or straight beam to introduce selected sound modes into the test object. A compression wave is formed when the transmitter is applied perpendicular (straight) to the surface. Encasing the transducer in a plastic housing that fixes the angle of incidence between the transducer and test component forms an angle-beam transducer. By appropriate specification of the transducer angle, one can introduce compression, shear, or surface waves into the component. A long, narrow beam is obtained from "paint-brush transducers" that are made from a mosaic pattern of small crystals matched to give a uniform-sound beam. These devices are used to scan large surfaces to discover areas that are then examined in finer detail with a more sensitive transducer.

Double transducers incorporate a separate transmitter and receiver into a single unit. The transmitter and receiver can be mounted side by side or stacked but are separated by a sound barrier to prevent interference. Different piezoelectric materials may be employed to optimize the send and receive modes, and the transducers can be obtained in straight-beam or angle-beam configurations.

Faced-unit or contour-focused transducers include frontal units mounted to the transducer for various reasons. Wear plates, for example, are frequently added for contact testing to protect the fragile piezoelectric crystals. Contour plates may also be added to direct the sound waves to be perpendicular to curved surfaces and introduce a more uniform signal into the test object.

Focused probes converge the sound beam to a narrow cross section, much in the same way that visible light is focused. These probes are frequently used for immersion ultrasonic systems. Although the piezoelectric material may be ground to the desired shape, a more common technique is to cement a "lens" (usually a polymer such as plexiglass) to the transducer. Focusing concentrates the sound beam in a small test area, reducing surface roughness effects and aiding detection in that area. The usable range of the beam is, however, shortened. Focused beams are useful for object thicknesses that range between 0.01 and 2.0 in (2.5 and 50 mm).

Linear array of ultrasonic crystals pulsed in
sequence so that individual sound waves add
to form composite sound beam with desired
direction of travel

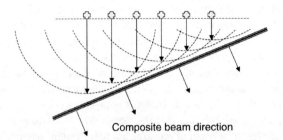

Composite beam direction

Figure 12.20. Selective pulsing of a linear array of ultrasonic crystals allows composite sound wave to be focused or transmitted in various directions. (After Reference 3.)

Electronically focused transducers, also called phased arrays, employ linear arrays of individual crystals arranged in a set pattern, as shown in Figure 12.20 [3]. Here the small individual crystals [typically 0.02 in (0.5 mm) in width] may be selectively pulsed in a desired sequence that allows the composite sound beam to be focused and/or swept without moving the part, thus decreasing scanning time. Linear arrays can be used for normal or angle transmission.

12.6.2   Electromagnetic–Acoustic Transducers

Electromagnetic–acoustic transducers (EMATs) are noncontacting probes that employ electromagnetic–acoustic phenomena to generate and receive ultrasound [1, 3]. Here, ultrasound is developed by vibratory Lorentz forces produced by the interaction of a static magnetic field and eddy currents generated in a coil by radio-frequency (RF) current. The specimen vibrates in sympathy with the RF current, forming either shear or compression waves. For reception purposes, the vibrating test object acts as a moving conductor in a magnetic field [an electromotive force (emf) is generated in the surface and causes current in the coil].

Since EMATs are noncontacting transmitter/receivers, they are especially suited for high-temperature inspections and/or applications with rough surfaces and moving specimens. In addition, they can generate shear, compression, surface, and Lamb waves. On the negative side, however, EMATs have a low signal-to-noise ratio that usually requires very large magnets and limits portability.

12.6.3   Laser-Generated Ultrasound

Repeated pulsing of the object surface with a laser causes local heating and cooling (and subsequent expansion and contraction) that can generate ultrasonic waves. At low energy levels the test object is heated directly by the laser to induce the mechanical cycling (thermoelastic approach). At higher energy levels, a small surface layer of the test member is vaporized, causing dielectric breakdown of the air around the test object and resulting in generation of sound waves by the "plasma" or "ablation" method. Although this process results in more efficient energy transfer, there is material loss and damage as pits are formed on the surface.

A significant advantage of laser-generated sound is the elimination of direct contact with the test object, avoiding many sound-coupling issues and permitting a simpler contour-following apparatus. In addition, hot surfaces can be inspected and rapid scanning can be accomplished since the laser beam has no inertia. It is also possible to generate large sound amplitudes that could be advantageous for some applications.

Disadvantages of this approach include poor energy transfer mechanisms that lead to low signal-to-noise ratios and the possibility for damaging the test object. In addition, reception of the signal requires another system (usually an optical beam that measures small surface displacements). Moreover, since laser-excited sound is impulsive, the resulting sound waves have a wide frequency spectrum (although this feature could be advantageous in some applications). Finally, relatively high capital costs are associated with laser-generated ultrasound systems.

12.7   ULTRASONIC PARAMETER SELECTION

This section discusses the influence of various ultrasonic test parameters on NDE capabilities.

Frequency   Frequency considerations often dictate the required ultrasonic method, with most ultrasonic NDE performed at frequencies in the range 200 kHz–25 MHz. Some composite materials, for example, attenuate high-frequency sound so much that lower test frequencies must be used to obtain satisfactory penetration. In addition, some pulse-echo methods cannot be used at low frequencies due to pulse-width reflection time considerations.

Ultrasonic test frequencies can be divided into two broad categories:

- the 1–10 MHz frequency range used for most NDE work and
- the low frequencies (<1 MHz) used for inspecting wood, concrete, and large-grain metals (the eddysonic and resonance techniques discussed earlier are in this category).

Penetration (depth of detection) decreases as frequency increases, and this effect is more pronounced with coarser grained materials that lead to more scattering of the sound beam. Although, in general, the highest frequency that gives adequate material penetration should be used for a given application, there are several competing issues that impact the frequency selection. *Sensitivity* (the ability to detect small discontinuities), for example, increases with frequency (i.e., sensitivity increases with smaller wavelengths). *Resolution* (the ability to distinguish separate discontinuities located near each other) increases with frequency and bandwidth but decreases with pulse length. Finally, *beam spread* (beam divergence from central axis) decreases as frequency increases.

Piezoelectric Transducer Selection   Selecting the appropriate transducer for a given NDE application involves a number of considerations that often require trade-offs for a given application. The type of transducer construction can be important, as discussed in Section 12.6.1. The size of the active area controls the amount of sound transmitted (power) and beam divergence, and large units provide more penetration. Moreover, increasing the transducer size for a given frequency gives a straighter beam but results in a loss of sensitivity.

The *characteristic frequency* is the resonant frequency for the transducer, and more efficient transmission and reception are obtained at this frequency. The transducer material and crystal thickness determine the characteristic frequency, and as the thickness decreases, the resonant frequency increases. Since crystals become too thin at higher frequencies, however, contact NDE is usually conducted at frequencies below 10 MHz.

The transducer bandwidth (i.e., the frequency band centered on the resonant frequency of the transducer) is controlled by the damping of the backing material that contacts the active element. A narrow bandwidth provides good penetration and sensitivity but poor resolution. A broad bandwidth affords good resolution but poorer sensitivity and penetration.

12.8 DATA PRESENTATION

As described in this section, ultrasonic signals may be displayed in several formats (or scans) to aid interpretation of the results.

A-Scan   The conventional A-scan is the most common form of display where the "raw" signal amplitude is plotted versus elapsed time, as shown in Figure 12.21. The A-scan format indicates flaw location on the time axis by relating its echo (FE) with respect to the back-surface echo (BE). When the anomaly results in complete reflection, however, the back-echo reference point is lost. Crack size can also be estimated by comparing the amplitude of the reflected signal with that from a calibrated crack length.

B-Scans   B-scans provide an improved presentation format by synchronizing the elapsed time for the signal to return to the receiver with transducer position along a given scan line, as shown in Figure 12.22. Here the time of travel is plotted versus transducer position. Since the reflection travel time is related to flaw depth, the B-scan effectively shows both anomaly presence and location. The flaw indication is based on the first received echo at the given probe location, however, and defects located under the flaw that is closest to the probe are not indicated. Although echo intensity is not measured directly as in the A-scan, it may be indicated by the relative brightness of the signals on the oscilloscope screen.

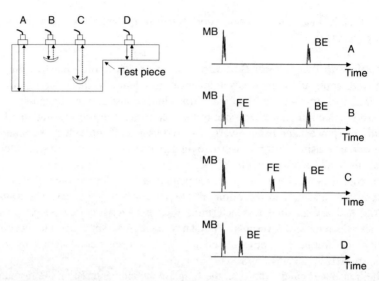

Figure 12.21. A-scan display of reflected signal amplitude versus elapsed time at various transducer positions.

B-scans are analogous to a cross-sectional view where the front and back surfaces are shown in profile and flaws correspond to a change in thickness. The main advantage of B-scans is their ability to reveal the distribution of flaws in a given cross section, as shown in Figure 12.22. They are, however, used more for medical applications than for industrial NDE. Figure 12.23, for example, shows an ultrasonic B-scan of a human fetus.

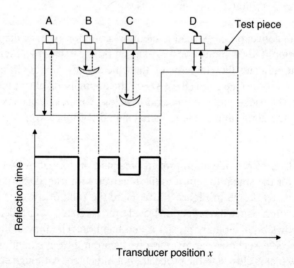

Figure 12.22. B-scan display of travel time of reflected ultrasonic signal (related to flaw depth) versus transducer position.

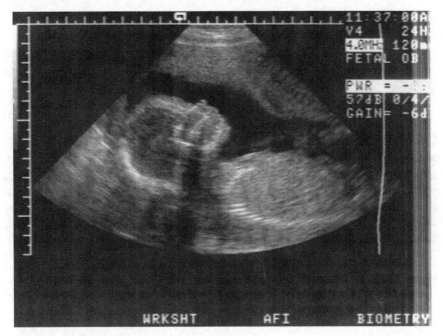

Figure 12.23. Sonogram of human fetus at 20 weeks of age.

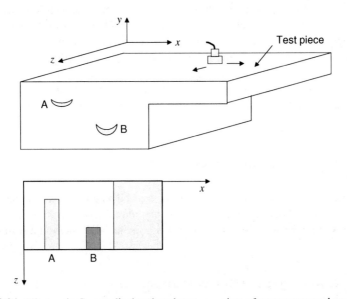

Figure 12.24. Ultrasonic C-scan display that shows x–z view of component and anomalies.

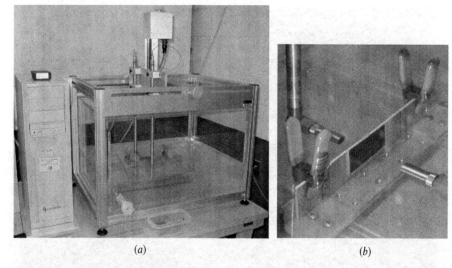

(a)   *(b)*

Figure 12.25. (*a*) Ultrasonic immersion apparatus with C-scan setup. (*b*) Transducers in through-transmission arrangement to measure disbonds under a composite patch that is bonded to an aluminum test specimen.

C-Scans As shown in Figure 12.24, C-scans display the reflected signal strength versus the x–z position of the transducer as it scans the component. These useful scans provide a plan view of the part that indicates flaw locations and shapes. The reflected signal strength is related to flaw depth and can be indicated by various colors or shades of gray on the monitor or printout. C-scans are very effective displays, particularly when examined in connection with part drawings. The schematic C-scan shown in Figure 12.24 indicates the relative depths, shapes, and locations of flaws A and B and the change in thickness of the test piece. Note, for example, that crack A occurs on a different plane than crack B and penetrates further into the edge of the component (i.e., z direction) than crack B. Figure 12.25 shows a C-scan apparatus with an immersion system arranged to examine the adhesive bond between a composite patch and a cracked aluminum test specimen. C-scans are a very powerful and popular display method and are used for many field- and depot-level inspections.

12.9 CONCLUDING REMARKS

Ultrasonic inspections are a most useful form of NDE that has may types of applications. Additional details of the ultrasonic method are given in References 1–5 and 9. The advantages and disadvantages of this popular NDE method are summarized below.

12.9.1 Advantages of Ultrasonic Inspection

1. Both surface and subsurface anomalies may be detected in relatively large and opaque objects.

2. A wide range of metallic and nonmetallic materials and thicknesses may be inspected.

3. Detailed information may be obtained about flaw size, orientation, and location. Component dimensions and material properties can also be measured.

4. There are a variety of techniques for obtaining and presenting the signal information that can be tailored to a particular application.

5. Portable instrumentation is available for field-level inspections.

6. Results are obtained in real time via electronic equipment that may be adapted for automatic data recording.

7. The technique is sensitive to relatively small defects.

8. Only single-surface access is needed.

9. Relatively little part preparation is required, although surfaces do need to be clean.

12.9.2 Disadvantages of Ultrasonic Inspection

1. At least one test surface must be accessible to the ultrasonic probe.

2. Rough surfaces cause problems, and a couplant is often required between the ultrasonic transducer and test object.

3. Inspections are highly sensitive to the sound beam and defect orientation, and "tight" cracks can be difficult to detect.

4. A high degree of skill and training are required to set up the equipment and interpret the results, increasing inspection costs.

5. Although flaw imaging is possible, these test procedures can be fairly complex.

6. Special scanning systems may be required for inspecting large surfaces, and it may be impractical to inspect complex shapes.

PROBLEMS

12.1 Use the Internet to locate the home page for a vendor that supplies ultrasonic equipment for NDE purposes and record this company's address for future reference. Examine the products and applications described on the web site and prepare a short report that describes the various capabilities and uses for the ultrasonic products sold by this company.

12.2 List and describe the various types of ultrasonic sound waves employed for nondestructive inspection and briefly discuss their NDE application.

12.3 Verify that the percentages of energy transmitted and reflected at an aluminum–water interface given in Table 12.3 are correct. Repeat your calculations for the steel–plexiglass interface.

12.4 Figure 12.7 was drawn with the assumption that the velocity of sound in material A is greater than that in material B. Redraw this figure if $V_B > V_A$.

12.5   Figure 12.8 was drawn with the assumption that the velocity of sound in material B is greater than that in material A. Redraw this figure if $V_A > V_B$.

12.6   Table 12.4 gives the first and second critical angles for *compression* waves that are incident to various material interfaces. Determine the first and second critical angles for these same cases for an incident *shear* wave.

12.7   A compression wave traveling in plexiglass strikes a plexiglass–aluminum interface at an angle of incidence of 10° (measured with respect to the normal to the interface). Draw a sketch that specifies the direction and speed of all subsequent sound waves that propagate in the *aluminum*.

12.8   Determine the first and second critical angles for ultrasonic immersion (water) testing of a mild steel part. What is the practical significance of these angles?

12.9   It has been decided to use an ultrasonic pitch–catch experiment to measure the elastic properties of a given block of material. To this end, a compression beam is applied with an immersion test to the top of the 2-in-thick member at point 1 with an angle of incidence $\theta = 10°$ as shown in Figure P12.9. Shear and compression waves are received at points 2 and 3. The results of these measurements indicate that the compression wave speed for the material $V_C = 248,000$ in/s and the shear wave speed $V_S = 121,000$ in/s. Determine Poisson's ratio for the test material. Where were the sound measurement (i.e., "catch") points 2 and 3 located for this experiment (i.e., what are distances X_2 and X_3)?

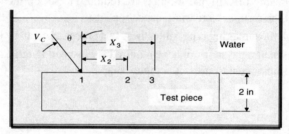

Figure P12.9

12.10   Shear wave transducers are often formed by incorporating a piezoelectric transducer into a plastic (i.e., plexiglass) wedge. The wedge transmits a *compression* wave from the transducer and sends a *shear* wave into the test object. Referring to Figure P12.10, determine the permissible range in wedge angle α if the ultrasonic probe assembly is to *only* transmit a *shear wave* into an *aluminum* test object.

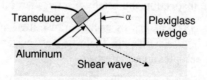

Figure P12.10

12.11 Briefly describe the pulse-echo and through-transmission ultrasonic inspection methods. Compare the advantages and disadvantages of these methods.

12.12 What types of ultrasonic NDE applications are best suited for the through-transmission procedure?

12.13 Briefly discuss the effects of ultrasonic *frequency* on ultrasonic inspections.

12.14 Briefly discuss NDE applications where it is advantageous to employ "low-frequency" ultrasonic inspections. Quantify "low frequency" in this context.

12.15 *Small* edge cracks have been observed to develop on the backside of an aluminum T-shaped component as shown in Figure P12.15. Since the backside is inaccessible, it has been decided to employ an ultrasonic inspection with the angle-beam pitch–catch arrangement shown. Geometric constraints require that the sending transducer be located at point A. You may assume that this transducer is mounted in a plexiglass block arranged so that a *compression* wave is applied at point A with an incident angle θ.

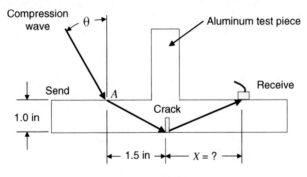

Figure P12.15

(a) Determine the incident angle θ that will give the *most sensitive* inspection for this case.

(b) Where should the receiving transducer be located in this case (i.e., specify the distance X in Figure P12.15)?

12.16 Briefly discuss the factors you would consider in deciding whether to employ radiographic or ultrasonic NDE for a particular component inspection. Compare/contrast the relative features which would cause you to select one inspection method over the other for a given application.

REFERENCES

1. Y. Bar-Cohen and A. K. Mal, "Ultrasonic Inspection," in *ASM Handbook,* Vol. 17: *Nondestructive Evaluation and Quality Control,* ASM International, Materials Park, Ohio, 1989, pp. 231–277.

2. "Ultrasonic Inspection of Materials," in D. E. Bray and D. McBride (Eds.), *Nondestructive Testing Techniques*, Wiley, New York, 1992, Chapter 11.

3. R. Halmshaw, "Ultrasonic Testing of Materials," in *Non-Destructive Testing*, 2nd ed., Edward Arnold, London, 1991, Chapter 4.

4. D. E. Bray and R. K. Stanley, *Nondestructive Evaluation—A Tool in Design, Manufacturing, and Service*, CRC Press, Boca Raton, Florida, 1997.

5. *Nondestructive Testing, Ultrasonic,* Classroom Training Handbook, CT-6-4, General Dynamics, 1981.

6. M. G. Silk, "Sizing Crack-Like Defects by Ultrasonic Means," in R. S. Sharpe (Ed.), *Research Techniques in Non-destructive Testing*, Vol. 3, Academic, 1977, Chapter 2.

7. N. Trimborn, "The Time-of-Flight-Diffraction-Technique." *NDTnet,* Vol. 2, No. 2, September 1997.

8. A. Hecht, "Time of Flight Diffraction Technique (TOFD)—An Ultrasonic Testing Method for all Applications?" *NDTnet,* Vol. 2, No. 2, September 1997.

9. A. S. Birks, R. E. Green, Jr., *Ultrasonic Testing—Nondestructive Testing Handbook,* 2nd ed., Vol. 7, American Society for Nondestructive Testing, Columbus, Ohio, 1991.

CHAPTER 13

EDDY CURRENT INSPECTION

13.1 OVERVIEW

Eddy current testing generates small electrical currents in the test piece by means of a varying electromagnetic field from a coil placed near the object of interest. These eddy currents develop, in turn, a reverse magnetic field that is sensed by the test coil. The eddy currents and the resulting inspection are sensitive to many factors, including electrical conductivity, magnetic permeability, specimen thickness, residual stress, microstructure, geometric factors, magnetic coupling, and discontinuities or cracks. This sensitivity to a wide range of parameters makes eddy current NDE a versatile method that can be tailored to many different applications. Suitable for field-level testing, the technique is, however, limited to materials that are electrical conductors and to examining surface or near-surface anomalies.

Although eddy current inspection equipment has been available since the 1940s, the basis for the technique dates to 1831 when Faraday discovered that varying the electrical current in one coil induces another current in an adjacent coil [1]. The mathematical foundation for eddy current NDE was established in 1864 when Maxwell provided the classic dynamic theory for the electromagnetic field. Hughes used eddy currents to measure electrical conductivity, magnetic permeability, and metal temperature in 1879, but further advances with the technique were slow to develop. In the 1920s, Kranz employed an eddy current device to measure wall thickness and later pioneered eddy current inspection of welded steel tubing (1930–1935). Forster and Zuschlag further developed inspection instruments in the early 1940s, leading to the sophisticated eddy current NDE equipment available today.

Faraday's Law   Briefly summarized, Faraday's law states that a magnetic field that "cuts" a conductor with a closed path (or a conductor that cuts a magnetic

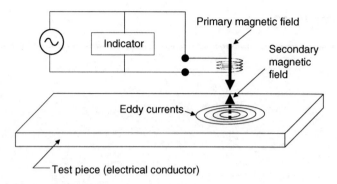

Figure 13.1. Schematic representation of basic eddy current NDE system showing primary magnetic field from test coil that forms eddy currents in test piece and subsequent secondary opposing magnetic field.

field) will generate an electric current in the conductor. The electric current can be induced in the conductor in two ways. First, the conductor may be moved across the magnetic lines of force. Or, the strength of the magnetic field may be cycled about a stationary conductor. This latter method is employed for eddy current NDE, where the alternating magnetic field generated by alternating test coil is brought near the inspection object.

A schematic representation of an eddy current inspection is given in Figure 13.1. Note that the alternating current in the coil develops eddy currents near the surface of the test piece. These alternating eddy currents induce a reverse secondary magnetic field that opposes the original magnetic field direction in the coil. The opposing magnetic field alters the coil impedance, which changes the current flowing in the coil, and is detected in the coil circuit. Anything that affects the eddy currents (e.g., cracks, porosity, inclusions) will modify the secondary magnetic field and be sensed by an impedance change in the test coil.

13.2   COIL TYPES

As shown in Figure 13.2, there are three basic types of test coils used in eddy current testing: surface coils, external coils, and internal coils [2]. Although these coils have different configurations, the eddy currents induced in the test piece always flow parallel to the plane of the coil. The induced eddy currents are strongest near the surface of the test object, and their strength quickly decreases with depth (see Section 13.4.6). Thus, for maximum detection, the discontinuity should be oriented perpendicular to the direction of eddy current flow and be located at or near the surface of the test piece.

Surface Coils   Surface coils are used to inspect relatively flat areas. They introduce eddy currents near the surface of the test member in a circular pattern with approximately the same diameter as the coil. The surface coils are frequently wound around a ferrite core that concentrates the magnetic flux into a smaller volume,

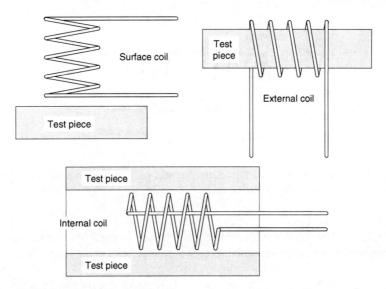

Figure 13.2. Schematic drawing of surface, external, and internal eddy current test coils.

resulting in eddy current distributions that are more sensitive to small discontinuities. The coils are often enclosed in hand-held probes, as shown schematically in Figure 13.3. Note that the hole probe shown in Figure 13.3*b* is a variation of a surface probe designed to scan the open hole bore for cracks or other discontinuities. Probe diameters may be as small as $\frac{1}{32}$ in (1 mm), and special probe holders can be designed for unique applications. Another common version of the surface coil is the

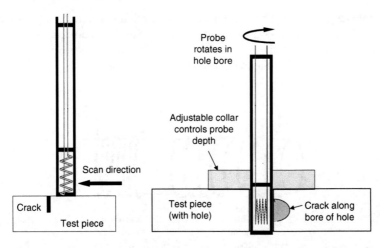

Figure 13.3. Examples of surface probes for eddy current examination of flat surfaces and open holes. Surface coil is oriented in probe on left to inspect flat objects. Surface coil on right is designed to be inserted into the bore of an open hole with collar adjustment to control depth location of probe in hole.

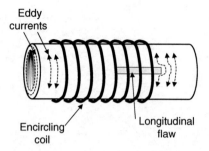

Figure 13.4. Encircling coil develops circumferential eddy currents in a tube or rod that are sensitive to longitudinal discontinuities.

"pancake" coil, in which the conducting elements are arranged in a circular spiral that is compressed to a single plane that is then placed parallel to the test surface.

Encircling Coils   Encircling coils are used to inspect tubes or bars for longitudinal flaws, as shown in Figure 13.4. Here the coil is placed around the object so that eddy currents are developed in a circumferential direction around the perimeter of the test piece. Again, the eddy current strength is greatest at the surface and decreases toward the center of the test piece. Since the eddy currents flow in the same direction as the coil current, this inspection is most sensitive to discontinuities located parallel to the length of the rod (a common type of flaw in tubes). As shown in Figure 13.5, the coil length can be adjusted to suit the application. Wide coils, for example, cover large areas and are mostly sensitive to bulk effects (i.e., conductivity changes), whereas narrow coils examine small areas and are more sensitive to small discontinuities.

Internal Coils   Internal bobbin-type coils are similar to external coils, except that they are placed inside the hollow test object (tube). Since the eddy currents are greatest at the inner diameter, these inspection coils will be most sensitive to longitudinal discontinuities located on the inner diameter of the test piece.

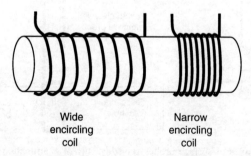

Figure 13.5. Schematic drawing showing wide encircling coils that are sensitive to bulk effects and narrow coils used to locate small anomalies near the surface of rods or tubes.

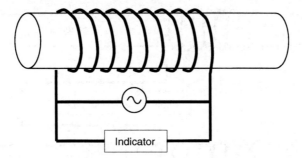

Figure 13.6. Single coil used for both generation and measurement of eddy currents in absolute arrangement.

13.3 COIL ARRANGEMENTS

A few of the many ways to arrange the test coils for eddy current inspection are described below [1, 2].

Single Coil (Absolute Arrangement) Here, as shown in Figure 13.6, the same coil is used to induce eddy currents in the test article and to measure the coil's reaction to the eddy currents. The measurement device (indicator) could be a voltmeter, ammeter, oscilloscope, or strip chart recorder. This arrangement is called "absolute" because the output is not compared with an external reference standard and can be used for all three classes of coils: encircling coil, inside coil, or surface coil.

Double Coil (Absolute Arrangement) Here one coil is used to establish the magnetic field that induces the eddy currents in the test object, whereas a second coil detects changes in the induced eddy currents. Although Figure 13.7 shows these two coils physically separated, the second coil is usually wound inside the first coil but is not connected to the alternating current (AC) source.

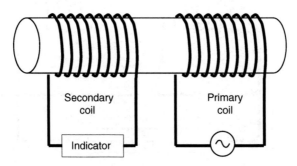

Figure 13.7. Double coils in an absolute arrangement with one coil used to generate and the other to measure eddy currents.

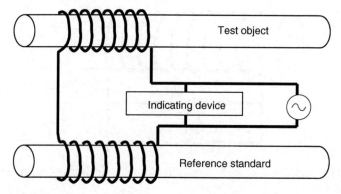

Figure 13.8. Differential arrangement of separate coils in a bridge to allow comparison of test object with a reference standard.

Differential Coil Arrangements   Differential arrangements connect coils around two objects in separate legs of a bridge circuit, allowing comparison of the test object with a reference standard. The bridge may be balanced in a manner that results in a simple go/no–go decision regarding component acceptance.

The differential coil arrangement shown schematically in Figure 13.8, for example, compares the test piece with an external standard. No signal is obtained from the sensing device if both objects have the same impedance. If the test object is flawed, however, its eddy current flow and subsequent magnetic field and impedance are altered, resulting in a no-go indication. The two coils may also be arranged to allow comparison of two areas of the same test object, as shown in Figure 13.9. This procedure compares the left and right sides of the test object and gives an indication if a crack develops in one section. Note, however, that when a discontinuity continues across both sections (such as a longitudinal seam), there would be no test indication. Several other multiple-coil arrangements are also possible for self- and external comparisons [2].

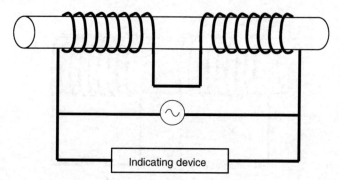

Figure 13.9. Differential arrangement of separate coils in a bridge to allow comparison of test object with itself.

13.4   PARAMETER SENSITIVITY

This section describes eddy current operating variables that may be exploited for inspection purposes. As indicated, eddy currents are sensitive to several parameters, allowing various types of NDE applications. This versatility is a two-edged sword, however, as proper interpretation of test results requires that the inspector correctly identify the parameter(s) leading to the indication at hand.

13.4.1   Electrical Conductivity

Recalling that the test object must be electrically conductive, it is not surprising that the degree of conductivity has a significant influence on eddy current formation. Higher conductivity of the test object leads to a more sensitive test but results in less component penetration (see Section 13.4.6). The electrical conductivity of metal varies with temperature (conductivity decreases as temperature increases), alloy composition, heat treatment and resulting microstructure (i.e., grain size), hardness, residual stress, and discontinuities in the metal. Variations in these parameters can be monitored with eddy currents, provided they influence the conductivity of the test piece. The relative conductivity of various metals and alloys are given in Tables 13.1 and 13.2.

Note in Table 13.2 that aluminum alloys with the same chemical composition (e.g., 2024-0, 2024-T3XX, 2024-T8XX) can have significantly different electrical conductivities. This difference depends on the thermal mechanical treatment and its

TABLE 13.1   Conductivity of Various Metals and Alloys

Metal or Alloy	Conductivity, % (IACS[a])
Silver	105
Copper, annealed	100
Gold	70
Aluminum	61
Magnesium	37
70-30 Brass	28
Phosphor bronzes	11
Monel	3.6
Zirconium	3.4
Zircalloy-2	2.4
Titanium	3.1
Ti–6A1–4V alloy	1.0
304 Stainless steel	2.5
Inconel 600	1.7
Hastelloy X	1.5
Waspalloy	1.4

Note: See References 1 and 2.

[a] The IACS (International Annealed Copper Standard) gives copper a reference conductivity of 100%.

TABLE 13.2   Electrical Conductivity Ranges for Aluminum Alloys

Alloy and Temper	Electrical Conductivity, % (IACS)[a]	
	Minimum	Maximum
1100 (all tempers)	57.0	62.0
2024-0	45.5	50.0
2024-T3XX	28.0	33.0
2024-T4XX	28.5	32.5
2024-T6XX	35.0	41.0
2024-T8XX	36.0	42.5
3003-0	44.5	50.0
6061-0	47.0	51.0
6061-T4XX	35.5	41.5
6061-T6XX	40.0	45.0
7075-0	44.0	48.0
7075-T6XX	30.0	35.0
7075-T73X	38.0	42.5
7075-T76X	36.0	39.0

Note: See References 1–3.

[a] International Annealed Copper Standard.

subsequent effect on microstructure and tensile strength. In general, an alloy has its largest conductivity when in the fully annealed condition. Some alloy tempers and, thus, strength levels are indicated in Table 13.2 by the -TXXX designation following the four-digit code (e.g., 2024) for alloy chemical composition. (Here the X's refer to several potential variations in thermal mechanical treatment.) The -0 designation specifies the fully annealed condition before the alloy is thermally and mechanically treated.

13.4.2   Magnetic Permeability

The magnetic permeability (μ) of the test object has an important impact on eddy current NDE. The permeability defined in Equation 13.1 measures the ease with which a magnetic flux can be established in the test material:

$$\mu = \frac{B}{H} \tag{13.1}$$

where B = strength of magnetic field

H = magnetic force in material

Since eddy currents are induced by the magnetic field from the test coil, the permeability of the test material strongly influences the eddy current response.

The magnetic permeability μ is 1 for nonmagnetic materials. For ferromagnetic materials, $\mu > 1$, and the induced magnetic field B is greatly intensified in the test object for a given magnetic force H. This fact poses a potential problem for eddy current NDE since variations in the magnetic field strength in the object can over-whelm changes in coil impedance caused by other factors of particular interest (e.g., flaws, microstructure, thickness, conductivity). The solution in this case is to magnet-ically saturate the ferromagnetic test piece so that small changes in magnetic flux do not cause large changes in eddy currents that mask other more interesting phenomena. In this case, a direct-current (DC) coil is first used to magnetically saturate the test part, and then the AC signal is applied to develop the eddy currents.

13.4.3   Lift-off Factor

Eddy current indications are very sensitive to the distance between the coil and the surface of the test object, with the eddy current density decreasing rapidly as the coil is removed from the component. Since variations in the coil–surface separation (i.e., magnetic coupling) can mask changes in impedance from other sources of primary interest, it is important to maintain a uniform distance between the coil and the test surface. Although this requirement complicates the NDE of intricate shapes, the eddy current lift-off phenomenon can be exploited to measure thickness of nonconductive coatings (e.g., paint, anodizing) on metal surfaces.

13.4.4   Fill Factor

The fill factor describes a problem with encircling coils that is similar to the lift-off phenomenon for surface coils. Here eddy current response is sensitive to how well the test part "fills" the coil. Since small changes in the diameter of the test object can cause large changes in eddy currents and mask other features of interest, it is important to keep the ratio between the diameters of the test object and the coil constant. The fill factor is defined by Equation 13.2, where D_i is the inner diameter of the test coil and D_o is the outer diameter of the test object:

$$N_{\text{fill}} = \left(\frac{D_o}{D_i} \right)^2 \tag{13.2}$$

In general, the fill factor should be close to 1.

An analogous phenomenon occurs for internal (bobbin) coils, where the fill factor is based on the ratio of the outer coil diameter to the internal diameter of the hollow test piece. Although eddy current NDE results are sensitive to the fill factor, this magnetic coupling effect can again be exploited to measure the outer diameter of circular rods with encircling coils or the inner diameter of tubes with internal coils.

13.4.5   Edge Effect

The edge effect deals with eddy current distortion caused when the test coil ap-proaches a free edge or junction between materials. Large variations in eddy currents can result at these locations and mask changes due to other anomalies. Although small coil diameters or special fixtures that hold the probe a set distance from the edge may minimize the problem, the edge effect cannot be completely eliminated. In general, one cannot inspect closer than $\frac{1}{8}$ in (0.32 cm) from the edge of the test object, although the particular limit depends on the coil size and test frequency.

13.4.6 Skin Effect

Figure 13.10 plots eddy current strength normalized by the value at the surface versus normalized depth of penetration. As shown, the eddy currents are strongest at the surface of the test object and rapidly decay below the surface. This limited depth of penetration, commonly known as the "skin effect," results from the secondary magnetic field created by the induced eddy currents. The secondary magnetic field opposes the original (primary) field in the test coil, reducing subsequent eddy current formation in the test piece. Thus, factors that increase the original eddy current strength (e.g., conductivity or magnetic permeability) also strengthen the opposing secondary magnetic field and limit the depth of penetration into the component thickness.

The standard depth of penetration δ is defined as the component depth where the eddy current density has decayed to 37% (i.e., $1/e$ precent) of the surface value. The standard depth of penetration δ is shown schematically in Figure 13.10 and is given by

$$\delta = (\pi \sigma \mu f)^{-1/2} \tag{13.3}$$

where σ = conductivity of test object (mhos)
    = 1/electrical resistivity of object
 f = inspection frequency (Hz)
 μ = magnetic permeability (1 for nonmagnetic materials, >1 for ferromagnetic materials)

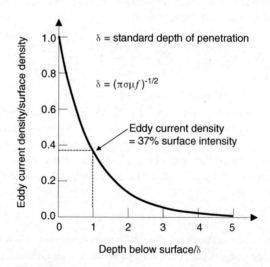

Figure 13.10. Normalized eddy current strength as a function of depth below specimen surface showing definition of standard depth of penetration.

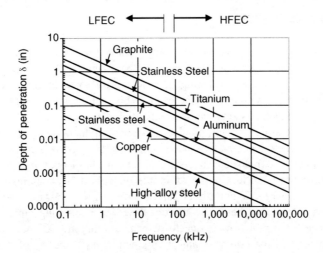

Figure 13.11. Standard depth of penetration versus frequency for various materials showing distinction between high- and low-frequency eddy current designations.

Figure 13.11 [2] shows the standard eddy current depth of penetration as a function of inspection frequency for several materials with different electrical conductivity and permeability. Note that the depth of penetration δ increases *inversely* with both frequency and conductivity, leading to more penetration in poorer conductors. In addition, the depth of penetration is less for ferromagnetic materials (i.e., penetration decreases as the magnetic permeability μ increases). As mentioned previously, ferromagnetic objects are usually magnetically saturated with a separate DC coil to minimize variations in the magnetic field (and the eddy currents) through the test piece.

The skin effect implies that there is a thickness factor for eddy current NDE. Although specimen thickness may be measured with eddy currents, the eddy current response is independent of thickness if the component thickness exceeds a few δ's for the given material and frequency. Thus, eddy current measurements are more accurate on thin rather than thick specimens, a trend that complements ultrasonic testing, where thickness measurements are more accurate for thick objects.

13.4.7 Other Factors

The coil current and temperature also influence eddy current formation and subsequent NDE. Increasing the coil current strengthens the induced magnetic field and, thus, strengthens the eddy currents in the test piece. Since conductivity decreases with increased temperature, eddy current strength also varies inversely with temperature. Temperature sensitivity must be kept in mind when employing differential coil arrangements that compare the test object with a reference standard (i.e., both members should be at the same temperature to produce the same output).

13.5   IMPEDANCE TESTING

As discussed in the previous section, eddy current formation depends on several geometric and material factors. Whereas this parameter sensitivity suggests many applications for eddy current testing, it is necessary to separate the influence of these factors to properly interpret test results. This section briefly summarizes some procedures that may be used to analyze eddy NDE.

Eddy current testing is based on electromagnetic induction of current in the test object and the subsequent development of an opposing magnetic field in the test coil. Thus, coil impedance (resistance to current flow) is the basic quantity detected during the eddy current inspection. Consider a simple AC circuit that contains a test coil, as shown in Figure 13.12. When direct current flows in the coil, the coil's electrical resistance R is the only barrier to current flow. For alternating current, however, the resistance to current flow is due to two factors that are 90° out of phase: the AC resistance R of the coil wire and the inductive reactance X_L of the coil (both expressed in ohms). The inductive reactance depends on the frequency as given by

$$X_L = 2\pi f L_0 \tag{13.4}$$

Here f is the AC frequency expressed in hertz and L_0 is the coil inductance given in henrys. One henry is the inductance that produces one volt across the coil when the current is changed by one ampere. The henry is a fairly large quantity, and inductances associated with eddy current testing are frequently expressed in millihenrys or microhenrys. Note that X_L depends on the coil parameters and test frequency. The magnitude of the total coil impedance Z (resistance to current flow) is given by

$$Z = \sqrt{R^2 + X_L^2} \tag{13.5}$$

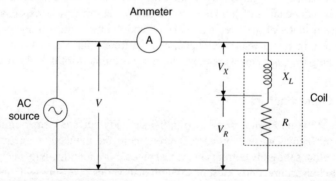

Figure 13.12. Schematic representation of an eddy current coil showing the inductive reactance X_L and resistance R in an AC circuit.

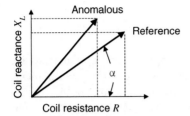

Impedance plane

Figure 13.13. Impedance plane diagram showing the relationship between coil reactance X_L and resistance R (note X_L lags R by the phase angle α). The original coil impedance measured with a reference object can change due to material and/or component anomalies.

This frequency dependent impedance resists current flow in the coil and decreases the strength of the coil's magnetic field, which, in turn, reduces the eddy currents developed in the test piece. The voltages V_X and V_R (see Figure 13.12) due to the coil reactance X_L and coil resistance R are 90° out of phase, with the reactance voltage lagging the resistance voltage. Since the current i is the same at any instant, the reactance and resistance impedances are also out of phase by an angle α, as shown in the "impedance plane" plot of Figure 13.13. Here the coil reactance X_L is plotted against the coil resistance R. These two impedances are given by

$$X_L = \frac{V_X}{i} \tag{13.6}$$

$$R = \frac{V_R}{i} \tag{13.7}$$

These two coil voltages and impedances may be shown as vectors in voltage or impedance plane diagrams, as indicated in Figure 13.13. Note that the reference coil impedance may be changed by anomalies that alter eddy current formation in the test piece and, thus, change the magnetic field that opposes current flow in the coil.

Several analysis methods may be used to relate various test parameters with changes in coil impedance resulting from the eddy currents in the test piece. Discussion here will focus on impedance-testing systems and phase analysis systems [2]. (See References 1–5 for additional details and other analysis techniques.)

13.5.1   Impedance-Testing Systems

Impedance-testing systems involve measuring changes in the total impedance of the test coil but do not record phase shift information (i.e., the angle α). One assumes that various test parameters (e.g., lift-off, thickness, conductivity) are held constant and that changes in the coil impedance are only due to the one parameter of interest. This simple, direct procedure works well when the test variable causes a sudden change in impedance (e.g., a crack or other geometric discontinuity). The procedure is not as effective for detecting subtle changes in parameters such as conductivity, thickness, and hardness. In those cases, it is difficult to isolate the cause(s) of the impedance change from the several possible sources, and additional measurement data are required to interpret the results.

13.5.2 Phase Analysis Systems

These systems record both impedance and phase angle information and provide the opportunity to separate the influence of several test variables. The following discussion is adapted from Reference 2, and although there are other useful formats for presenting impedance plane data, the examples shown here demonstrate how one may discriminate between various eddy current test parameters.

Conductivity Effect   Consider eddy current measurements from a series of identical, thick blocks. The test frequency, thickness, and geometry, for example, are all held constant, and the coil is placed on the object so that the lift-off is zero. Only the material conductivity is changed in these experiments. The eddy current coil impedance is measured for the various test pieces and the results are plotted in the impedance plane, as shown in Figure 13.14. Here the inductive reactance (X_L) and resistance (R) of the coil impedance are plotted versus each other to form a "conductivity locus" for the different materials. These schematic results are obtained at two different frequencies ("high" and "low") from tests of identical "thick" blocks made from materials with different conductivity. The two limits in conductivity shown in Figure 13.14 are "air" (0% IACS) and annealed copper (100% IACS). Other materials (1, 2, . . . , 6) with increasing conductivities fall between these limits (i.e., conductivity of air < material 1 < 2 < 3 < 4 < 5 < 6 < copper).

Note that the relative values of the coil reactance and resistance depend on test object conductivity, giving each material a different position on the master conductivity locus. In general, increasing test object conductivity results in a greater resistive component of impedance, whereas decreasing conductivity leads to a greater reactive component. Also note that the relative positions of materials on the conductivity locus change with frequency. At the low frequency, for example, highly conducting materials are "spread out" on the conductivity curve whereas poor conductors are grouped together. At the high test frequency, however, this trend is reversed, with good conductors grouped together and poorer conductors more separated. Thus, frequency plays an important role in the ability for eddy current NDE to discriminate between materials of different electrical conductivity.

Lift-off Effect   Now, repeat these experiments with a single highly conductive material such as material 5 (same frequency, thick specimen, etc.). In this case, however, the lift-off factor is gradually changed by removing the coil from the test piece. The "lift-off" line (i.e., the dashed line) in Figure 13.15 shows that the measured impedance for a given conductivity depends on the distance separating the coil and the test piece. When the lift-off is zero (i.e., the coil contacts the test object), one obtains the original location for material 5 on the conductance curve. When the lift-off is large, the conductance curve result for air (0% IACS) is reached. Different lift-off values lie between these extremes on the lift-off curve. A family of such lift-off curves

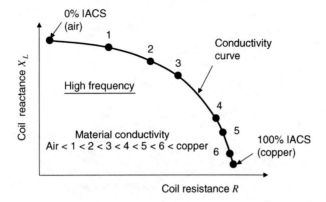

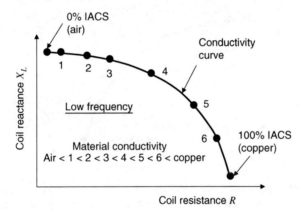

Figure 13.14. Conductivity locus of reactance versus resistance components of coil impedance shown in an impedance plane diagram. These results are obtained from tests of identical thick blocks made from materials with different conductivity. Note relative positions of materials when the test frequency is changed.

can be obtained for all of the materials used to generate the original conductance curve (e.g., note the lift-off curve for material 3).

Since the relative material location along the original conductance curve changes with frequency, the angle between the lift-off and conductivity curves also depends on frequency. Compare, for example, the high- and low-frequency lift-off curves for material 3 in Figure 13.15. In this case a more "shallow" intersection occurs at the lower frequency than at the higher frequency. Thus, it would be easier to distinguish between impedance changes caused by lift-off or conductivity at the higher frequency in material 3, since that lift-off curve deviates more quickly from the conductivity curve. For materials with other conductivity, however, a different frequency might give more separation between the lift-off and conductivity curves. Thus, frequency

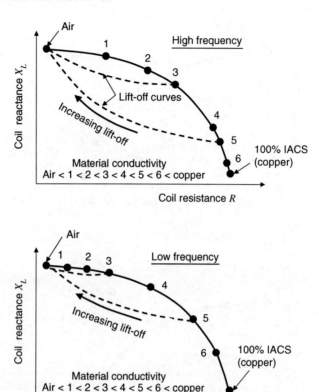

Figure 13.15. Effect of lift-off on the impedance plane behavior for two materials with different conductivity. Here specimen thickness and eddy current frequency are fixed.

selection again plays a key role in separating the influence of various parameters in eddy current testing.

Thickness Effect An indication of how eddy current tests may be used to determine component thickness may be seen by another series of tests with specimens of various thicknesses. Figure 13.16 compares impedance measurements with various thickness specimens (designated by points $A, B, \ldots, E$) made from material 5 (dotted line). Note that as thickness increases, the original thick-specimen point on the conductivity locus is reached (recall the original conductivity line was obtained from tests with thick specimens). The thickness curve approaches the air point on the conductivity locus for extremely thin specimens (i.e., air corresponds to zero thickness). Thus, a family of thickness curves could be established for various materials and used to measure component thickness.

Due to the limited depth of eddy current penetration discussed in Section 13.4.6, however, eddy current NDE is only able to distinguish between thicknesses of relatively thin specimens (i.e., on the order of two or three δ's per Equation 13.3). Again,

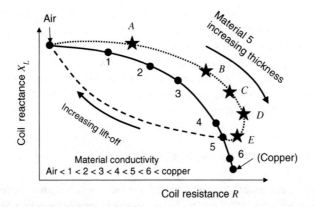

Figure 13.16. Effect of specimen thickness on the impedance plane behavior for a particular material and test frequency.

since depth of penetration depends inversely on conductivity, frequency, and permeability as shown in Equation 13.3, the relative positions along the thickness curves depend on test frequency and material. Thus, there will be an optimal test frequency for the particular material and range of specimen thickness of interest.

13.6 OPERATING FREQUENCY

As seen in the previous section, test frequency has a large role in eddy current testing. Typical operating frequencies range between 200 Hz and 6+ MHz, and the subsets of eddy current NDE may be defined as follows:

- Low-frequency eddy current (LFEC): $0.1 \text{ kHz} < f < 50 \text{ kHz}$.
- High-frequency eddy current (HFEC): $f > 100 \text{ kHz}$.

Frequency selection for a given application often involves compromise. Recall, for example, that the depth of penetration increases as f decreases but that sensitivity decreases. Whereas detecting surface flaws in nonferromagnetic materials may suggest a relatively high frequency (~ 5 MHz), a lower test frequency (~ 1 kHz) would be required for ferromagnetic materials due to the loss in penetration associated with increased permeability (μ). Low-frequency eddy current (LFEC) is employed for maximum depth of penetration, such as for locating subsurface cracks or cracks in multilayer stack-ups. Low-frequency applications require larger coils, however, and have less sensitivity than high-frequency eddy current (HFEC). Due to this trade-off between depth of penetration and sensitivity, one usually picks the highest frequency consistent with the desired penetration.

Multifrequency techniques operate at two or more frequencies. This procedure allows one to perform simultaneous tests at different frequencies to enable multiparameter measurements. In addition, one can employ multiple frequencies to separate the influence of various test parameters (e.g., lift-off, thickness, cracks, dimensional changes). Continuous-eddy-current methods involve the application of several frequencies at one time. Here, bandpass filters are tuned to specific frequencies that separate the outputs. Sequential test methods involve switching between frequencies during the NDE test. Often referred to as multiplexed systems, they employ phase/amplitude processing of the test data. It is also possible to Fourier transform a pulse to acquire multiple-frequency data.

13.7   EDDY CURRENT APPLICATIONS

Eddy current applications are based on the various inspection parameters that influence the eddy currents in the test piece. In principle, any feature that alters the eddy current flow can be detected. Some examples of eddy current parameter sensitivity to NDE applications are given below.

Component Discontinuities   Eddy currents are particularly sensitive to discontinuities that are located perpendicular to the direction of current flow (e.g., surface cracks, seams, pits, weld flaws), and eddy current testing is a powerful NDE method for those types of anomalies. The technique is, however, less sensitive to discontinuities located parallel to the eddy current direction. As stated before, HFEC testing is the most sensitive technique and is capable of finding quite small fatigue cracks. As reported in Reference 6, for example, 200-kHz eddy current measurements with simulated Boeing 737 fuselage lap splice joints had 90% probability-of-detection crack lengths ranging from 0.071 to 0.088 in (1.80–2.24 mm). Since HFEC has limited depth of penetration, however, low-frequency measurements are needed for cracks that do not penetrate the surface or for second-layer inspections. A 0.5-in- (1.3-cm-) diameter LFEC coil operating at 500 Hz, for example, can find 1-in- (2.5-cm-) long cracks in aluminum splice plates that are covered by a 0.2-in- (0.5-cm-) thick aluminum skin [7].

Electrical Conductivity   Sensitivity of eddy currents to electrical conductivity can be used for alloy sorting (by composition), to detect residual stresses, to measure heat treatment or hardness, and to measure the thickness of *metallic* cladding or plating.

Magnetic Permeability   Sensitivity to magnetic permeability can also be used for alloy sorting (magnetic materials), sensing heat treatment conditions (magnetic materials), case hardening depth measurements, or plating thickness measurement.

Geometric and Material Homogeneity   Metal thickness (for thin sheets) can be measured by the skin effect. Changes in material homogeneity can indicate the presence of cracks, segregation, seams, inclusions, pits, corrosion, and so on.

Magnetic Coupling   Also known as the lift-off or fill factor, magnetic coupling can indicate the thickness of *nonmetallic* coatings or insulation thickness. This effect can also be used for proximity gages, for diameter measurements, and to determine gaps between metal sheets. The lift-off effect can also characterize surface roughness.

Example 13.1   The ability of the eddy current method to measure surface roughness as well as specimen thickness is demonstrated in Figure 13.17 [8, 9], where eddy current methods were used to examine geometric changes associated with corrosion. Here one side of 1-in- (2.5-cm-) wide 2024-T3 bare aluminum specimens were exposed to a 3.5% NaCl environment over a 2-in- (3.1-cm-) gage length through the ASTM G-44 alternate emersion corrosion test. This accelerated corrosion protocol causes local pitting along with a general loss in specimen thickness. References 8 and 9 propose that an increase in local stress caused by the general thickness loss combined with an equivalent initial crack size distribution (see Section 9.5) associated with local pitting or surface roughness could be used to estimate the influence of prior corrosion on subsequent fatigue life. Thus, the goal here was to employ eddy current NDE to measure thickness loss and surface roughness caused by various degrees of corrosion.

The actual thickness of 40 nominal 0.09-in- (2.3-mm-) thick specimens were measured before and after corrosion using the impedance plane concepts discussed in Section 13.5.2. Here calibration blocks with steps of approximately 0.0005 in (0.013 mm) were used as a reference for subsequent eddy current measurements of average specimen thickness. Note in Figure 13.17a that the average thickness following exposure to the corrosive media decreased as the corrosion period increased. These reductions

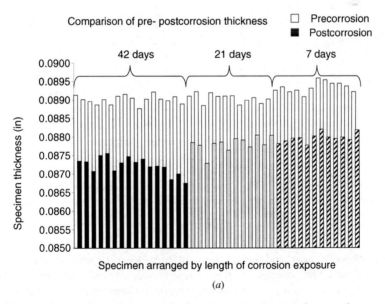

Figure 13.17. Use of eddy current to measure thickness loss and surface roughness caused by various levels of ASTM G44 alternate emersion corrosion on 2024-T3 bare aluminum [8]. (a) Comparison of specimen thickness before and after corrosion showing general loss of material thickness. (*continued*)

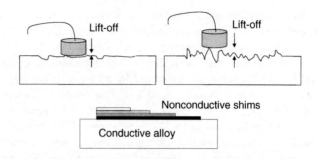

Roughness of corroded surface measured by
eddy current lift-off factor

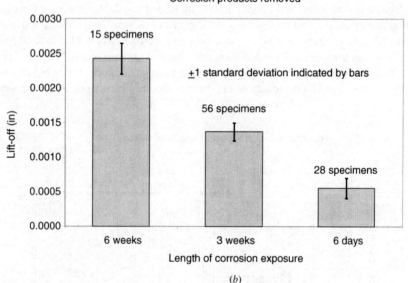

Lift-off for various exposure periods
Corrosion products removed

Figure 13.17. (*continued*) (*b*) Eddy current lift-off measurements showing average increase in surface roughness with longer exposure to corrosion. Corrosion products were removed before eddy current testing [8].

in specimen thickness, and thus decrease in cross-sectional area, would correspond to 0.8, 1.8, and 2.8% increases in remote tensile stress for the respective 7-, 21-, and 42-day corrosion exposures. In this case, corrosion was only allowed to occur on one side of the thin specimens, and the thickness measurements were obtained by touching the eddy current probe to the uncorroded side of the specimen.

The local increase in surface roughness caused by corrosion pitting was measured by the eddy current lift-off effect, as shown in Figure 13.17*b*. Here the rough pits prevented uniform contact between the probe and specimen surface. In this case, the lift-off was calibrated by placing a series of nonconducting shims on an uncorroded specimen. In addition, corrosion products were first removed from the specimens so that only surface roughness associated with pitting of load-bearing material was assessed. As shown in Figure 13.17b, the average lift-off measurement (and thus specimen roughness) also increased with the length of exposure to the corrosive environment, giving another eddy current indication of the extent of corrosion damage. (Problems 17.17–17.20 in Chapter 17 indicate how those types of NDE measurements might be used to evaluate the severity of corrosion damage on subsequent fatigue life.)

13.8   CONCLUDING REMARKS

Eddy current testing is a most versatile and powerful NDE tool for examining components made from electrical conducting materials. The reader is encouraged to consult References 1–2, 4, 5, and 10 for additional details of this versatile inspection method. The advantages and disadvantages of eddy current NDE are briefly summarized below.

13.8.1   Advantages of Eddy Current Testing

1. Eddy current NDE is sensitive to several structural and metallurgical variables, leading to many types of applications. Special probes can be designed for specific purposes.
2. Eddy current NDE is a moderately fast technique that gives an instantaneous indication.
3. Portable instrumentation is available for field use.
4. The technique is adaptable to simple go/no-go testing.
5. The method can detect small surface and near-surface discontinuities. Low-frequency eddy current testing can be used for inspection of multiple stack-ups of thin sheets.
6. Eddy current NDE has little chance of harming the test object.
7. Minimal part preparation is needed (nonconductive paint layers that are less than 0.015 in thick may be readily penetrated).

13.8.2   Disadvantages of Eddy Current Testing

1. The test object must be an electrical conductor.
2. The test surface must be accessible to the eddy current probe.
3. There is only a limited depth of penetration [$< \frac{1}{4}$ in. = (6 mm) in standard cases].
4. Skill and training are required to suppress noninteresting test variables.
5. There can be problems with ferromagnetic materials and with rough surfaces.
6. The method can be time consuming for inspecting large areas.

PROBLEMS

13.1 Use the Internet to locate the home page for a vendor that supplies eddy current devices for NDE purposes and record this company's address for future reference. Examine the products and applications described on the web site and prepare a short report that describes the various capabilities and uses for the eddy current products sold by this company.

13.2 Briefly summarize the principles behind the eddy current inspection method.

13.3 Briefly discuss the various parameters that influence a metal's conductivity. How does sensitivity to conductivity influence eddy current NDE?

13.4 What special considerations are needed for eddy current NDE on ferromagnetic materials?

14.5 Describe the lift-off and fill factors associated with eddy current NDE. How may these factors be utilized for inspection purposes?

13.6 Define *high-frequency eddy current* and *low-frequency eddy current* in the context of NDE. What are the advantages and disadvantages of each technique?

13.7 Would LFEC or HFEC eddy current be better for the following cases? Briefly explain your answer.
  (a) Sorting aluminum alloys by their composition
  (b) Sorting titanium alloys by composition
  (c) Verifying that a high-alloy steel has received the proper heat treatment
  (d) Inspecting for cracks in a titanium member that is underneath a thin aluminum sheet

13.8 How might eddy current NDE be used to verify that the thickness of a nonconductive paint meets specification?

13.9 An eddy current inspection method is being used to measure the thickness of 7075-T6 aluminum sheets. The test frequency is 1 kHz. It is discovered that it is possible to accurately discriminate between sheet thicknesses that are less than 0.4 in but that thicker specimens cannot be resolved as accurately. Briefly explain why eddy current thickness measurements lose accuracy for the thicker specimens. What would the maximum resolvable thickness be if the test frequency were changed to 100 kHz?

13.10 It is desired to use eddy current NDE for quality control measurements of the thickness of sheets produced from a certain alloy. Laboratory tests indicate that suitable accuracy is obtained for 0.01-in-thick sheets when the eddy current frequency is 60 kHz. What frequency would you employ if you wished to perform similar quality control measurements on 0.05-in-thick sheets made from the same material? What changes to the test frequency, if any, would you expect to make if you wished to design a similar test for a different material that is a *better* electrical conductor?

REFERENCES

1. ASM Committee on Eddy Current Inspection, "Eddy Current Inspection," in *ASM Handbook,* Vol. 17: *Nondestructive Evaluation and Quality Control,* ASM International, Metals Park, Ohio, 1989, pp. 164–194.

2. *Nondestructive Testing, Eddy Current,* 2nd ed., Classroom Training Handbook CT-6-5, General Dynamics, 1979.

3. *Non-Destructive Inspection Methods*, Technical Manual TM55-1500-335-23, Royal Australian Air Force, June 25, 1992.

4. R. Halmshaw, *Non-Destructive Testing,* 2nd ed., Edward Arnold, London, 1991, Chapter 6, pp. 230–261.

5. E. Bray and D. McBride, *Nondestructive Testing Techniques,* Wiley, New York, Chapter 18, 1992.

6. W. D. Rummel and G. A. Matzkanin, *Nondestructive Evaluation (NDE) Capabilities Data Book,* Nondestructive Testing Information Analysis Center, Aging Aircraft NDI Validation Center at Sandia National Laboratories, Albuquerque, New Mexico, May 1996.

7. G. Ansley, S. Bakanas, M. Castronuovo, T. Grant, and F. Vichi, *Current Nondestructive Inspection Methods for Aging Aircraft,* Report No. DOT/FAA.CT-91/5, FAA Technical Center, Atlantic City International Airport, New Jersey, June 1992.

8. J. Scheuring, and A. F. Grandt, Jr., "A Fracture Mechanics Based Approach to Quantifying Corrosion Damage," *SAE 1999 Transactions, Journal of Aerospace,* Sec. 1, Vol. 108, pp. 270–281, 1999.

9. J. Scheuring, and A. F. Grandt, Jr., "Quantification of Corrosion Damage Utilizing a Fracture Mechanics Based Methodology," in *Proceedings of The Third Joint FAA/DoD/NASA Conference on Aging Aircraft,* Albuquerque, New Mexico, September 20–23, 1999.

10. M. L. Mester, *Electromagnetic Testing—Nondestructive Testing Handbook*, 2nd ed., Vol. 4, American Society for Nondestructive Testing, Columbus, Ohio, 1986.

CHAPTER 14

MAGNETIC PARTICLE INSPECTION

14.1 OVERVIEW

This chapter deals with magnetic particle inspection, a practical subset of the more general class of magnetic inspection techniques. In a magnetic inspection, the ferromagnetic test piece is first magnetized by one of several methods employing magnets or an applied electrical current. Anomalies in the test piece cause magnetic flux "leakage" fields that are then detected by a variety of visual or electronic sensors.

The magnetic particle technique relies on visual detection of small particles that migrate to the flux leakages associated with flaws in the test member. These magnetic particles may be applied dry or wet and come in a variety of colors (including fluorescent). The technique is sensitive to surface or near-surface cracks and, although limited to ferromagnetic materials, has many practical applications in industry.

In addition to magnetic particles, other visible flux sensors include magnetic rubber, magnetic paint, magnetic printing, and magnetic foils (see Section 14.5). Additional magnetic inspection methods (e.g., magnetic flux leakage, electric current injection, Barkhausen noise) are briefly summarized in Section 14.6.

14.2 PRINCIPLES OF MAGNETISM

This section overviews some principles of magnetism that are related to magnetic particle NDE. More detailed discussions of the theory and applications of magnetism are presented in References 1–5.

The theory of ferromagnetism is based on the Bohr–Sommerfeld atomic model where the basic magnetic particle is an electron spinning about its axis. As the electron spins, its moving electrical charge generates a magnetic moment whose polarity is determined by the spin direction. Since many of the electrons spin in opposite directions in demagnetized material, there is no net magnetic moment. A magnetic field exists, however, when there are uncompensated spins. Ferromagnetism results when a magnetic force increases the number of uncompensated spins and, thus, the strength of the magnetic field.

Ferromagnetic materials have small regions (domains) where the magnetic moments are parallel and add rather than cancel. These domains are smaller than the material grain size and are separated by boundaries (domain walls) that allow the moments in a given area to remain parallel. Magnetic fields cause the domain walls to move in ferromagnetic materials so that the randomly oriented domains align with the direction of the magnetic field. The result is a change in overall magnetization of the sample.

In 1919 Barkhausen demonstrated that the magnetization process is not continuous but consists of small abrupt steps when the magnetic domains move under an applied magnetic field. When all the electrical signals caused by domain movement are added, they result in a noiselike signal known as Barkhausen noise. (Barkhausen noise leads to another distinct NDE method, which is discussed in Section 14.6).

Figure 14.1 shows the magnetic field associated with a bar magnet. Note the continuous flow through the magnet from the south to the north pole and the return around the outside of the magnet. If the bar is broken into pieces, separate poles are formed at the ends of each piece. When the pieces are brought together, the original magnetic field returns, except for a local perturbation around the air gap between the broken ends. The basic premise for magnetic NDE is that such interruptions

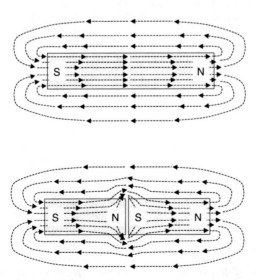

Figure 14.1. Schematic representation of the magnetic lines of force around a bar magnet. Note local distortion of magnetic field around gap between two adjacent magnets that are in close proximity. Similar distortions are caused by discontinuities in magnetized test piece.

in the local magnetic field caused by discontinuities can be sensed and related to component anomalies. In magnetic particle inspections, these "flux leakages" attract small magnetic particles that are visually detected.

Electromagnetism involves magnetizing materials with an electric current. As shown in Figure 14.2, a circular magnetic field develops when electric current flows through a conductor. Current flow in a circular coil forms a longitudinal magnetic field inside and outside the coil. The coil increases the magnetic flux density for a given current and facilitates magnetization of ferromagnetic material. Note that the directions of the magnetic field are determined by the right-hand rule in both cases. In addition, discontinuities that are oriented perpendicular to the field directions will be most disruptive and sensitive to detection.

The magnetic flux density and field intensity are related through the permeability of the material by the equation

$$\mu = \frac{B}{H} \tag{14.1}$$

Here H is the magnetizing force (field intensity), B is the flux density, and μ is the magnetic permeability (a measure of the ease with which a material can be magnetized). The SI units of these quantities are amperes per meter for H, the tesla

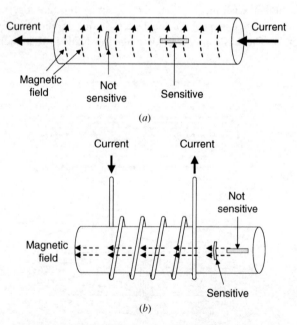

(a)

(b)

Figure 14.2. Magnetization obtained by current flowing through a rod or by a circular coil. (a) Current flow through the rod results in a circular magnetic field around the rod that is sensitive to longitudinal discontinuities. (b) Longitudinal magnetic field that results from a circular coil placed around the test piece is sensitive to circumferential anomalies.

(webers per square meter) for B, and henrys per meter for μ. [The centimeter–gram–second (cgs) units for H and B are the oersted and gauss, respectively.] The term *lines of force* is often used to identify either lines of constant flux density (B) or lines of magnetic field intensity (H). Flux density and magnetic field intensity lines only coincide, however, in nonmagnetic materials or in a vacuum.

Magnetic and electric circuits are analogous. In electrical circuits, Ohm's law, given by Equation 14.2, relates the electrical current I, the voltage E, and the resistance R:

$$ I = \frac{E}{R} \tag{14.2} $$

The corresponding relation in magnetic circuits given by Equation 14.1 may be rewritten as

$$ B = \frac{H}{R'} $$

where the reluctance $R' = 1/\mu$. Note that both equations indicate that "energy" flow (I or B) is related to a "force" (E or H) divided by the "resistance" to flow (R or R', which is equal to $1/\mu$). Also note that the electrical resistance R and magnetic reluctance R' are inversely proportional to electrical conductivity and to magnetic permeability, respectively.

14.3 MAGNETIC FIELD GENERATION

As indicated previously, detected perturbations in the "normal" magnetic fields are evidence of material anomalies. The "continuous" inspection approach examines the component while the magnetizing current is applied and can be used for most parts. The "residual method" studies the component *after* the magnetizing current is turned off. Successful inspection in this case depends on the component retaining some residual magnetization (retentivity). The residual approach cannot be used, for example, on low-alloy steels that have poor retentivity.

The remainder of this section discusses various methods to introduce magnetic fields into the test object. Several of these approaches are readily accomplished with the apparatus contained on a modern magnetic particle inspection unit (e.g., see Figure 14.3), whereas other techniques require additional equipment. In most cases, it is necessary to demagnetize the component following the inspection (see Section 14.4).

14.3.1 Coils

Electromagnetic coils are used to provide longitudinal magnetic fields in the test object, as shown in Figure 14.4. The strength of the longitudinal magnetic field inducted in the test part is proportional to the coil current (I) and the number of

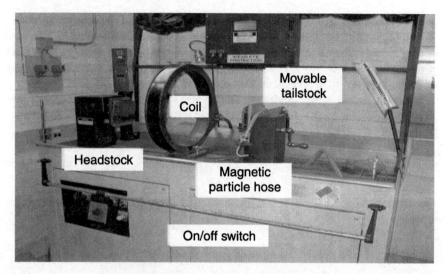

Figure 14.3. Typical magnetic particle test machine.

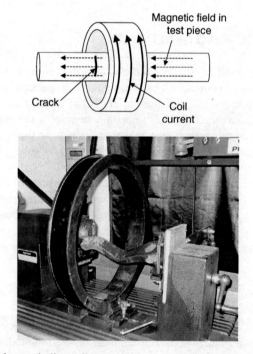

Figure 14.4. Use of an encircling coil to establish a longitudinal magnetic field in test object.

turns (N). Note that in this case discontinuities should be oriented perpendicular to the longitudinal magnetic field (i.e., in the plane of the coil windings) for maximum indications.

The ease of magnetism depends on the component length (L) and diameter (D). An empirical expression for the product of the coil current (amperes) and the number of coil turns needed for adequate NDE flux is given by Equation 14.3 in terms of the member's L/D ratio [4]:

$$NI = \frac{45,000}{L/D} \tag{14.3}$$

This empirical expression applies when the component is located near the edge of the test coil. If $L/D < 10$, there can be a demagnetizing effect at the ends of the part, and it is sometimes necessary to add material to increase L/D. Again, discontinuities perpendicular to the longitudinal magnetic fields will give the strongest indications.

Electromagnetic coils provide a relatively simple means to develop longitudinal magnetic fields and do not involve direct contact with the test piece. On the negative side, the component needs to be centered in the coil to obtain a uniform field. If centering is not possible with complex parts, multiple processing may be required to readjust the component in the coil. As noted above, the L/D ratio can be problematic for small parts (desire large L/D). It is desirable to have $L/D > 20$, if possible, and rods may be added to the end of the component to achieve this length.

14.3.2 Central Conductors

Central conductors may be used to introduce circular magnetic fields in tubes and rings or in short parts with holes, as shown in Figure 14.5. Here current flowing through the conductor develops a circular magnetic field in the test piece that is placed around the conductor. The central conductors can be either magnetic or nonmagnetic and be used with direct or alternating current.

The magnetic flux density in the test piece varies inversely with the distance from the central conductor, as shown in Figure 14.6. Here the variation in flux density (B) and the magnetic force (H) are given along a radial line from the center of a nonmagnetic central conductor containing direct current. Note that the magnetic flux is largest at the inner diameter of the test ring and decreases through the wall thickness. It may be necessary to offset the central conductor to one side of large rings that have weak magnetic fields when the conductor is located at their center. Although placing the conductor near one wall increases the magnetic field at that location, it results in a nonuniform magnetic field in the rest of the piece. It would be necessary to rotate the ring around the conductor and to make several measurements to obtain complete coverage in this case.

One of the main advantages of central conductor magnetization is that direct electrical contact is not made with the test piece, minimizing the potential for spark damage in the component. On the negative side, the central conductor must be large enough to carry the amount of electrical current needed to generate a sufficient

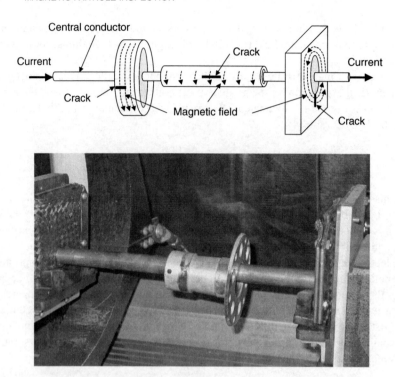

Figure 14.5. Use of central conductors for the circular magnetization of hollow tubes and rings or short parts with holes.

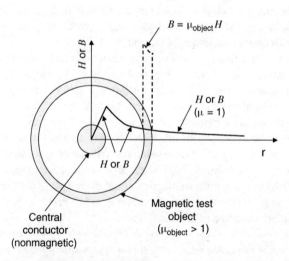

Figure 14.6. Flux density and field intensity along a radial line from the center of a DC central conductor (nonmagnetic) through the wall of a hollow magnetic cylinder.

magnetic field in the test part. In addition, there is less sensitivity to anomalies at the outer diameter than at the inner diameter of the test piece, and the inspection is not sensitive to transverse flaws.

14.3.3 Direct Contact

Electric current can also be applied directly though the test piece to develop magnetic fields in small components, as shown in Figure 14.7. Here the ends of the part are clamped in a fixture, and current is sent directly through the member, producing a circular magnetic field along the length of the test object. One empirical rule of thumb suggests that suitable magnetic fields can be obtained when the applied current is 1000 A per inch (2.5 cm) diameter of the part being magnetized. This "head shot" magnetizes the entire object at once, so that large areas can be inspected in a short time. The uniformity of the magnetic field will, however, depend on the complexity of the component. Large woven copper pads are frequently inserted between the object and the testing machine in order to provide better electrical contact and to remove heat generated by the current flow.

Disadvantages of the direct-contact approach are that large currents (8000–20,000 A DC) are needed for large components, the effective magnetic field is limited to the outer surface of the test piece, and the method is not sensitive to transverse flaws. In

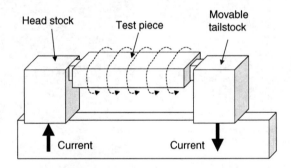

Figure 14.7. Circular magnetization obtained by applying current directly though the test piece with a headshot.

addition, the ends of the component must be shaped so as to permit good electrical contact between the headstock and tailstock, and there is danger of arcing at these contact locations.

14.3.4 Yokes

The test object may also be magnetized by connection to a yoke magnet, as shown in Figure 14.8. Here the primary inspection area is the uniform region of magnetic flux located between the ends of the yoke. Discontinuities perpendicular to the flux lines connecting the yoke will be most sensitive to detection. Either permanent or electromagnets can be used for the magnetizing yoke. Permanent magnets are especially useful when electric power is unavailable, and since there are no electrical connections, there is no danger of arcing. The strength of permanent magnets cannot be adjusted, however, and they are not strong enough to magnetize large areas. It may

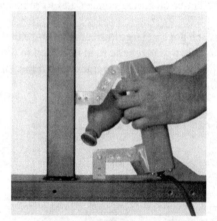

Magnetic particle inspection of weld
with Magnaflux Y-6 AC Yoke

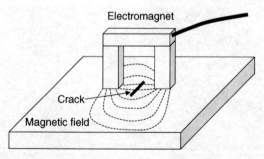

Figure 14.8. Magnetic yoke used to generate magnetic field in test piece. Strongest field is between yokes, and most sensitivity is obtained when discontinuity is perpendicular to magnetic field. (Photographs courtesy of Magnaflux Corporation, a Division of ITW.)

also be difficult to separate the test piece from permanent magnets, and particles can cling to the magnet following the inspection.

Electromagnetic yokes consist of a coil wound around a soft iron U-shaped core. The yoke shown in Figure 14.8 has adjustable legs that may be adapted to different piece contours. (Here dry particles are being dusted on the weld joining two perpendicular components.) Either alternating or direct current can be used for yoke magnetization. A DC yoke gives more magnetic penetration into the test object, whereas AC yokes concentrate the magnetic flux near the surface (skin effect). Advantages of the electromagnetic yoke are its ability to be turned on or off and the opportunity to adjust magnetic strength by controlling the electric current. In addition, there is no direct electrical contact with the test object. Yokes are highly portable devices that can be oriented in various directions. Moreover, as discussed in Section 14.4, AC yokes can also be used to demagnetize the test object. On the negative side, AC yokes have poor sensitivity to subsurface flaws. Finally, it should be noted that working with either permanent or electromagnetic yokes can be a time-consuming process with large test pieces that require frequent repositioning of the yoke to locate random discontinuities.

14.3.5 Prod Contacts

Prod contacts are used to pass a high-amperage, low-voltage current directly through a localized area of the test object, as shown in Figure 14.9. Note that circular magnetic fields are formed where the ends of the prods contact the component. Maximum detection occurs for surface discontinuities that are located perpendicular to the circular magnetic fields between the two prods, which are usually located within 2–8 in (3–20 cm) of each other.

Prod contacts are widely employed for magnetic NDE since they are portable and easy to use for large tanks, welded structures, and so on. (Note, e.g., the vintage photograph shown in Figure 14.8, where two men are using prods to conduct a magnetic

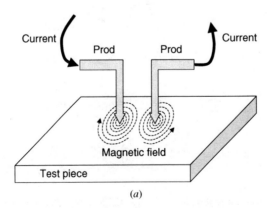

(*a*)

Figure 14.9. (*a*) Schematic representation of how electric prods generate magnetic field in test object. (*continued*)

(b)

Figure 14.9. (*continued*) (b) Vintage photograph showing use of prods for magnetic particle inspection of large test object. (Photograph courtesy of Magnaflux Corporation, a Division of ITW.)

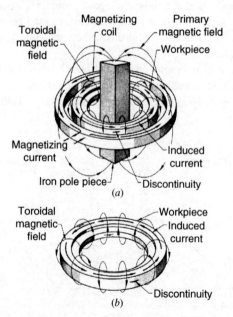

Figure 14.10. Induced-current method of magnetizing a ring-shaped part (from p. 98, Ref. 1). (a) Ring being magnetized by induced current. Current direction corresponds to decreasing magnetizing current. (b) Resulting induced current and toroidal magnetic field in test ring.

particle inspection on a large steel component [6].) Prod magnetization allows one to selectively apply circular magnetic fields to various locations in the test object and to incrementally inspect large surfaces with nominal electric currents. Since suitable magnetic fields only occur between the prods, however, time-consuming multiple coverage is required to inspect large areas. In addition, interference between the external magnetic field around the prods and that induced in the test piece can be a problem. Care must also be taken to avoid damaging the test object by sparking from the prods.

14.3.6 Induced Current

Circular magnetic fields may also be induced in ring-shaped specimens by use of magnetized coils, as shown in Figure 14.10 [1]. First the test piece is placed in a magnetizing coil to generate a circular magnetic field around the part and coil. (An iron pole piece is often placed inside the test ring to increase the strength of the magnetic fields.) Then, rapidly cycling the primary magnetic field in the outer coil induces an electric current in the test object that, in turn, induces a circular (toroidal) magnetic field in the test part. This induced magnetic field is well behaved with uniform, predictable characteristics and can be used to detect circumferential flaws in the component. Note that there is no direct electrical or physical contact with the test piece during the magnetization process.

14.4 MAGNETIC HYSTERESIS/DEMAGNETIZATION

An important issue for magnetic NDE lies in the fact that many ferromagnetic materials do not return to a completely unmagnetized state when removed from external magnetic fields. This residual magnetism may require the component to be demagnetized following inspection.

14.4.1 Magnetic Hysteresis

Assume that a virgin piece of magnetic material is subjected to a slowly applied, completely reversed magnetic field H and the magnetic flux density B is continuously measured, as shown in Figure 14.11. Three plots are shown in this figure: the applied magnetic force H versus time t, the measured magnetic flux B versus time t, and the B-versus-H *hysteresis loop*. (Note these three plots are arranged in a manner similar to the cyclic stress–strain results discussed in Section 9.3.) Common points on these three curves are designated by the subscripts 0, 1, . . . , 6.

Follow the measured flux density B on the hysteresis curve as the applied magnetic force goes from a value of zero (point 0) to positive and negative maximum H (points 1 and 4) and back to positive maximum H (point 1 again). Initially, in the virgin material, B increases rapidly as H is applied but eventually reaches saturation (line 0–1). Now, when the magnetizing force H is returned to zero (point 2), the flux density B also decreases but follows a different path on the B–H curve (1–2) than for the initial loading. Note that there is a residual (*retentive*) magnetic flux B_r at point 2 when the

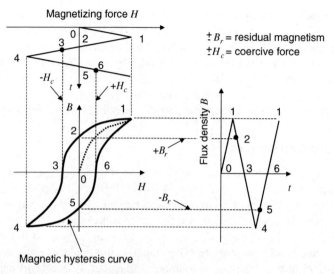

Figure 14.11. Development of magnetic hysteresis curve showing the flux density B resulting from application of positive and negative magnetizing force H.

magnetizing force H is removed. Continued decrease of H (i.e., applying negative H) eventually returns the flux B to zero (point 3), although a *coercive* magnetic force H_c is required to achieve this demagnetized condition.

Now, continue to decrease H to its completely reversed value (point 4) and then increase it back to the original positive maximum H (point 1). Repeated cycling between these $\pm H$ limits (points 1 and 4) results in the closed B-versus-H hysteresis loop shown in Figure 14.11. (This hysteresis loop is analogous to the cyclic stress–strain behavior discussed for metal fatigue in Section 9.2.3.) Note that if the maximum H is large enough, saturation in B occurs at the $\pm H$ peaks. In addition, residual magnetism of $\pm B_r$ results at $H = 0$ (points 2 and 5), and the coercive forces $\pm H_c$ exist when $B = 0$ (points 3 and 6).

The shape of the magnetic hysteresis loop depends on the value of the applied magnetic force H and the material in question. Soft steels, for example, have a "narrow" loop, whereas harder materials result in a "wide" H–B trace. Figure 14.12 [7] compares actual hysteresis loops for AISI 410 stainless steel treated to two different hardness levels and subjected to cyclic magnetic forces of $H = \pm 20$ Oe or $H = \pm 80$ Oe. Note the dependence on the applied H and the fact that the harder steel (Rockwell C42) has a wider hysteresis loop than the softer (Rockwell C22) steel. (These two hardness values correspond to yield stresses of 85 ksi = 586 MPa and 50 ksi = 345 MPa, respectively, in the AISI 410 stainless steel [7].)

14.4.2 Demagnetization

It is often necessary to demagnetize components following inspection (especially "hard" steels, which have greater magnetic retention). This demagnetization is an extra step that adds to the cost of the inspection.

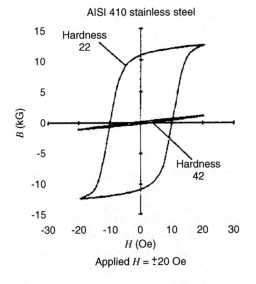

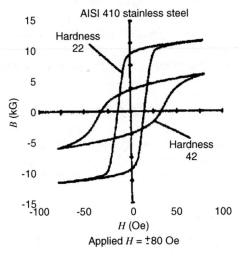

Figure 14.12. Hysteresis curves for AISI 410 stainless steel heated treated to two Rockwell C hardness levels and subjected to two different cyclic magnetic force fields [7].

When to Demagnetize   Demagnetization is necessary if the residual magnetization interferes with the intended function of the test member. Magnetization could, for example, influence subsequent operation of electronic controls. Residual magnetization may also attract particles that could clog drain holes or keyways or cause excessive wear between members that are in sliding contact. Problems may also arise if residual magnetization affects subsequent machining of the component by attracting filings and chips that could interfere with cutting operations, cause excessive tool wear, or result in poor surface finishes. Magnetization could also interfere with arc welding of parts or attract particles that impede subsequent painting or plating. Finally, remagnetization for future inspections could be affected by the residual magnetization.

Demagnetization is not required if the residual magnetism does not result in the undesired consequences mentioned above. It may also be unnecessary if the component is subsequently heated above its Curie point (i.e., the temperature where ferromagnetic ordering of the atomic moments is destroyed) and loses its magnetism. (Such heating of the test piece could, however, alter its heat treatment and mechanical properties.)

Demagnetization Procedure Although there are several techniques to remove residual magnetization, all are based on subjecting the component to a completely reversed cyclic magnetic force that slowly decreases in magnitude. By gradually decreasing the amplitude of the applied magnetic cycle, one can slowly "work down" the hystersis loops and, in principle, return to a state of zero magnetism (i.e., $B = 0$ when $H = 0$). The applied cyclic H may be decreased by gradually reducing the current (either AC or DC) in an electromagnetic coil or by slowly removing the test piece from the coil. Small components may employ ordinary 60-Hz AC line current for this purpose, although larger components may require a slower frequency to obtain sufficient magnetic penetration. The demagnetizing field may also be applied by yokes or by through-current techniques. It should be noted, however, that it can be very difficult to completely demagnetize irregularly shaped components.

14.5 MAGNETIC FLUX LEAKAGE SENSORS

Several methods may be used to locate the magnetic flux leakages caused by anomalies in the test piece. This section briefly describes magnetic particles along with other visual and nonvisual sensors employed for magnetic NDE applications.

14.5.1 Magnetic Particles

Magnetic particles are the most common visual "sensor" used for magnetic inspections. Here, magnetic particles are spread over the magnetized test piece and "flow" to component discontinuities that perturb the magnetic field. These small particles can be applied in wet or dry form and are detected visually by the inspector with normal or ultraviolet lighting.

Desired Particle Properties The particles must be ferromagnetic and have a high magnetic permeability so as to be magnetized by low-level leakage fields. They should also have low retentivity (low residual magnetism) so as not to be attracted to the test piece before it is magnetized.

Particle size and shape have important consequences. If the particles are too large, they will not be held by weak flux leakages. If too small, they will adhere to adjacent surfaces, resulting in a poor background for identifying discontinuities. The desired particles include a blend of elongated and spherical shapes. The elongated particles form north and south poles and arrange nicely in strings but do not flow well. The spherical or equiaxed particles flow properly but tend to clump up. Thus, multifacet nugget-shaped particles are the best compromise.

Since the particles will be detected visibly, their color and contrast are important. Pigments (e.g., dayglo, yellow, red, black, gray) are used to give the best contrast for dry particles, although wet-particle colors are limited to black and red oxide. Fluorescent particles may be used for both wet and dry applications but are more common for wet applications.

Dry particles are larger than the wet version, with an approximate diameter of 6 mils (150 μm). They are applied to the magnetized test piece by an air stream from rubber spray bulbs or small blowers with the goal of obtaining a uniform cloud of particles with minimum motion. Dry particles work well with AC magnetization on rough surfaces and for subsurface flaws but are not as sensitive to very fine cracks. Dry particles work best when the magnetized surface is vertical or overhead but tend to settle without much mobility when applied to horizontal or sloped surfaces.

Wet particles are smaller than their dry counterparts, ranging from 0.04 to 1 mil (1–25 μm) in size, with a mean diameter of approximately 0.2 mil (6 μm). They are applied with various liquids such as water or light petroleum distillates (e.g., kerosene or other light oils). Although a water bath is the cheaper choice and is not flammable, one must be careful about electrical grounds. Uniform bath strength is a major factor in the quality of inspection, requiring continuous agitation of the bath–particle mixture. It is also important to keep the bath clean from impurities. Wet particles are best suited for detecting fine surface discontinuities (e.g., fatigue cracks) and can be used with DC fields to detect subsurface cracks. They are used for smooth surfaces, oily or wet surfaces, and can be applied under water.

14.5.2   Other Visual Sensors

Although magnetic particles are perhaps the most common sensor, there are several other visual techniques that can be used to locate anomalies associated with magnetic flux leakages.

Magnetic Rubber   Magnetic rubber is another form of visual magnetic inspection. Here magnetic particles are dispersed in a room temperature curing rubber, and the mixture is flowed onto the magnetized test piece. The rubber is allowed to harden, and after approximately 1 hour the rubber impression is removed for examination under a microscope. Since rubber does shrink, however, measurements should be made within the first 72 hours after hardening.

Magnetic rubber has several distinct advantages. The process provides a permanent record of the inspection. The rubber impressions are quite useful for hard-to-see locations and for complex test shapes. They are especially valuable for indications that require magnification or for applications with coated surfaces. Although rubber impressions can record surface topography in nonmagnetic materials, there is no particle migration to indicate flaw locations. Magnetic rubbers are, however, slower to use than magnetic particle inspections since additional time is required for the rubber to set and cure.

Magnetic Painting   This process uses a visually contrasting paint with a magnetic particle slurry that is first brushed or sprayed on the test surface. The test object is then magnetized, and particles migrate to discontinuities. The visual contrast of the paint requires no special lighting, and the paint covers both light and dark surfaces well. The nondrying paint is not affected by wind currents or by wet surfaces and is good for field applications. The paint can also be applied under water. The paint does, however, require some minimal surface preparation.

Magnetic Printing   Magnetic printing is another visual sensing technique for magnetic inspections [4]. Here the test piece is sprayed with a white plastic coating and then dusted or sprayed with dry magnetic particles. A flat coil (or printer) is placed against the surface and causes a pulsating effect that marks (or "prints") the coating in the area of discontinuities. This technique is reported to be sensitive but complex to apply in practice.

Magnetic Foils   Magnetic foils are visual sensors that employ sheets impregnated with magnetic powder and a viscous fluid [4]. The foil is placed against the surface of the magnetized test object and pressure is applied to the foil in order to mix the fluid and powder. The powder then flows to the discontinuities. After 1–2 minutes have elapsed, the fluid–powder mixture hardens and the foil can be removed for examination. The foils provide a permanent record of the inspections and are reported to work well under water or in other areas of poor visibility.

14.5.3   Other Magnetic Sensors

In addition to the magnetic particles and other visual sensors described above, various additional techniques may be employed to locate flaw-induced changes in the test object's magnetic field. Several of these methods are briefly summarized below (see Chapter 21 of Reference 5 for more detailed discussion). These flux leakage sensors may be classified as active or passive devices. The visual methods (except for magnetic printing) discussed previously are passive sensors, whereas active sensors require some form of external excitation.

Hall Elements   These are active semiconductor devices that develop a voltage when subjected to a current and a magnetic field. They can be made quite small, effectively becoming point sensors. The output of Hall sensors is independent of the velocity used to move them through the magnetic field, so they may be readily scanned about the test piece.

Magnetodiodes   These are also solid-state devices whose electrical resistance depends on the magnetic field strength and the ambient temperature. Constructing a circuit that places the diodes in series electrically but in opposite directions magnetically can minimize the temperature sensitivity. The sensitivity of the magnetoresistance of the diodes is relatively high and, coupled with their flat frequency response, makes them quite useful for AC magnetic tests.

Magnetic Recording Tape   Magnetic recording tapes developed to inspect steel billets have the ability to distinguish between surface and subsurface flaws. Here, the neoprene tape is embedded with finely divided ferrite particles that become magnetized by the flux leakage fields. Different portions of the tape are magnetized to different levels by the varying magnetic field about the test piece. Subsequent electronic filtering of the tapes makes it possible to discriminate between surface and subsurface flaws.

Pick-up Coils   Pick-up coils are passive devices that involve moving a straight wire or coil through the magnetic field surrounding the test piece. The strength of the induced current and resulting voltage depends on the magnetic field and the speed of the moving coil. Pick-up coils can be designed to operate parallel or perpendicular to the magnetic field directions and may be shaped to a particular application. Coils are robust and easy to manufacture but must be shielded to prevent eddy current effects.

Förster Microprobes   Förster microprobes are active devices that consist of coils that are tightly wound about a ferrite core. The ferrite core has a nonlinear $B–H$ relationship that exhibits little magnetic hystersis and saturates easily. Coils may be connected in a differential circuit and excited at various frequencies. Due to the nonlinear permeability of the ferrite core, it is possible to develop probes that are sensitive to the magnetic field strength.

14.6   OTHER MAGNETIC NDE METHODS

Although this chapter focuses on magnetic particle and other visual and nonvisual methods for detecting magnetic perturbations in the test piece, there are other magnetic-based NDE methods.

14.6.1   Electric Current Injection

The electric current injection (ECI) approach examines the magnetic fields that result when an electric current is applied to a *nonmagnetic* test object. Again, the current flow causes magnetic fields around the conductor (i.e., test piece), and one searches for flaw-related perturbations in these magnetic fields. Although the test material is nonferromagnetic in this case, it must still be an electrical conductor. Since the basic phenomenon examined here is still a magnetic field, ECI is considered to be a magnetic NDE method.

The electric current is applied to the test object by mechanical contact with electric probes. The magnetic fields in this case are several orders of magnitude smaller than in ferromagnetic materials, however, and are much harder to detect. Thus, special techniques are required to sense the magnetic field changes. Coils of some type are usually employed for this purpose. (Hall effect probes are not very effective.) Although ECI methods are not widely used, they hold promise for nonferromagnetic, electrical conducting materials such as aluminum, magnesium, and titanium [4].

14.6.2   Barkhausen Effect

As discussed in Section 14.2, Barkhausen discovered that the ferromagnetism process produced sound that could be amplified. These sound signatures can be measured and related to grain size, orientation, and material inhomogeneities. There are two particular material characteristics of interest to NDE that affect the intensity of the Barkhausen noise signals: metallurgical structure and residual stresses.

Since the Barkhausen noise intensity decreases when the material hardness increases, sound emissions can be related to changes in hardness (and the subsequent metallurgical microstructure). It is also known that elastic stress influences the orientation of the magnetic domains through a process known as *magnetoelastic interaction*. In materials with positive magnetic anisotropy (e.g., iron, most steels, cobalt), the Barkhausen noise is decreased by compressive stresses and increased by tensile stresses. This phenomenon can be exploited to measure residual stresses and/or to determine the direction of principal stresses.

Thus, Barkhausen noise methods can evaluate residual stresses provided the microstructure is reasonably controlled and vice versa. They may also locate defects that involve changes in residual stresses and/or microstructure (e.g., grinding burns, soft spots, edges on hardened surfaces, decarburized areas). Although measurement depth is limited to near the surface [< 0.04 in = (1 mm)], the technique is sensitive to surface finish, stress relief, and so on. Whereas Barkhausen noise methods are limited to ferromagnetic materials, mechanical contact is not required with the test object.

14.7   CONCLUDING REMARKS

Magnetic NDE methods are valuable tools for examining ferromagnetic materials. The magnetic particle approach is particularly convenient to use and is one of the major inspection methods employed in industry. Additional details of this NDE method may be found in References 1–5 and 8. The advantages and disadvantages of magnetic particle inspections are summarized below.

14.7.1   Advantages of Magnetic Particle NDE

1. Complex shapes are readily tested.
2. Magnetic particle equipment is fairly portable and relatively inexpensive.
3. The technique is sensitive to small surface (or near-surface) defects.
4. The method requires moderate skill levels and is fairly easy to learn.
5. The inspection results are obtained relatively quickly.

14.7.2   Disadvantages of Magnetic Particle NDE

1. Magnetic particle inspections are limited to ferromagnetic materials.
2. Component surfaces must be accessible.
3. Large electric currents can be required to magnetize large parts.
4. The test object may need to be demagnetized after the inspection.
5. The method is limited to surface or near-surface defects.
6. Rough surfaces can cause problems.
7. Surface finishes or sealants may need to be removed in some cases.
8. Flaw detection is sensitive to the magnetic field orientation.

PROBLEMS

14.1   Use the Internet to locate the home page for a vendor that supplies materials for magnetic particle inspection purposes and record this company's address for future reference. Examine the products and applications described on the web site and prepare a short report that describes the various capabilities and uses for the magnetic particle products sold by this company.

14.2   Nondestructive evaluation magnetization methods can be classified as techniques that bring the test piece into the presence of an external magnetic field and those that run an electric current through the test piece.

   (a) Briefly discuss the magnetizing techniques that employ an external magnetic field to magnetize the test object.

   (b) Briefly discuss the methods that apply an electric current to the test piece in order to magnetize it.

14.3   It is desired to inspect thin-walled steel tubes for longitudinal cracks by the magnetic particle method. Briefly discuss the techniques you could use to induce magnetic fields in the tubes for these inspections.

14.4   Discuss the relative merits of DC versus AC yokes for NDE purposes.

14.5   What parameters influence the strength of the magnetic field induced by a coil?

14.6   Discuss the influence of flaw orientation on magnetic particle inspection. What magnetization methods would be appropriate for a component that is expected to have flaws located in a longitudinal direction? What methods would be appropriate for detecting transverse flaws?

14.7   Sketch a magnetic hysteresis loop for a low-strength steel and label the following points: locations of magnetic saturation, residual retentivity, and coercive force. Repeat this problem for a high-strength steel.

14.8   Discuss the factors that would determine whether or not demagnetization would be required following a magnetic particle inspection. Briefly describe how the demagnetization process would be accomplished.

14.9 With Figure 14.11 as a model, sketch how demagnetization would be accomplished by working down the hysteresis loop. Briefly explain your answer.

14.10 Discuss the relative merits of wet versus dry magnetic particles for NDE purposes. Give examples when one type of particle would be preferred over the other.

14.11 Discuss the similarities and differences between magnetic particle and liquid penetrant NDE. Give examples where one method would be preferred over the other.

14.12 Discuss the similarities and differences between magnetic particle and eddy current NDE. Give examples where one method would be preferred over the other.

14.13 Briefly summarize other *visual* methods besides magnetic particles that can be used to locate flux leakages in the test piece.

14.14 Briefly summarize *nonvisual* methods for locating flux leakages in the test piece.

REFERENCES

1. A. Lindgren, "Magnetic Particle Inspection," in *ASM Handbook,* Vol. 17: *Nondestructive Evaluation and Quality Control,* ASM International, Metals Park, Ohio, 1989, pp. 89–135.

2. R. Halmshaw, *Non-Destructive Testing,* 2nd ed., Edward Arnold, London, 1991, Chapter 5.

3. *Nondestructive Testing, Magnetic Particle,* 2nd ed., Classroom Training Handbook CT-6-3, General Dynamics, 1977.

4. D. E. Bray and D. McBride, *Nondestructive Testing Techniques,* Wiley, New York, Chapters 15 and 16, 1992.

5. D. E. Bray and R. K. Stanley, *Nondestructive Evaluation, A Tool in Design, Manufacturing, and Service,* rev. ed., CRC Press, Boca Raton, Florida, 1997.

6. Magnaflux Corporation, Chicago, Illinois, personal communication, 2002.

7. H. Kwun and G. L. Burkhardt, "Effects of Grain Size, Hardness, and Stress on the Magnetic Hysteresis Loops of Ferromagnetic Steels," *Journal of Applied Physics*, Vol. 61, p. 1576, 1987.

8. J. T. Schmidt and S. Kermit, *Magnetic Particle Testing—Nondestructive Testing Handbook,* 2nd ed., Vol. 6, American Society for Nondestructive Testing, 1989.

CHAPTER 15

OTHER INSPECTION METHODS

15.1 OVERVIEW

The most common inspection methods employed in industry have been discussed in Chapters 10–14. This chapter will present an overview of several other specialized and emerging NDE techniques available to assess structural integrity. Particular attention is given to acoustic emission, thermal imaging, and advanced optical methods.

15.2 ACOUSTIC EMISSION INSPECTION

Acoustic emission (AE) has been defined as "a transient elastic stress wave generated by the rapid release of energy from a localized source within a material" [1, page 611]. NDE applications of AE involve loading the test object, sensing the AE, and correlating it with a source of damage [2–4]. Forms of AE have been employed since ancient times. Early potters, for example, associated damaged earthenware with cracking sounds given off during the cooling process, and cracking noises from timbers in mines or ships were an obvious warning of imminent failure.

Modern utilization of AE began in the 1950s, however, and has had success with "noisy" materials such as composites, concrete, and the like. There are many sources for AE that can be recorded for NDE purposes [2]: crack formation and growth, magnetomechanical realignment or growth of magnetic domains (the Barkhausen effect discussed in Section 14.6.2), microstructural changes such as dislocation movement, twinning or phase changes, fracture of brittle inclusions or surface films, or even chemical activity resulting from corrosion. Sources of AE in composite materials include fiber breakage, crazing, or delaminations, whereas emissions in concrete

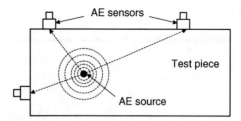

Figure 15.1. Schematic representation of acoustic emission NDE showing how discrete bursts of sound that emanate from a local inhomogeneity in the test piece are sensed and used to locate the source of an anomaly.

originate at aggregate or reinforcement separations from the cement matrix and from microcracks.

Nondestructive evaluation applications of AE involve "listening" for sound generated by growing "damage" in the test piece, as indicated in Figure 15.1. Although the *location* of AE damage can be determined by triangulation techniques based on the time for the signal to reach strategically placed sensors, damage *size* cannot be found directly by AE. (The ASME Boiler and Pressure Vessel Code dealing with fiber-reinforced plastic vessels requires, for example, that anomalies found by AE must also be examined by other NDE methods and then be repaired and reexamined before entering service [5].) It should also be pointed out that since each loading event emits unique bursts of sound, AE inspections are not directly repeatable.

Although AEs are generated by the material itself and not by an external source, a stimulus is required to trigger the original AEs. These stimuli include mechanical, thermal, or magnetic induced stress or chemical activity. Whereas special test loads can be applied for NDE measurements, AE inspections may also be conducted in conjunction with a proof test or a full-scale fatigue test or perhaps during in-service loading of the structure.

Many (but not all) materials exhibit suitable AEs for NDE purposes [2]. Material factors that *increase* AE include anisotropy, brittleness or cleavage failures, large grain sizes, martensitic phase transformations, twinning, thick sections, cast structures, high strain rates, and low temperatures. Factors that *reduce* AE activity include isotropy, ductile failure mechanisms, wrought structures, small grain sizes, low strain rates, and high temperatures. Acoustic emission has become important for NDE of composite materials, and AE use is covered by the ASME Boiler and Pressure Vessel Codes [5]. Acoustic emission is, in fact, one of few practical ways to inspect some specialized composites.

15.2.1   Acoustic Emission Signals

Although AE signals cover a wide range of energy levels and frequencies, they consist of the two basic types shown in Figure 15.2: *burst emissions,* which correspond to a particular emission event, and *continuous emissions,* which give a sustained signal

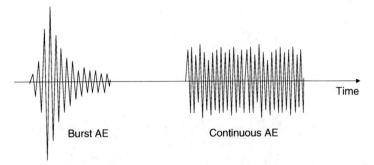

Figure 15.2. Schematic examples of burst AEs corresponding to a particular emission event and continuous emissions resulting from a series of rapidly occurring events.

associated with rapidly occurring events. Both types of emissions may propagate by combinations of the standard ultrasonic modes (shear, longitudinal, or surface waves) and may undergo reflection and refraction (including mode changes) at boundaries (recall Section 12.3). The AE signals may vary in amplitude and frequency and may be above or below the range of human hearing. While signals as high as 50 MHz have been detected, AE monitoring is typically performed in the 20–1200-kHz frequency range.

Attenuation of AEs as they travel through the structural component is an important NDE consideration that influences the number and location of transducers needed for the inspection. Sound-suppressing mechanisms include scattering at small discontinuities such as precipitates or grain boundaries, absorption by magnetic or thermoelastic damping, and interactions with dislocations. The AE signals may also be diminished by diffraction or coupling losses at the sensor locations and may follow several different sound paths as they travel between the source and sensor. This latter phenomenon can confuse AE analysis when multiple signals from the same event arrive at the sensor at different times.

Early applications of AE often gave inconsistent or misleading results due to inadequate standards for quantifying and reporting the recorded emissions and led to an unfavorable assessment of the technique in some circles. This problem has been addressed in recent years, and there are several generally accepted parameters used to quantify AE signals, as shown in Figure 15.3 [2–4]:

1. The peak amplitude A is the highest peak attained by the AE waveform and characterizes the detectability of the received signal.
2. The adjustable threshold level is a key parameter that is selected to eliminate low-level AE signals and/or background noise from consideration.
3. The rise time R to meet the peak amplitude is measured with respect to the given threshold level.
4. The total signal duration time D is defined with respect to the given threshold.
5. The frequency of the AEs is an important characteristic.

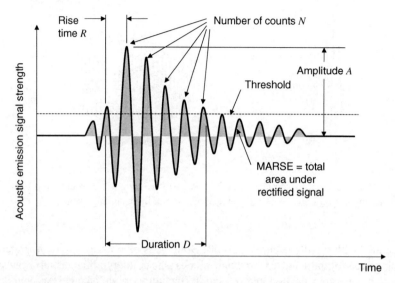

Figure 15.3. Definition of terms used to characterize the AE signal.

6. The number of crossings across the specified threshold ("ring down" counts N) is frequently used to quantify AE activity.

7. The AE energy, as measured by either the square of the amplitudes or the area under the rectified signal envelope (MARSE), is another important measure of AE signal amplitude. The MARSE is a preferred measure because it is sensitive to both amplitude and duration and is less dependent on threshold setting and operating frequency. More complex circuitry is, however, required to measure MARSE.

A few of the many formats used to display AE data are shown schematically in Figure 15.4. Here history plots are given in cumulative or rate form versus test time or applied load. Cumulative plots are convenient for representing the total AE activity, whereas the rate format shows changes in AE during the test. A cumulative history of events or counts is often plotted versus applied load (stress or strain), as shown in Figure 15.4c, and is quite useful in that it relates cause (load) and effect (AE counts). These plots are especially helpful for separating poor- from high-quality components. Other formats for presenting AE data are described in References 2–4.

Load history plays an important role in acoustic emission NDE, as shown in Figure 15.5, where cumulative AE behavior is presented for several load cycles. Note, for example, the AE activity as the load is initially increased from point 0 to 1. Although activity may continue briefly after stopping at maximum load, AE will generally cease and not continue further when the load is decreased to point 2 and then returned to the prior peak. Known as the *Kaiser* effect, this phenomenon indicates that no new damage is created during the unload/reload portion of the cycle until the previous maximum applied load is exceeded (points 3–4). Indeed, many of the initial AEs that occur during the first load cycle are associated with initial stability of the structure

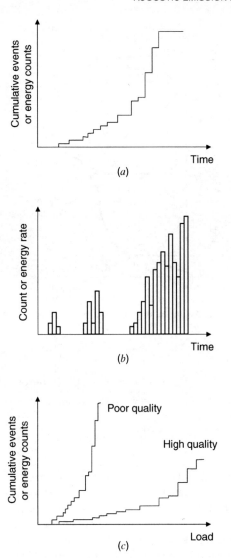

Figure 15.4. Various ways of presenting AE data: (*a*) cumulative AE activity versus time, (*b*) AE rate versus time; (*c*) cumulative AE activity versus applied load.

through local yielding or residual stress relief rather than with component damage per se [4].

It is possible, however, to get new AE during the reload cycle in some cases, as shown by the 4–5–7 loading event. Note here that new acoustic emissions begin at load 6, which is below the prior load peak 4, suggesting that new damage (i.e., crack propagation) has formed during reloading. This latter occurrence is described by the *felicity ratio* FR and is defined as

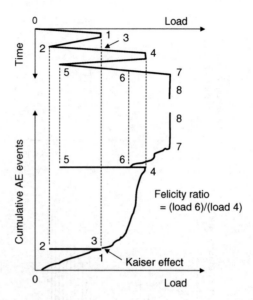

Figure 15.5. Acoustic emission activity during cyclic loading showing Kaiser effect and definition of the felicity ratio.

$$FR = \frac{\text{load at subsequent AE onset}}{\text{previous maximum load}} \qquad (15.1)$$

$$= \frac{\text{load 6}}{\text{load 4}} \quad \text{in Figure 15.4}$$

The Kaiser effect describes the situation where FR $=$ 1. Note that FR $<$ 1 indicates the presence of damage caused by prior loading and can be a reason for component rejection. Fiber-reinforced polymer tanks and pressure vessels are to be rejected, for example, if FR $<$ 0.95 [4]. Although AEs often continue to occur for a short time after the load is stopped and held at a fixed value, continued emissions during a sustained hold time (e.g., event 7–8 in Figure 15.5) are evidence for growing flaws in the test component and suggest the presence of significant structural damage. Thus, load history has an important influence on interpretation of AE results and needs to be carefully planned for a valid AE inspection.

15.2.2 Acoustic Emission Sensors

Acoustic emissions are ultrasonic waves that can be measured by a number of techniques similar to those employed for ultrasonic testing [2]:

- *Piezoelectric* transducers are used for much of current AE monitoring. As discussed in Chapter 12, there are many types of piezoelectric materials with a wide range of properties that provide sensitive and reliable sensors that are suitable for wide ranges of frequencies. The main disadvantage of piezoelectric sensors for AE applications is their limited bandwidth, which prevents accurate signal reproduction outside their range of response.

- *Accelerometers* are devices that measure accelerations and are a rugged form of sensor that was used for early AE research. They have a limited frequency range (100 kHz maximum), however, which limits their use for many applications.

- *Air-gap capacitive* transducers monitor surface displacements via changes in the capacitance between the probe and the surface of the test object. These sensors have a relatively flat frequency response that limits their application to analysis of signal frequencies and wave shapes associated with specific AE generation mechanisms. They have not found other wide applications.

- *Optical/laser* transducers that employ laser interferometry are effective for measuring the true wave shape and frequency of AE. Their advantages over piezoelectric transducers include a wider frequency response, improved coupling, and the ability to integrate signals over large area.

- *Magnetostriction* transducers are more rugged than piezoelectric transducers and maintain transducer characteristics at high temperatures and/or in the presence of nuclear radiation. They are, however, limited by high-frequency response (maximum frequency $\sim$1 MHz).

15.2.3 Acoustic Emission Techniques

Practical AE inspection techniques involve consideration of several factors, including the frequencies to be monitored, sensor location, and separation of the AE signal from other ambient noise. Acoustic emission frequencies typically range from subsonic to above 50 MHz. A growing crack, for example, emits a wide range of frequencies, any of which, in principle, can be monitored for NDE. The monitored frequency is usually selected to be above ambient noise but below those attenuated by the test object. (Recall from Section 12.4 that ultrasonic signal attenuation increases with frequency.)

Acoustic emission sensors are located on the test piece in a particular pattern, so that the times various sensors record the detected signal can be used to locate the source of AE (e.g., the crack of interest). Triangular arrays of AE detectors are usually used for flat surfaces, although linear arrays of sensors can be used for detecting growing damage along a straight weld. Complex structural shapes might involve other sensor locations. Sensor placement is also influenced by economic considerations, material attenuation, mode conversions, and so on. As mentioned previously, the possibility for multiple sound paths between the AE source and sensor must also be considered when interpreting the results.

The acoustic emissions are separated from ambient noise and electrical interference by shielding of cables, filtering power supplies, and so on. Amplitude and frequency discrimination is often used to remove unwanted signals [i.e., blocking low-frequency (audio) sound removes equipment noise]. Selection of the threshold level is key to removing unwanted sounds and requires careful characterization of AE from known damage sources in order to know what sounds to block.

15.2.4   Summary of Acoustic Emission NDE

Acoustic emission is a specialized form of NDE that differs from other inspection methods [4]. It and proof testing (recall Section 4.3) are unique, for example, in that both measure the *growth* (and, thus, the presence) of existing damage, whereas other NDE methods detect damage size or shape. Acoustic emission is not repeatable in the sense that each event is unique, and AE is especially sensitive to the material rather than the geometric characteristics of the test piece. (Additional details of AE testing are given in References 2–4 and 6.) As summarized below, AE has several practical applications and limitations.

Advantages of Acoustic Emission NDE

1. Acoustic emission signals locate the source of growing damage in a loaded structure.
2. Monitoring equipment is relatively simple and inexpensive.
3. Signals may be monitored on-line or recorded and stored for later analysis.
4. Acoustic emission can detect internal discontinuities hidden below the surface of the structure.
5. Wide-area coverage is possible as the entire structure can be monitored at one time.
6. Acoustic emission can be used for long-term structural monitoring and for use in hostile environments with remote sensing.

Disadvantages of Acoustic Emission NDE

1. There must be an applied stress or chemical activity for emissions to occur, and as a result, stabilized damage is not detected.
2. Not all damage will emit AE, and the procedure is not applicable to all materials.
3. Although discontinuities can be located by AE, crack sizes cannot be readily determined. Thus, AE inspections are best used to complement other NDE techniques when damage size is required.
4. Since the AEs for each loading of the test object are unique, the method is not directly repeatable.
5. Electrical interference and ambient noise must be filtered for proper interpretation of AE results.
6. Although improving, the lack of standard methods of data interpretation has hampered development and acceptance of the technique.
7. Multiple sound paths between the emitting damage and the sensor complicate signal identification in complex structures.

15.3 THERMAL INSPECTION

Thermal inspection (or thermal imaging) is a collection of NDE techniques that involve heating (or cooling) the test object and then seeking changes in local surface temperature associated with component anomalies [7–10]. Thermal NDE often employs remote sensing and can rapidly evaluate large surface areas. The procedure is based on the fact that all bodies emit electromagnetic radiation when their temperature exceeds absolute zero, with the amount of radiation increasing with temperature.

The wavelength of thermal radiation varies between 4×10^{-6} and 4×10^{-4} in (0.1 and 100 μm), depending on component temperature. Infrared wavelengths lie in the 2.8×10^{-5}- to 4×10^{-3}-in (0.7–100-μm) range and are between visible light and microwave energy on the electromagnetic spectrum. Since visible radiation is in the 1.6×10^{-5}- to 2.8×10^{-5}-in (0.4–0.7-μm) wavelength range, the temperature of an extremely hot body can be visible to the human eye (e.g., "red hot" steel). Although cooler members continue to radiate in the infrared regime, this radiation is not visible to the naked eye and must be sensed by some other means. Most thermal NDE procedures employ infrared heating for the inspection stimulus.

Thermal inspection is dependent upon applying a thermal gradient to the test member and sensing temperature changes caused by component anomalies. Thus, defects must be located near the surface and oriented in a manner that results in local changes in surface temperature. The heat transfer characteristics (e.g., thermal conductivity and specific heat) of the flaw must differ from that of the parent member in order to be detected. Voids and inclusions can be easy to locate, for example, whereas tight cracks are harder to detect. The defect size and orientation are also important considerations. Flaws that lie in planes parallel to the component surface are more sensitive than those that are perpendicular, and in general, the larger areas are more detectable.

15.3.1 Temperature Measurement

Measurement of surface temperature is the key step for thermal NDE and may be accomplished by a variety of methods. A simple example of thermal imaging found in nature is shown in Figure 15.6, where the uneven melting of an early morning frost reveals outlines of the internal roof support structure in the author's office building.

Although infrared imaging was discovered in 1800, it was not used for NDE purposes until after World War II and only became more popular after high-speed temperature measurements became possible in the 1960s. There are many forms of infrared sensing devices, ranging from the simple to the complex, from large to small fields of view, and from laboratory to field applications. Thermal measurements may be classified as *thermometric* (measurement of temperature) or as *thermographic* (mapping of equal temperature contours or isobars). Temperature measurements may also be classified as *noncontact* or *contact* methods.

Noncontact temperature measurement methods include radiometers and pyrometers—devices that are limited to measuring heat at a point or along line—as well

Figure 15.6. Example of thermal imaging in nature, as melting frost leaves image of internal roof support structure.

as infrared cameras and microscopes that determine temperatures over larger areas. Some cameras employ film, and others are electronic devices.

Contact methods include material coatings and thermoelectric devices such as those summarized below:

- *Thermocouples* employ two pairs of bimaterial junctions arranged in a bridge circuit that produces an electric voltage when the temperature of one junction is changed relative to the other.
- *Thermopiles* consist of multiple thermocouples arranged in series.
- *Thermistors* sense changes in electrical resistance caused by temperature changes.
- *Known melting temperature materials* (e.g., waxes, lacquers) can be coated on the test piece and their calibrated melting temperature used to map surface temperatures.
- *Paints and inks* that change colors with temperature can be used to coat large areas of a component.
- *Thermally sensitive papers* and labels can also be attached to the test piece to indicate surface temperatures.
- *Liquid crystals* are materials that give a reversible color change when subjected to a given temperature. It is possible to control the color–temperature range and to produce liquid crystals that can measure temperatures to within 0.18°F (0.1°C) [8].
- *Thermal phosphors* are materials that emit visible light when exposed to ultraviolet radiation. Since the brightness depends on temperature, they can be used as temperature-sensitive coatings.

15.3.2 Heating Methods

Heating (or cooling) of the test piece is another critical step that can be accomplished by several "active" or "passive" or methods. An *active* test object generates the heat needed for inspection by itself. Examples here include temperature rises from faulty electrical components, from friction generated by moving machinery, or by blocked or leaking energy transfer devices (i.e., heat exchangers). Changes in operating temperature may also provide suitable thermal stimuli. Cold fuel in an aircraft that just landed from high altitude would warm at a different rate than the surrounding structure, creating a thermal gradient throughout various portions of the aircraft. Note, however, that these examples often imply a limited window of opportunity for the thermal inspection.

A thermally *passive* test object must be heated or cooled externally for the inspection. Since temperature detection is often more sensitive at higher temperatures, NDE usually involves heating the test object, rather than cooling it by various radiant, convective, conductive, or energy conversion methods. *Radiant* heating includes natural sunlight as well as the use of heat lamps and lasers. Radiant heating can be focused on local areas or scan larger surfaces and is relatively easy to control by simply blocking the source. *Convective* heating can be accomplished with ovens or blowers, but it may be difficult to uniformly heat the entire test piece. *Conductive* heating may be obtained with hot plates or heating blankets but is seldom used because of the difficulty in maintaining uniform thermal contact with the test object. *Energy conversion* is another classification of the heating required for passive objects and may result, for example, from the operation of electronic components or from heat emitted by plastic deformation associated with yielding or mechanical vibrations. Thermal energy released during chemical activity is another example of this form of heating.

There are several other ways in which one can classify heating of a thermally passive test object for NDE testing [11]. In *pulsed thermography*, for example, a short-duration heat cycle is applied, and one measures the decay of surface temperatures. *Step heating* employs a longer duration thermal cycle (a "long pulse"), and variations in the increase in surface temperature are related to component anomalies. *Lock-in thermography* employs continuous sinusoidal heating to generate oscillating thermal waves within the test object that are monitored with respect to both amplitude and phase. Finally, *vibrothermography* uses mechanical vibrations to heat the test object.

15.3.3 Typical Applications of Thermal NDE

Thermal imaging is especially useful for real-time, remote measurements of large structures and/or components that are in service. Several general types of applications are given below [9].

Plant condition monitoring is an important application of thermal NDE used to determine if a device is operating correctly or if a process is being properly performed. Electrical equipment, for example, may emit excessive heat due to poor connections, shorts, large resistances, open circuit, or faulty components. These faults may all lead to local heat buildup that can be detected by thermal imaging. Mechanical devices can also be examined during service for evidence of extraneous heat due to friction caused by bad bearings and/or poor lubrication. Heat transfer systems may also be readily examined by thermal imaging for evidence of blocked pipes or faulty heat exchangers.

Buildings and structures may be examined for heat loss caused by bad insulation in walls and roofs or through poorly sealed windows, doors, or construction joints. It may also be possible to monitor heating and air-conditioning distribution systems and moisture buildup in the roof of a building. Other construction applications include

Figure 15.7. Thermal imaging of structures; (*a*) image showing liquid level in storage tank; (*b*) local hot spot in wall of tank suggesting damage to wall; (*c*) images of building comparing heat loss through double- and single-pane windows. (Photographs/images courtesy of FLIR Systems, www.flirthermography.com.)

location of interior support members (i.e., studs in walls or rebar in concrete) as well as detection of cracks or spalling in concrete runways or roads. It is also possible to remotely examine storage tanks. Several of these applications are dramatically demonstrated in Figure 15.7. (Note that these black-and-white photographs are much more dramatic in their original color.)

Process monitoring and control may often be accomplished by performing temperature measurements at a particular point in a manufacturing process. These measurements can then be used to monitor how well the process is proceeding as well as to adjust operating parameters through open- or closed-loop control. Examples here include determining temperature gradients in molten metal processing, ensuring proper temperatures of ovens and molds, as well as verifying correct thermal cure cycles.

Other applications include traditional NDE assessment of component damage, particularly that associated with laminar materials (e.g., delaminations), joining of thin sections (e.g., corrosion between lap joints), and water absorption in a honeycomb structure. There are many medical and veterinary applications of thermal imaging. Thermal methods may also be employed for experimental stress analysis using the thermoelastic effect that relates small temperature changes to stress gradients in a loaded component [12].

15.3.4 Summary of Thermal NDE

Thermal imaging is a powerful inspection tool in its own right and provides a valuable complement to other traditional NDE methods. Continued advancements in heat-sensing equipment ensure that it will remain an increasingly important NDE tool. The advantages and disadvantages of this inspection method are summarized below.

Advantages of Thermal Inspection

1. Thermography is a noncontact method that can be used for remote sensing.
2. It is capable of a wide field of view that allows large areas to be monitored quickly and accurately.
3. Components can be monitored during actual operation in service.
4. The results of the visual display are often intuitive and easy to interpret.
5. Signal processing and enhancement can be applied to the output from electronic sensors.
6. No special safety precautions are required.

Disadvantages of Thermal Inspection

1. It can be difficult to apply uniform thermal energy to large surface areas, and low or variable surface emissivity may require application of a surface coating to ensure proper thermal radiation.
2. Discontinuities must be located near the component surface and have anomalous heat transfer characteristics that perturb surface temperatures.

3. Test equipment can be relatively expensive.

4. Outdoor measurements can be significantly influenced by the weather conditions or time of day (e.g., best results are often obtained after the sun has set).

5. Although precise temperature measurements are possible, many applications are limited to a qualitative nature.

6. Thermal measurements are subject to drift, and frequent calibrations are needed if absolute temperature measurements are desired.

7. Sensitivity is affected by reflected or other background radiation.

8. Glass is opaque to infrared wavelengths, requiring that windows be open when viewing from a vehicle.

15.4   ADVANCED OPTICAL TECHNIQUES

Visual inspection is discussed in Chapter 10 and, as indicated there, includes the aid of light sources, low-power lenses, borescopes, and the like. This section briefly summarizes several more advanced optical NDE methods. Strictly speaking, these techniques are regarded as strain analysis methods in that they locate anomalous displacements associated with surface or near-surface discontinuities. Since a detailed review of basic optics is beyond the scope of this work, the reader is referred to References 13–15 for summaries of the underlying optics principles behind these methods.

15.4.1   Optical Interference

Several inspection methods employ optical interference to obtain precise measurements of surface displacements. The concept is shown schematically in Figure 15.8, where two monochromatic light rays are incident on a block of polished glass with two perfectly flat, parallel surfaces. This "optical flat" is placed a distance d from the free surface of the test piece. The two rays have the same wavelength and are in phase as they leave the source but travel different path lengths to the observer. One ray, for example, penetrates the flat and is reflected off of the test piece back through the optical flat, whereas the other is reflected off of the bottom surface of the flat.

Note that as the two reflected rays travel to the observation point, they may be out of phase depending on the difference in path length ($2d$). The intensity at a given observation point is the sum of the two waves and will result in a series of light and dark fringes depending on the phase relation between the rays. These fringes depend on the distance d between the optical flat and the test surface and are essentially contour maps of equal separation between the two surfaces. If the wavelength is known, simple equations are available to quantitatively relate the fringe patterns with the distance between surfaces [13, 15–17]. Since the fringe patterns are related to multiples of the wavelength of light, very precise displacement measurements are possible by interferometric methods. (This interference fringe formation is the basis for the crack opening displacement measurements shown previously in Figures 7.32 and 7.33.)

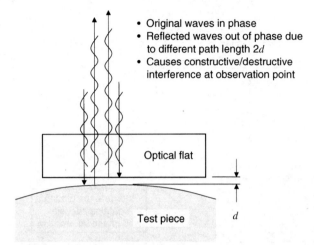

- Original waves in phase
- Reflected waves out of phase due to different path length $2d$
- Causes constructive/destructive interference at observation point

Optical flat

Test piece d

Figure 15.8. Schematic representation of optical interference that results when two monochromatic light waves reach same point by different path lengths.

Interferometry has many uses in optics, and one common NDE application is to determine the surface roughness of a machined surface. This inspection would employ an optical flat as discussed previously with respect to Figure 15.8. Here variations in the flatness of the test piece result in a series of interference fringes when the optical flat is placed on the surface. Although one can quantify the variation in flatness through analysis of the fringe patterns, it is often sufficient to simply note fringe aberrations associated with deformations caused by surface anomalies (i.e., strain concentrations).

15.4.2 Holography

Three-dimensional virtual images may be reconstructed from holograms that are created by recording both amplitude intensity and phase information from light reflected from the test object [13–15, 17, 18]. In a normal photograph, only amplitude is recorded by the negative, and phase information data are lost in the subsequent print. Holograms also contain phase data that give the reconstructed image its three-dimensional character. The key to creating the hologram lies in the use of coherent light to illuminate the test piece. As shown in Figure 15.9, the original coherent beam is split, with part going directly to the hologram (reference beam) and part reflecting off the test object. Although these two beams originally have the same phase, they travel different path lengths to the recording plane and are not necessarily in phase when meeting again on the film. Thus, they create interference patterns that are recorded along with the normal variation in light intensity on the film negative. When this hologram is again examined with a coherent light source, the reconstructed image has many of the three-dimensional characteristics of the original object.

Holographic NDE involves creating a three-dimensional virtual image of the test object in an unloaded state that is then used as a template to compare with the

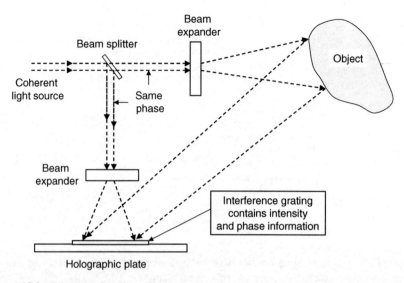

Figure 15.9. Schematic creation of holographic plate that contains both phase and amplitude (intensity) information from light reflected by test object.

deformed object. Interference fringe patterns that result from superposition of the original hologram image and the deformed test object provide precise measurements of surface deformations that can be used to locate flaws or other strain concentrations. Holograms can, in principle, be created with any type of wavelike radiation, including electromagnetic and other types of waves. The two most common forms for NDE are optical holography that uses visible light and acoustical holography that uses ultrasonic waves. Holograms could also be formed with neutrons.

There are a number of different holographic NDE techniques that use continuous or pulsed lasers. Pulsed lasers provide the opportunity to "freeze" motions in time. *Real-time holography* superimposes the hologram with the original object in its deformed position to create the interference fringes. This procedure requires precise position of the holographic plate (to approximately $\frac{1}{4}\mu$m) and is sensitive to rigid-body motions. *Double-exposure holography* exposes part of the hologram in the original state and part in the deformed state. Reconstruction of the hologram image then shows the interference fringes directly without the need for aligning the virtual image with the original object. *Time-average holograms* are made of vibrating surfaces where bright fringes appear at nodal points where the vibrations are small.

Holographic NDE may be used for detecting debonds in sandwich structures such as tires and other laminates. It has also been used for crack detection in hydraulic fittings and for qualitative evaluation of turbine blades. The advantages and disadvantages of this method are given below.

Advantages of Holographic Inspection

1. The method can be applied to any type of solid material, and the test object can have almost any size/shape, provided it can be stressed.

2. The inherent sensitivity of the inspection is associated with displacements of one-half wavelength of light ($\sim$125 nm), and it is possible to increase sensitivity 1000 times by special analysis methods.

3. Holographic NDE is a full-field rather than a point inspection method, allowing large areas to be quickly examined.

4. The readout can be qualitative and/or quantitative providing flexibility for different applications.

5. Pulsed-laser techniques allow inspection in hostile environments.

6. It is possible to obtain relative comparisons of a single object at two stress levels, rather than employing an absolute frame of reference or relying on a similar member for reference.

7. Interferograms can be reproduced at a later time.

Disadvantages of Holographic Inspection

1. The requirement for stable fixtures to hold and load the test object complicates field use, and the method can be impeded by spurious fringes associated with rigid-body motions.

2. The object must be deformed at stress levels that do not cause structural damage or rigid-body motions, placing limits on the maximum skin thickness that can be examined.

3. Anomalies must be located so as to perturb surface displacement.

4. The object needs diffusely reflecting surfaces with high reflectivity, sometimes requiring coatings.

5. The technique *locates* flaws but does not quantify their size, shape, or depth.

6. The test object is compared with itself rather than with some given standard.

7. Holographic NDE requires well-trained personnel and sophisticated equipment, and the use of lasers presents safety requirements.

15.4.3   Moiré

Moiré interferometry is a standard strain analysis technique that involves obtaining surface displacements by interference of two fine gratings with slightly different pitch and/or orientation [14]. The creation of Moiré fringes is shown in Figure 15.10 by overlapping two grids of evenly spaced, parallel lines. In Figure 15.10a, the top and bottom grids have the same number of lines and are parallel to each other but have slightly different pitch (lines in the top grid are closer together). In the area of overlap, the difference in pitch leads to a regular pattern of light and dark fringes corresponding to the recurring alignment and misalignment of individual lines in the respecting grids. In Figure 15.10b, the two overlapping grids are identical, with the same pitch between individual lines, but are rotated relative to each other. Again, note the series of light and dark fringes formed by the intersection of individual grid lines.

   In Moiré strain analysis, the fringes caused by differences in pitch are related to normal strains, whereas those associated with angular misalignment describe shear

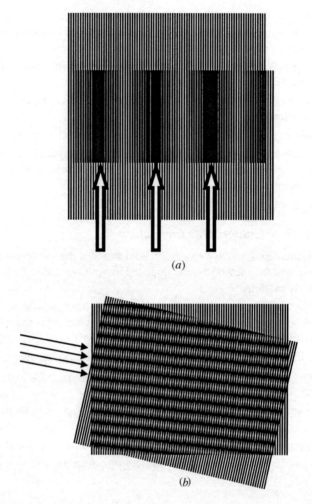

Figure 15.10. Moiré interference fringes patterns: Fringes formed by overlapping parallel grids with (*a*) slightly different pitch and (*b*) identical pitch but different angular orientation.

strains. In an experimental Moiré strain analysis, a grid of finely spaced lines (on the order of 50–1000 lines/in) is mounted to the test piece, which is then subjected to a small load that deforms the test grid. Next, a reference grid is superimposed over the deformed specimen grid, resulting in fringe patterns caused by the different spacing and/or orientation of lines in the two grids. These fringes patterns may then be analyzed by well-established equations [14] to determine the normal and shear strains in the test piece. There are a number of variations to this basic concept, but Moiré strain analysis is a relatively simple and powerful method to measure small surface strains.

For NDE purposes, however, one is usually not interested in quantitative measurements of strains but simply seeks strain *concentrations* caused by local anomalies on the surface of the test piece. These strain concentrations lead to sharp discontinuities that are readily apparent in the normal Moiré fringe pattern. The Moiré fringes result from very small surface strains and, depending on the grid patterns employed, are quite sensitive to small surface defects. Although the test grid may be physically mounted to the test piece (e.g., bonded, etched, photodeposited), one is interested in obtaining the fringe patterns as quickly as possible for NDE purposes, so noncontact techniques are often employed. Shadow Moiré and projection Moiré are two such methods that optically project the grid pattern onto the test object without physical contact [19]. Moiré NDE can be applied quickly and easily in the field and can locate anomalies associated with both in-plane and out-of-plane deformations.

15.4.4   Other Optical Methods

As described in References 13, 19, and 20, there are a variety of other optical techniques that can be used for inspection purposes. These methods include speckle pattern interferometry, shearography, photoelasticity, the use of structured light, Diffracto-Sight, optical profilometry, and video enhancement. Some of these techniques can be readily applied in the field, whereas others require specialized laboratory apparatus (e.g., stable optical benches, specialized optical equipment). Some can be employed with normal lighting, whereas others may require monochromatic and/or coherent wavelengths obtained from a laser. Some may require extensive surface preparation, whereas others need little surface preparation. In general, these methods are full-field displacement analysis techniques that locate strain concentrations associated with surface or near-surface anomalies. They are capable of detecting in-plane and/or out-of-plane deformations.

15.5   OTHER METHODS

Since it is not possible to review all the many other forms of specialized and emerging inspection methods here, the reader is encouraged to consult the technical literature (e.g., References 17, 19, and 21–23) for descriptions of other NDE methods. Leak testing [17, 22], for example, senses escaped fluids or gases from a pressure vessel to locate component damage. The leaks may be detected by pressure gages, trace-sensitive devices, mass spectrometers, or the sound of escaping fluids or gases. Measuring a component's dynamic behavior provides another indication of inhomogeneities or changes in material properties [17]. In this case the dynamic response of a vibrating component is compared with a reference standard. Other inspections employ microwaves [23] to locate moisture in dielectric materials and to measure component thickness.

15.6   CONCLUDING REMARKS

In addition to the traditional "big six" NDE methods (visual, liquid penetrant, radiography, ultrasonic, eddy current, and magnetic particle), there are a host of other inspection methods that can be used to assess structural integrity. Although this chapter overviews acoustic emission, thermal imaging, and several advanced optical approaches, many other techniques are available and/or under development. Indeed, only the inspector's imagination limits what forms of stimuli can be used to interrogate structures for evidence of anomalies.

PROBLEMS

15.1 Compare the acoustic emission and the ultrasonic inspection methods. How are they similar? How are they different?

15.2 Define the Kaiser effect and the felicity ratio for acoustic emission testing. What is their significance for NDE?

15.3 One of the major advantages of thermal inspection methods is the ability to inspect a component while in service. Give two examples of how thermal NDE could be used for real-time monitoring of a structure while in use. What types of anomalies would you expect to discover in these cases?

15.4 Use the Internet to find information about the leak testing method of NDE. Write a short report on the results of your search.

15.5 Use the Internet to find information about the Diffracto-Sight (commonly known as D-Sight) method of NDE. Write a short report on the results of your search.

15.6 Use the Internet to find information about the dynamic measurement method of NDE. Write a short report on the results of your search.

15.7 Use the Internet to find information about the proof test method of NDE. Write a short report on the results of your search. How does proof testing differ from the other major inspection methods?

15.8 Suppose that you have been asked to evaluate a new inspection method for a particular application. What criteria do you believe a new and/or nontraditional inspection method should satisfy before being accepted for field use?

15.9 The many different inspection techniques discussed in this text have various advantages and disadvantages. Often a particular inspection problem might require two *complementary* methods that work *together* to meet the desired NDE requirements. Give an example of two complementary inspection methods and briefly discuss how the two techniques would work together to ensure structural integrity for a particular application.

15.10 Assume that you work for a firm that provides NDE services and that a client has asked you to design an NDE "system" for a particular manufacturing facility. The client's product performs adequately provided that appropriate quality control standards are met, but manufacturing-induced damage has led to several premature failures in service. The client wants to install an inspection system that will guarantee that the components produced will "never fail" during the product's warranty period. Briefly discuss the information you would need to design such an inspection system. What data do you need to know about the component and its use? What information is needed about potential inspection systems and NDE personnel?

15.11 Which of the NDE methods considered in this text is, in your opinion, the most versatile technique (i.e., capable of being adapted for the most different uses). Explain (with examples) the versatility of your chosen method. Note that there may be several possible correct answers to this question, but you are to select one NDE method and then defend your answer.

Assume for the following questions that you are a maintenance supervisor with a tight schedule and a limited budget but with access to all of the inspection methods discussed in this text. Choose and describe an NDE method for the specific applications given below. Although there may be more than one appropriate method for a given situation, you only need to select one technique but must clearly justify and explain your choice.

15.12 You need to detect fatigue cracks that develop in the threaded section of steel bolts used to join two components that are subjected to large vibratory forces. You may assume that the bolts may be removed for inspection.

15.13 You need to detect fatigue cracks that develop in the threaded section of titanium bolts used to join two components that are subjected to large vibratory forces. You may assume that the bolts may be removed for inspection.

15.14 You need to inspect thin-walled aluminum hydraulic tubes that develop longitudinal discontinuities such as laps or seams during the extrusion process.

15.15 You need to ensure that a uniform thickness of paint is applied to the surface of aluminum panels.

15.16 You need to ensure that a uniform cladding thickness has been applied to a sheet of aluminum alloy (the clad is a thin layer of pure aluminum plating on the parent alloy).

15.17 You need to locate fatigue cracks along the bore of a fastener hole in the middle layer of a three-row stack-up of aluminum plates:
(a) Assume that the fasteners may be removed for inspection.
(b) Assume that the fasteners may *not* be removed for inspection.

15.18 You need to routinely examine the frozen produce in a grocery store to ensure that the refrigeration unit is working properly and the food product is thoroughly frozen.

15.19 You need to locate the wooden studs behind a plaster wall in your home in order to secure a hook for a wall hanging.

15.20 You are a medical doctor who needs to monitor the development of a human fetus.

15.21 You are a medical doctor who needs to examine a patient for evidence of colon cancer.

REFERENCES

1. "Standard Terminology for Nondestructive Examination," ASTM Designation E 1316-95a, in *1995 Annual Book of ASTM Standards, Volume 03.03, Nondestructive Testing,* American Society of Testing and Materials, West Conshohocken, Pennsylvania, 1995, p. 611.

2. D. E. Bray and D. McBride, *Nondestructive Testing Techniques,* Wiley, New York, 1992, Chapter 12.

3. R. Halmshaw, *Non-Destructive Testing,* 2nd ed., Edward Arnold, London, 1991, Chapter 8.

4. A. Pollock, "Acoustic Emission Inspection," in *ASM Handbook,* Vol. 17: *Nondestructive Inspection and Quality Control,* ASM International, Materials Park, Ohio, 1989.

5. "Acoustic Emission Examination of Fiber-Reinforced Plastic Vessels," in *Boiler and Pressure Vessel Code, Article 11, Subsection A, Section V,* American Society of Mechanical Engineers, 1983.

6. R. K. Miller, *Acoustic Emission Testing—Nondestructive Testing Handbook,* 2nd ed., Vol. 5, American Society for Nondestructive Testing, Columbus, Ohio, 1987.

7. G. Hardy and J. Bolen, "Thermal Inspection," in *ASM Handbook,* Vol. 17: *Nondestructive Evaluation and Quality Control,* ASM International, Materials Park, Ohio, 1989, pp. 396–404.

8. D. E. Bray and D. McBride, *Nondestructive Testing Techniques,* Wiley, New York, 1992, Chapter 22.

9. H. Kaplan, *Practical Applications of Infrared Thermal Sensing and Imaging Equipment,* Tutorial Texts in Optical Engineering, Vol. TT 13, SPIE Optical Engineering Press, Bellingham, Washington, 1993.

10. X. Maldague and P. Moore, *Infrared and Thermal Testing—Nondestructive Testing Handbook,* 3rd ed., Vol. 3, American Society for Nondestructive Testing, Columbus, Ohio, 2001.

11. X. Maldague, "Applications of Infrared Thermography in Nondestructive Evaluation," in P. Rastogi (Ed.), *Trends in Optical Nondestructive Testing,* Elsevier, Oxford, U.K., 2000, pp. 591–609.

12. N. Harwood and W. Cummings, *Thermoelastic Stress Analysis,* Adam Hilger, Bristol, 1991.

13. D. D. Duncan, J. L. Champion, K. C. Baldwin, and D. W. Blodgett, "Optical Methods," in P. J. Shull (Ed.), *Nondestructive Evaluation: Theory, Techniques, and Applications,* Marcel Decker, New York, Chapter 10, 2002.

14. J. W. Dally and W. F. Riley, *Experimental Stress Analysis,* McGraw-Hill, New York, 1978, Chapters 11 and 12.

15. J. D. Trolinger, "Fundamentals of Interferometry and Holography for Civil and Structural Engineering Measurments," *Optics and Lasers in Engineering,* Vol. 24, pp. 89–109, 1996.

16. P. F. Packman, "The Role of Interferometry in Fracture Studies," in A. S. Kobayashi (Ed.), *Experimental Techniques in Fracture Mechanics,* Vol. 2, Society for Experimental Stress Analysis, Westport, Connecticut, 1975, pp. 59–87.

17. D. E. Bray and D. McBride, *Nondestructive Testing Techniques,* Wiley, New York, 1992.

18. J. W. Wagner, "Optical Holography," in *ASM Handbook,* Vol. 17: *Nondestructive Evaluation and Quality Control,* ASM International, Materials Park, Ohio, 1989, pp. 405–431.

19. A. Beattie, L. Dahlke, J. Gieske, B. Hansche, G. Phipps, R. Shagam, and K. Thompson, *Emerging Nondestructive Inspection Methods for Aging Aircraft,* Report No. DOT/FAA/ CT 94-11, FAA Technical Center, Atlantic City, New Jersey, October 1994.

20. M. W. Allgaier and N. Stanley, *Visual and Optical Testing—Nondestructive Testing Handbook*, 2nd ed., Vol. 8, American Society for Nondestructive Testing, Columbus, Ohio, 1993.

21. R. K. Stanley, *Special Nondestructive Testing Methods—Nondestructive Testing Handbook*, 2nd ed., Vol. 9, American Society for Nondestructive Testing, Columbus, Ohio, 1995.

22. C. N. Jackson and C. N. Sherlock, *Leak Testing Methods—Nondestructive Testing Handbook*, 3rd ed., Vol. 1, American Society for Nondestructive Testing, Columbus, Ohio, 1997.

23. A. Bahr, R. Zoughi, and N. Qaddoumi, "Micro Wave," in P. J. Shull (Ed.), *Nondestructive Evaluation: Theory, Techniques, and Applications,* Marcel Decker, New York, 2002, Chapter 9.

PART IV

APPLICATIONS

The objective of the three chapters in Part IV is to demonstrate applications of the damage tolerance analysis concepts and nondestructive inspection methods developed in Parts II and III. Chapter 16 discusses practical aspects of designing a new structure to be resistant to initial manufacturing and service-induced damage. Chapter 17 focuses on issues related to continued or extended service of older structures that near or exceed their original design life goals. Chapter 18 treats special concerns associated with some specialized types of structures and materials and provides final concluding comments regarding the structural integrity concepts and goals developed in this book.

CHAPTER 16

DESIGN CONSIDERATIONS

16.1 OVERVIEW

The objective of this chapter is to discuss practical issues associated with the initial design and construction of damage tolerant structures. Emphasis here is on matters that should be considered for new designs, in contrast to maintaining the integrity of existing structures. (The latter topic is the subject of Chapter 17.) The chapter begins with an overview of certification requirements and the structural integrity process followed by a discussion of structural loads. Next, attention is focused on use of crack resistant structural configurations and selection of damage tolerant materials. Implications of nondestructive inspection on structural design are then reviewed. The chapter concludes with issues that should be considered when incorporating new technology into structural design.

16.2 CERTIFICATION REQUIREMENTS

Historical background for the development of modern damage tolerant design criteria is reviewed in Chapter 1 and discussed in more detail for aircraft in References 1 and 2. There were, for example, a number of structural failures during the 1950s and 1960s that pointed out serious shortcomings in the structural design and certification procedures in practice at that time. Chief among the deficiencies was the failure to formally account for the possibility of initial manufacturing or service-induced damage during the design process. Recognition of the drastic consequences of preexistent damage on structural integrity led to development of the fracture mechanics analysis and nondestructive inspection methods that are the focus of this book. The fact that these requirements are formally incorporated in many certification procedures further attests to their importance as well as to the fact that practical methods are available for their implementation.

16.2.1   FAA and USAF Damage Tolerance Criteria

Tiffany and Masters described an early use of fracture mechanics as a design tool for rocket motor cases at a 1964 conference [3]. Kaplan and Lincoln [2] assert that this was the first time a major manufacturer directly discussed the consequences of initial manufacturing defects and credit Tiffany and Masters with a major change in design philosophy regarding preexistent damage. The USAF became the first organization to formally require damage tolerance with the July 1974 issuance of MIL-A-83444, Airplane Damage Tolerance Requirements [4]. Quoting from MIL-A-83444 [5]:

> This specification contains the damage tolerance design requirement applicable to air-plane safety of flight structure. The objective is to protect the safety of flight structure from potentially deleterious effects of material, manufacturing and processing defects through proper material selection and control, control of stress levels, use of fracture resistant design concepts, manufacturing and process controls and the use of careful inspection procedures. . . . The analyses shall assume the presence of flaws placed in the most unfavorable location and orientation with respect to the applied stresses and material properties.

Note the clear mandate to consider worst-case *initial* damage. The choice of the particular design approach to achieve damage tolerant structure is, however, left to the manufacturer. All USAF aircraft have been designed to the MIL-A-83444 requirements since 1974, and all active fleets in service prior to that time have been subsequently reanalyzed with respect to that damage tolerant philosophy.

On the commercial aircraft side, Swift [1] cites concerns in the 1970s by the British Civil Aviation Authority (CAA) and by the U.S. Federal Aviation Administration (FAA) about the continued operation of older aircraft as key factors leading to adoption of damage tolerance. The CAA was worried that the redundancy of fail-safe designs could be compromised by widespread fatigue damage and proposed to limit the service life of Boeing 707 commercial jet transports to 60,000 hours. The FAA was also disturbed by the potential sale of older fail-safe aircraft to inexperienced operators who might not follow the rigorous maintenance and inspection procedures needed to ensure structural integrity. Both organizations held a series of meetings with industry during the 1970s that culminated with a December 1978 FAA decision that requires all commercial transport aircraft be designed to expanded damage tolerance requirements specified in Federal Airworthiness Requirement (FAR) 25.571 [6].

That section of the transport airplane airworthiness standards begins with the following statement regarding initial damage: "An evaluation of the strength, detail design, and fabrication must show that catastrophic failure due to fatigue, corrosion, manufacturing defects, or accidental damage, will be avoided throughout the operational life of the airplane." These requirements are further expanded in paragraph

(b) of FAR 25.571: "The evaluation must include a determination of the probable locations and modes of damage due to fatigue, corrosion, or accidental damage. Repeated load and static analyses supported by test evidence and (if available) service experience must also be incorporated in the evaluation. Special consideration for widespread fatigue damage must be included where the design is such that this type of damage could occur." The possibility for discrete source (i.e., accidental) damage requires additional consideration in paragraph (e) of FAR 25.571 as follows: "The airplane must be capable of successfully completing a flight during which likely structural damage occurs as a result of (1) impact with a 4-pound bird . . . (2) Uncontained fan blade impact; (3) Uncontained engine failure; or (4) Uncontained high energy rotating machinery failure."

Thus, by the end of the 1970s, both commercial and military aircraft certification authorities had rigorous requirements that aircraft be tolerant of preexistent and service-induced damage. Although these documents have been modified from time to time to account for special circumstances, they remain firm in their mandate to prevent failure by unforeseen damage. Similar philosophies are also incorporated in the design procedures for other types of structures (e.g., spacecraft, pressure vessels, nuclear reactors).

16.2.2 Inspection Requirements

Inspection requirements for various structures are discussed in a number of federal documents. Inspection of spacecraft hardware is, for example, the subject of MSFC-STD-1249 [7]. The scope (Section 1.0) of this document states:

> This standard provides guidelines for selection and prescription of NDE (Nondestructive Evaluation) techniques required to fulfill the demands of a Fracture Control Program. Also provided are the minimum crack-like flaw sizes deemed reliably (90% probability, 95% confidence) detectable via the traditional NDE techniques, Penetrant, Eddy Current, Magnetic Particle, Ultrasonics, and (manual) Radiography, when implemented as prescribed herein.

The general requirements for a fracture control plan that entails inspection then follows:

> The Fracture Control Program concept dictates that fracture mechanics design assessments of aerospace structures and pressure vessels assume the potential existence of crack-like flaws in crucial, higher stressed zones of hardware. Mission fulfillment expectations are thus confirmed by fracture mechanics calculations of acceptable maximum size initial crack-like flaws (Critical Initial Flaw Size—CIFS) that would *not* grow to catastrophic-failure size within the number of cyclic exposures dictated for mission assurance. NDE then is obligated to assure that the CIFS does not exist in the newly fabricated hardware.

With respect to commercial aircraft, FAR 25.571 [6] states the following regarding inspections:

Inspections or other procedures must be established, as necessary, to prevent catastrophic failure, and must be included in the Airworthiness Limitations Section of the Instructions for Continued Airworthiness. . . . Inspection thresholds for the following types of structure must be established based on crack growth analyses and/or tests, assuming the structure contains an initial flaw of the maximum probable size that could exist as a result of manufacturing or service-induced damage: (i) Single load path structure, and (ii) Multiple load path "fail-safe" structures and crack arrest "fail-safe" structure, where it cannot be demonstrated that the load path failure, partial failure, or crack arrest will be detected and repaired during normal maintenance, inspection, or operation of an airplane prior to failure of the remaining structure.

Although some have interpreted the initial damage tolerant crack size assumption to permit structural cracking, nothing could be further from the truth [1]. These regulations do not allow continued operation with known cracks in primary structure. Once detected, cracks are to be repaired or the components replaced. The intent of damage tolerance requirements is to specify the *minimum* crack sizes that must be detected at a given time. One must then determine the *frequency* of repeated inspections required to ensure that no crack at or below the NDE threshold can grow to fracture before the next inspection opportunity.

16.3   OVERVIEW OF STRUCTURAL INTEGRITY PROGRAM

The general tasks of the USAF aircraft structural integrity program (ASIP) are overviewed by Kaplan and Lincoln [2] and are summarized in Table 16.1. This "cradle-to-grave" procedure involves five comprehensive tasks. The first is to assemble the information needed for preliminary design and to develop strategies and approaches for meeting the structural life goals. The second task involves consideration of the anticipated service loads and operating environment and results in more detailed design analyses, materials selection, and development testing. Task III entails full-scale ground and flight tests to demonstrate the ability of the airframe to withstand static and dynamic loads with the desired degree of damage tolerance and durability. Another goal of the flight tests is to confirm that the original design load assumptions are consistent with the actual measured flight loads. Task IV determines the information needed to ensure that appropriate maintenance actions are performed in a timely manner during the life of the fleet. This task involves final analyses of loads, stresses, damage tolerance, durability, and so on; collecting data on actual aircraft usage; formation of a structural maintenance plan; and development of an individual airplane tracking program to prioritize maintenance actions for particular aircraft. Task V is the final management of all actions needed to ensure the safety and operational readiness of the fleet. (See Reference 2 for a more complete discussion of these five tasks.)

Ignoring the pun, the key point here is that structural integrity is not an accident. Preventing structure failures requires initial consideration of all possible contingencies, development of a plan to protect the structure from damage through design and nondestructive inspection, updating that plan when new information becomes available, and most of all, strict monitoring to ensure adherence to all aspects of the fracture control plan.

TABLE 16.1 Summary of USAF Aircraft Structural Integrity Program Tasks

Task I: design information	• ASIP master plan • Structural design criteria • Damage tolerance and durability control plans • Selection of materials, processes, and joining methods • Design service life and design usage
Task II: design analyses and development tests	• Materials and joint allowables • Loads analysis • Design service load spectra • Design chemical/thermal environment spectra • Stress analysis • Damage tolerance analysis • Sonic analysis • Vibration analysis • Flutter analysis • Nuclear weapons effects analysis • Nonnuclear weapons effects analysis • Design development tests
Task III: full-scale testing	• Static tests • Durability tests • Damage tolerance tests • Flight and ground operations tests • Sonic tests • Flight vibration tests • Flutter tests • Interpretation and evaluation of test results
Task IV: force management data package	• Final analyses • Strength summary • Force structural maintenance plan • Loads/environment spectra survey • Individual airplane tracking program
Task V: force management	• Loads/environment spectra survey • Individual airplane tracking data • Individual airplane maintenance times • Structural maintenance records

Note: See Reference 2.

16.4 LOADS

As indicated in the previous section, critical tasks in any structural design are to estimate, and then confirm, the forces that will be encountered during service. Clearly, the fatigue life is a function of the cyclic stresses experienced during various missions, whereas residual strength requirements depend on the maximum applied load. From a static strength standpoint, one is concerned with the limit and ultimate design loads [6]. The *limit load* is the maximum force that is expected to occur during service, and all components must resist this load without undergoing permanent deformation or other structural damage. In order to provide an additional margin of safety, however, aircraft are designed to withstand an *ultimate load* equal to the limit load times a safety factor (typically assumed to be 1.5). Although components may yield, they must not fracture at ultimate load.

Determining design forces is a daunting task, as the actual service loads are usually not known until after construction and initial performance tests have been completed. Moreover, service usage often varies during the life of the structure, so that the actual load history may differ significantly from that anticipated during the design stage. Although a detailed analysis of loads methodology is beyond the scope of this text (see References 8–10), it is hoped that the following discussion gives an appreciation for the many factors that influence the severity of structural loading.

Aircraft loads, for example, may be divided into several categories: ground, air, powerplant, and "other." Ground loads include forces encountered during taxi, take-off, and landing, whereas air loads result from maneuvers under the pilot's control or from gusts caused by atmospheric conditions that the pilot cannot control. Maneuver loads may be further subdivided into level flight, symmetric or nonsymmetric maneuvers, dynamic maneuvers, or loads resulting from autopilot control. The powerplant forces result from the thrust and torque exerted by the engine on the airframe. Other types of aircraft loading include fuselage pressurization, towing forces, or hoisting forces. In addition, the actual stresses caused by these types of loads depend on many other system factors (e.g., vehicle weight, speed, direction, acceleration, weather).

Although service loads may differ for other types of structures, they must also be estimated during the design process. A bridge, for example, must resist its own dead weight along with live loads applied by pedestrians and vehicles that cross its span. One must also consider the dynamic forces resulting from heavy winds, the potential vibrations of an earthquake, any special types of loading that may occur during construction and assembly, and, perhaps, accidental collision with another object.

16.4.1 Load Factors

Structural forces are often represented in terms of load factors. Consider, for example, an aircraft in equilibrium during straight and level flight, as shown in Figure 16.1*a*. Summing forces in the vertical direction indicates that the lift force L equals the aircraft weight W. The load factor n may be defined as the ratio of lift divided by weight, and in this case is simply given by

$$n = \frac{L}{W} = 1 \tag{16.1}$$

Now consider the case given in Figure 16.1*b*, where there is an acceleration a perpendicular to the flight path resulting from a maneuver or turbulence. Summing forces in the vertical direction gives

$$\sum F = L - W = ma \rightarrow L = W + ma = W + \frac{W}{g}a$$

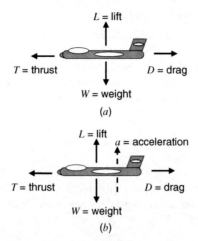

Figure 16.1. Definition of load factors: (*a*) aircraft in equilibrium during straight and level flight; (*b*) aircraft with acceleration *a* perpendicular to flight path.

Here m is the mass of the object and g is the acceleration of gravity. Now, computing the load factor n gives

$$n = \frac{L}{W} = \frac{W + Wa/g}{W} \rightarrow n = 1 + \frac{a}{g} \tag{16.2}$$

Similar load factors can also be obtained for the horizontal directions. Since the actual weight of the aircraft depends on the fuel, cargo, and so on, the load factor n gives a convenient scaling factor that combines weight and inertia effects. Load factors depend on the type of aircraft and mission and typically vary from 2–3 for a transport aircraft to 6–9 for high-performance fighters. These load factors are used to determine shear and moment diagrams, and ultimately stress distributions, at key structural locations.

Limit load factors at a given altitude may be presented as a function of indicated air speed V to specify the flight envelope (i.e., the $V-n$ diagram) shown in Figure 16.2. Flight conditions to the left of the stall lines are not possible, for example, since the aircraft cannot maintain lift at those slow speeds. The upper and lower portions of the flight envelope result in limit load and cause structural damage, whereas the right-hand side is defined by the maximum attainable velocity. The actual design envelope is derived from airworthiness requirements and specific design considerations and can differ greatly for individual aircraft types.

The corners of the simplified flight envelope of Figure 16.2 represent four basic flight conditions that often lead to critical structural loads in a given portion of the aircraft:

- PHAA (positive high angle of attack)
- PLAA (positive low angle of attack)

- NHAA (negative high angle of attack), and
- NLAA (negative low angle of attack).

As indicated in Figure 16.3 [11], the locations of the maximum tension and compression bending stress in a wing change with these four basic flight conditions. The resultant force **R** acting on the wing (in the plane of the cross section) is given by Equation 16.3 or 16.4:

$$\mathbf{R} = \mathbf{L} + \mathbf{D} \qquad (16.3)$$

Here **L** is the lift force and **D** is the drag measured with respect to the relative wind direction. One can also resolve **R** in terms of components **C** and **N** measured parallel and perpendicular to the chord direction, which makes an angle α to the wind direction:

$$\mathbf{R} = \mathbf{N} + \mathbf{C} \qquad (16.4)$$

Note that the directions of **N** and **C** change relative to **R** with the angle of attack α. The bending moments (and subsequent stresses) caused by **N** and **C** also depend on α, resulting in the location of the maximum tensile and compressive stresses moving from one quadrant of the wing to another. The maximum tensile stress, for example, occurs in the upper left quadrant of the wing for NLAA but moves respectively in a clockwise direction to the other three quadrants for the NHAA, PHAA, and PLAA conditions. Thus, the designer must consider several possible flight conditions in order to properly size the structural components. Similar scenarios could also be developed for other types of structures, further confirming that the designer must consider many different configurations and applications in order to ensure all worst-case loadings are adequately analyzed.

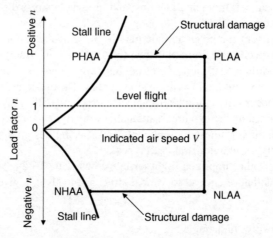

Figure 16.2. Idealized V–n diagram defining limits of flight envelope.

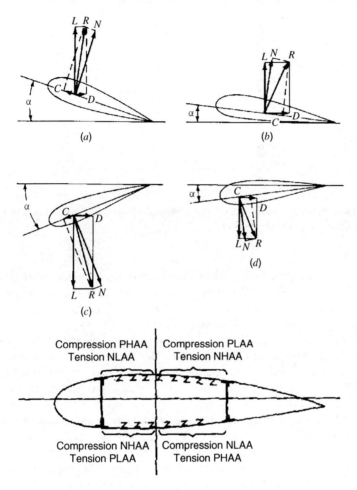

Figure 16.3. Four basic flight conditions showing how location of maximum stresses in wing depend on angle of attack: (*a*) PHAA; (*b*) PLAA; (*c*) NHAA; (*d*) NLAA. (Parts *a* and *b* from Ref. 11, copyright 1982 by McGraw-Hill, Inc. Reproduced with permission from The McGraw-Hill Companies.)

16.4.2 Load Spectrum Issues

Applied loads are rarely constant but usually cycle in a complex manner during service. Although cabin pressure loads, landing gear impact, or flap loads only occur once per flight, gust and maneuver loads, engine vibrations, or taxi loads are repeated many times during a given flight. This cyclic loading may be expressed in several ways. The exceedance curve shown in Figure 16.4, for example, plots the number of times a given load factor is "exceeded" during a given period of time versus the load factor *n*. This schematic describes the typical situation where there are relatively few large-load applications but many small-load exceedances in a given mission. The

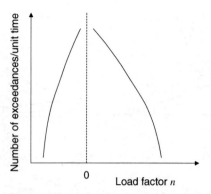

Figure 16.4. Schematic exceedance curve showing the number of times a particular load factor is exceeded in a given period of time.

actual exceedance data for a particular application are based on statistical analyses that combine laboratory and/or service measurements, analytical predictions, design specifications, or extensions of prior results. There will, in general, be several different exceedance curves for various types of missions.

Although these data describe the load spectrum content and can be converted to stresses by an appropriate calibration curve for the structural location of interest, exceedance information does not indicate the *loading sequence*. Since fatigue damage is quite sensitive to the order in which loads are applied (recall the load interaction effects discussed in Sections 7.4 and 9.2.5), the forces must be put in a logical sequence for fatigue testing and analysis. In addition, practical considerations may require that the large numbers of small, frequently applied loads be reduced for fatigue testing. Ordering the exceedance curve loads into a reasonable flight spectrum with defined peaks and valleys, as well as any truncation (or "clipping") of the curve for fatigue testing, requires considerable judgment and experience [9].

Some typical aircraft spectra are given in Figures 16.5 and 16.6 [12]. Figure 16.5, for example, shows the lower wing-skin stresses for a simulated high-altitude tanker aircraft mission followed by several practice landings and take-offs ("touch-and-go's"). Although greatly simplified, the general nature of this typical stress history begins with an initial compressive peak while the aircraft is on the ground and the wing is only loaded by its own weight. After take-off, there are numerous cycles about a large tensile mean stress. The tensile mean is the main lift load during flight, whereas the many smaller excursions result from turbulence and flight maneuvers. (In actuality, there would be many more small cycles than are shown here.) Upon landing, the lift is released, and the lower wing surface goes into compression. The large mean stress cycle, which begins and ends with compression peaks upon take-off and landing, is known as the ground–air–ground (or GAG) cycle. The touch-and-go GAG cycles are similar in nature to the main flight but are shorter in length. Figure 16.6 shows the stress history for a vertical fin for the same tanker aircraft

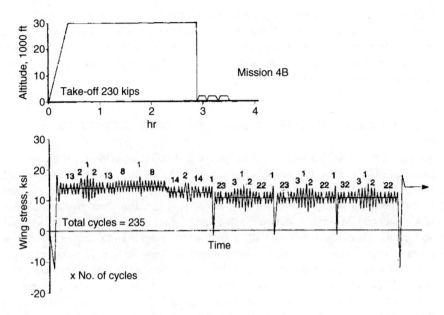

Figure 16.5. Simulated mission of aircraft tanker showing lower wing skin stress history for 3-hour flight followed by several touch-and-go landings [12].

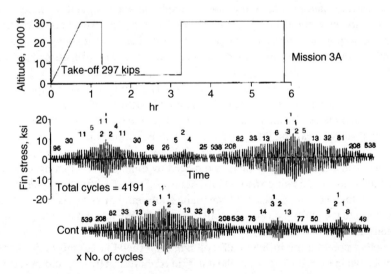

Figure 16.6. Simulated mission of aircraft tanker showing vertical fin stress history for extended periods of high- and low-altitude flight [12].

[12]. Note here that the fin spectrum is characterized by cycling about a zero mean stress and results mainly from turbulence and maneuver loads. Similar stress spectra must be developed for other structural locations and missions. The designer's task is to select the worst-case missions that are then used to size structural members and select component materials.

16.5 DAMAGE TOLERANT STRUCTURAL CONFIGURATIONS

Two key results from a structural design are a detailed description of the desired shape and size of all components (e.g., drawings that detail component dimensions, surface finish, location of fastener holes, manufacturing processes) and specification of the materials used to fabricate each member (e.g., alloy, heat treatment, grain flow direction). Material selection for damage tolerance is discussed in Section 16.6 and entails choosing materials with high fracture toughness to resist catastrophic fracture, superior fatigue crack growth resistance to delay the time for small cracks to grow to fracture, and resistance to corrosion. In addition to selecting damage tolerant materials, however, one may provide another line of defense by employing geometric features that impede or arrest subsequent crack growth. This section discusses the role that such "structural features" can play in preventing catastrophic failure.

16.5.1 Crack Arresting Structures

It is often possible to achieve damage tolerance by providing structural mechanisms that arrest or retard individual cracks before they grow to a size that causes total fracture. Multi-load-path structures, for example, provide damage tolerance by allowing individual members to fail without total system loss. Also known as fail-safe design, this approach guarantees safety by ensuring that the remaining structure carries the additional load from a failed member. Two generic multi-load-path examples are shown schematically in Figure 16.7. The simplest case shown in Figure 16.7*a* is that of separate members loaded in parallel. The failure of one component does not lead to total fracture provided the remaining structure withstands the additional load originally borne by the broken member. Since the remaining components carry additional stress, however, their fatigue lives will be shortened.

Recall that the thickness dependence of K_c leads to another potential damage tolerance benefit when a thick plate is replaced by thinner sheets that have the same total thickness. As shown previously in Figures 6.5 and 6.6, fracture toughness decreases as thickness increases until the plane strain minimum K_{Ic} is reached. Thus, in addition to the multi-load-path benefits discussed here, parallel members with thin sections have the extra advantage of larger fracture toughness.

Another common multi-load-path configuration found in aircraft is the stiffened panel shown in Figure 16.7*b*. Here the thin skin is reinforced by longitudinal stringers that are either attached or produced integral with the skin (e.g., by extrusion or machining from thick plate). Although the stiffeners help resist buckling from compression or torsional loads, they also provide a key mechanism to arrest skin fracture.

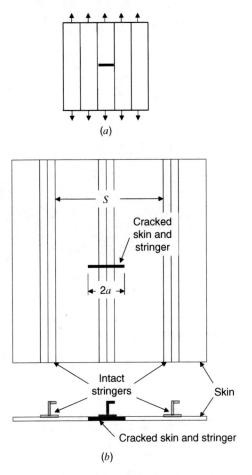

Figure 16.7. Schematic representation of multi-load-path structures: (*a*) separate members loaded in parallel; (*b*) stringer–skin configuration.

Here a stiffened panel has developed a crack that has broken a central longeron and extends into the skin growing toward the other two stringers. This typical "two-bay" crack situation is often assumed as a worst-case scenario resulting from the discrete-source damage incident discussed in Section 16.2.1.

The stress intensity factor solution for the skin crack is shown schematically in Figure 16.8. Note that K initially increases with crack length for a given applied stress but then decreases as the crack tip approaches a stringer [13]. (The broken central stringer is omitted for clarity in Figure 16.8.) This decrease in K is due to the fact that load is shifted from the skin to the stringer as the crack extends. If the remaining stringers are able to carry the entire additional load, the skin crack will arrest and the configuration is damage tolerant.

A schematic residual strength diagram for this simple stringer–skin configuration is shown in Figure 16.9. Here the residual strengths (i.e., fracture stresses) for both

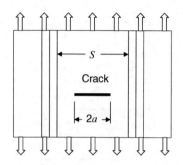

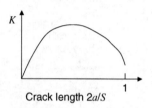

Figure 16.8. Schematic stress intensity factor solution for stringer–skin tension panel showing decrease in K as skin crack tip reaches stringer.

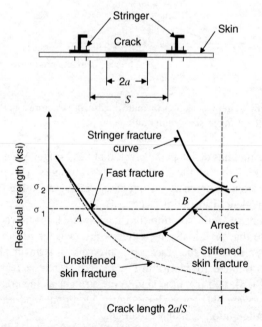

Figure 16.9. Schematic residual strength curve for stringer–skin cross section showing crack arrest when skin crack approaches stringer.

the skin and stringer are shown as a function of dimensionless skin crack length $2a/S$. Suppose, for example, that σ_1 is the maximum tensile stress applied to the stiffened panel. The skin crack will begin fast fracture at point A on the skin fracture curve but will arrest at point B as it grows toward the stringer. Now, total fracture of the reinforced panel is controlled by the stringer residual strength curve and will occur at an applied stress of σ_2 when the skin crack length $2a/S = 1$. For comparison purposes, the residual strength for the unstiffened panel also shown in Figure 16.9 indicates that complete loss of the structure occurs much more readily when there are no stiffeners.

The stress intensity factors and residual strength curves for the skin and stringers depend on the particular materials and geometry in question (see References 1 and 13–19 for more details). Note that the simple scenario discussed here assumes that a single skin crack exists in the structure. If the skin and/or stringer contains additional cracks, the residual strength can be significantly reduced, and the crack may not arrest at the stringers. This "widespread fatigue" issue is of considerable concern for older structures that could develop multiple cracking late in life and is discussed in more depth in Chapter 17. It should also be noted that if the stringer and skin are integral structures (i.e., one piece), the crack remains in the unit and can continue to propagate from the skin into the thicker stiffener portion. Although there may be economic advantages for the integral skin–stiffener structure, separate skins and stiffeners do provide a damage tolerant advantage in that a single crack cannot automatically propagate from the skin to the stiffener (or vice versa) but would need to reinitiate in the remaining secondary structure.

Pressurized cylinders provide another case where structural features can help arrest propagating cracks before catastrophic fracture occurs. Recall, for example, the leak-before-break criterion discussed in Section 6.6. In that instance, the goal was to select a pressure vessel wall thickness and material with sufficient fracture toughness to prevent surface cracks from causing rapid fracture. The concept was that if the crack penetrated the wall in a stable manner, without achieving the conditions for rapid fracture, the vessel would leak before fracturing and release the internal pressure and applied stresses. Thus, the member could only leak, and catastrophic fracture was not possible.

Two other crack arrestment mechanisms for pressure vessels are also instructive here. Consider, for example, the thin-walled cylinder that is reinforced with circumferential stringers (or frames) shown in Figure 16.10. (Longitudinal stringers would also be present, but are omitted here for clarity.) This structure is typical of an aircraft fuselage, and again, the frames reinforce the skin while also providing a means for crack arrest. If the pressurized fuselage is suddenly impacted with foreign object damage (e.g., an engine burst or collision with another object), a longitudinal crack could form and propagate due to the pressurized hoop stress. If the crack grows unimpeded, a catastrophic explosion would occur and the entire structure would be lost. If, on the other hand, the circumferential frame blocks the crack's path and causes it to turn 90° in direction, a "flap" would result, allowing the internal pressure to escape and unload the structure. A similar scenario could also be developed for circumferential skin cracks that are arrested by longitudinal stiffeners. This flapping mechanism has,

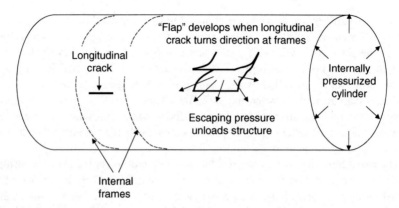

Figure 16.10. Schematic view of crack formed when longitudinal crack in a pressurized cylinder turns direction at internal frames and releases pressure.

in fact, proven to be a key feature that has saved several aircraft from catastrophic fracture.

The other crack arrestment scenario of interest here is that of a long pipeline subjected to an internal pressure. If a longitudinal crack were to develop as shown schematically in Figure 16.11, it is possible for a rapid fracture to propagate faster than the pressure could be released through the crack surfaces, resulting in many feet (miles?) of fractured pipe. One way of slowing the crack in order to relieve the pressure would be to employ periodic sections of pipe made from a tougher material. Although it may not be economically feasible to build the entire pipeline from this high-toughness material, intermittent segments could slow the rapid fracture enough to allow the internal pressure to escape through the crack surfaces and unload the pipe, resulting in crack arrest.

Another example of a "tough" material being inserted at strategic locations is the use of tear straps to reinforce the stiffeners and frames in the aircraft structure, as shown in Figure 16.12 [16, 17]. This figure demonstrates how shear clips provide clearance between intersecting frames and longerons, avoiding the necessity of

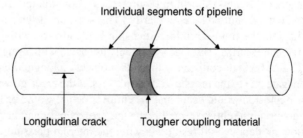

Figure 16.11. Schematic representation of a pipeline with "tough" coupling segments designed to arrest a longitudinal crack.

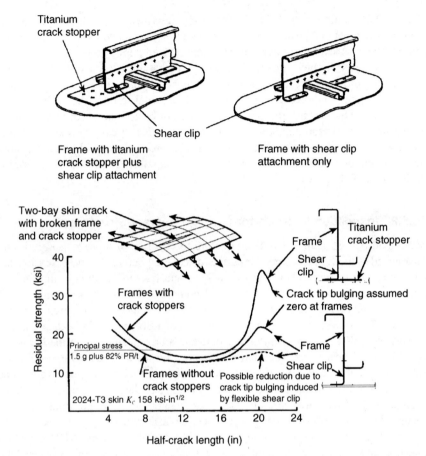

Figure 16.12. Two types of circumferential frame designs employed in commercial transport aircraft. The use of crack stoppers provides an additional barrier to cracks that attempt to propagate between the skin and frame [16, 17].

weakening either of those load-bearing members with cutouts. Although the shear clip may be fastened directly to the skin, riveting and/or bonding an additional strip or "tear strap" provides an effective "crack stopper" (see References 1 and 14–18). The increase in residual strength for a two-bay skin crack obtained by titanium crack stoppers is shown in Figure 16.12. Swift [17] reports that titanium crack stoppers have been employed for the Douglas DC-8, DC-10, and L-1011 aircraft whereas aluminum tear straps are used on the Boeing 727, 737, 747, 757, and 767. He also warns against a design trend to eliminate these crack stoppers for economic reasons.

Use of compressive residual stresses at key structural locations provides another practical means to delay and/or arrest the growth of preexisting cracks. These residual stresses may be introduced by shot peening, fastener hole coldworking, or interference fit fasteners and are quite effective in extending the fatigue lives associated with local stress concentrations. This application is discussed in more detail in Section

17.6.1, where compressive residual stresses are described as a method to extend the lives of structures that have developed cracks through extended service or unanticipated high stress levels. Although these beneficial compressive residual stresses may also be employed for new construction, some feel that this approach should be saved as a "trump card" to solve local cracking problems that may develop in service. It has also been suggested, however, that compressive residual stresses should be used for new structure but their beneficial effects should not be accounted for in the original design life analysis and component sizing (e.g., nominal stress levels should not be increased by reducing component thickness). This procedure would provide an additional margin of safety, and was, for example, successfully employed in the original design of the F-15 [20].

16.5.2   Inspection Intervals for Damage Tolerant Structures

Multi-load-path structures can delay fracture for a period of time, but safety is only guaranteed if the initial member failure is discovered before the increased loads in the secondary member also lead to cracking and fracture. Although the flapping failure and resulting controlled fuselage decompression described in Figure 16.10 would certainly be noticed, other forms of primary member fracture might not be so apparent. There was, for example, the tragic loss of a Boeing 707 in Lusaka, Zambia, in May 1977 that resulted from failure to detect a broken spar cap in the horizontal stabilizer [1]. The accident occurred 100 flights *after* the fatigue fracture of this fail-safe component, and although the secondary structure provided the desired additional resistance for a significant period of time, ignorance of the original failure led to the eventual loss of the entire aircraft. Thus, one cannot be lured into a false sense of security by fail-safe design and ignore in-service inspections.

The frequency of inspection depends on the capabilities of the NDE method in question and the damage tolerant strategy employed. The trade-off between NDE capability and inspection interval is shown by the schematic fatigue crack growth curve given in Figure 16.13 [1]. Here, a "long" inspection period is obtained if one is able to reliably detect "small" initial crack sizes. Such cracks are, however, likely to be hidden by adjacent structure and would require component disassembly and/or sophisticated (i.e., "expensive") inspection methods to locate. A larger crack would be easier to find but would result in shorter inspection periods and more frequent inspections during the life of the structure. The manner in which inspection intervals depend on the particular damage tolerant structural configuration may be demonstrated by examining the following five general classes of aircraft structure described by Swift [1]:

- safe life;
- multiple load path, externally inspectable;
- multiple load path, not inspectable for less than load path failure;
- multiple load path, inspectible for less than load path failure; and
- single load path, damage tolerant.

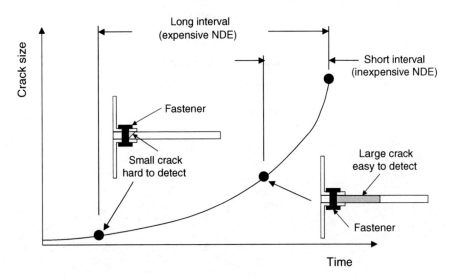

Figure 16.13. Trade-off between detectable crack size and inspection interval [1].

The safe-life approach is permitted by FAR 25.571 [5] only if damage tolerance is impractical (e.g., landing gear). In that case, it must be shown by test and analysis that the component can withstand the variable-amplitude fatigue loads expected in service without developing cracks. In addition, appropriate safe-life scatter factors must be applied for this type of structure. Since the safe-life approach assumes that components will be retired before cracks develop, in-service inspections are usually not performed for this type of design.

The externally inspectable multi-load-path structure shown in Figure 16.14 is the configuration recommended for major areas of aircraft structure, including wings, stabilizers, and fuselage. In this particular scenario, the stringer is the primary member and is assumed to fail first. Since the stringer is located internally, it is difficult to inspect without major disassembly. The goal, then, is to ensure safety by only inspecting the outside skin, which is readily accessible for examination. Thus, one must determine the safe inspection interval required to ensure that the secondary member (skin) does not develop cracks that grow to failure after the stringer fractures. The stringer and skin crack growth curves are shown in Figure 16.14. Here the undetected stringer fracture does not immediately cause catastrophic failure but is saved by the additional load picked up in the skin. The remaining life of the structure is now controlled by the skin crack growth curve (i.e., the secondary member). Although a small skin crack is assumed to be initially present and may not be immediately detectable, at some point it becomes so large that it is easily found (e. g., by a visual inspection) and the structure is repaired. The "safe" inspection period shown in Figure 16.14, then, is the additional time it would take for the "easily detected" skin crack to grow to fracture and is the margin of safety in this case (i.e., repeated inspections at this interval would guarantee that a catastrophic factor could not result from this crack

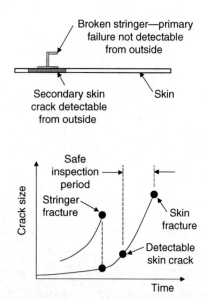

Figure 16.14. Multi-load-path failure scenario with easily detectable secondary-member crack serving as starting point for safe inspection period.

growth scenario). The easily detected crack size depends, of course, on the particular inspection method under consideration. It is important to note that the safe inspection interval shown in Figure 16.14, as well as the other inspection intervals discussed in this section, is normally divided by an appropriate safety factor to give an additional margin of safety. A safety factor of 2, for example, would provide two opportunities to detect cracks before they grow to fracture.

The second multi-load-path category occurs when the primary member cannot be easily inspected but its *failure* is readily discovered. An example of this configuration is the lug failure shown in Figure 16.15. Although internal surface cracks in the first lug are not easily found without component disassembly, fracture of that member is readily detected. After the primary member fails, member 2 carries the additional load, resulting in the schematic fatigue crack growth curves given in Figure 16.15. Here, the safe inspection period is the secondary crack growth life that begins with primary member failure. The secondary member is assumed to have an initial crack based on some appropriate criterion (i.e., initial manufacturing inspection limits) that will have grown to some predictable size at the time of the primary failure. The safe inspection period is now the time that it takes for this assumed secondary-member crack to grow to fracture at limit load after the primary member fails (and increases the stress in the second member).

Swift's third multi-load-path failure scenario is shown in Figure 16.16, where subcritical cracks can be detected in the primary member *before* it fractures. The example shown here consists of separate integrally stiffened lower wing planks that are loaded in parallel. The fatigue crack growth curves for the first-member failure

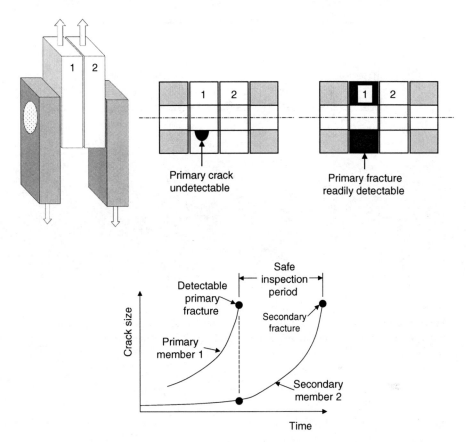

Primary crack
undetectable

Primary fracture
readily detectable

Safe inspection period

Detectable
primary
fracture

Secondary
fracture

Crack size

Primary
member 1

Secondary
member 2

Time

Figure 16.15. Multi-load-path failure scenario with easily detectable primary-member failure serving as starting point for safe inspection period.

and the subsequent crack growth in the secondary member are shown schematically in Figure 16.16. Note that in this case the internal stiffener crack can be readily detected *after* it penetrates the outer skin. This detectable crack could then grow and fracture the primary panel without catastrophic loss, since the remaining panels would take the additional load. The safe inspection period in this case, then, is the time for the *detectable* primary crack to grow to fracture plus the additional crack growth period in the secondary member (with its increased stress). As before, the secondary member is assumed to have a preexistent crack whose size can be determined at the time of the primary-member fracture.

Now consider the single-load-path, damage tolerant structure where the goal is to prevent an initial crack from growing to fracture without the protection of the redundant fail-safe members discussed previously. Known as *slow-crack-growth* structure in USAF terminology [4], a single-load-path lower wing skin scenario is shown schematically in Figure 16.17. Two crack growth periods are shown here. From the FAA standpoint, the "safe inspection period" would be based on the time it would take

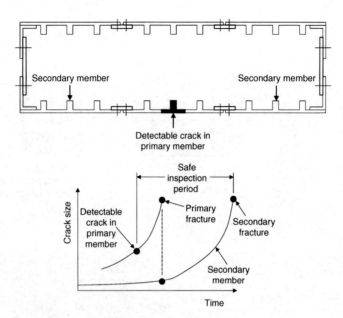

Figure 16.16. Multi-load-path failure scenario with easily detectable subcritical crack in primary-member failure serving as starting point for safe inspection period.

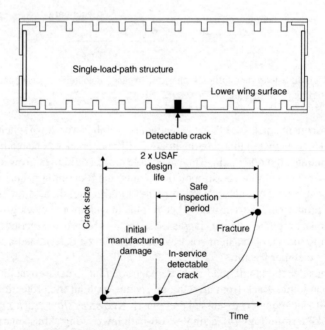

Figure 16.17. Single-load-path failure scenarios with in-service detectable crack or initial manufacturing damage serving as starting point for safe inspection period.

for an *in-service detectable* crack to grow to fracture at limit load. The other USAF crack growth period shown in Figure 16.17 begins with the *initial manufacturing* crack sizes specified in MIL-A-83444 and ends with fracture at the maximum load expected to occur in 20 lifetimes. The USAF would not require an in-service inspection for this type of structure if the component were retired before reaching half of that crack growth period (i.e., a safety factor of 2 on the crack growth life). Although the initial-crack-size assumptions specified by the USAF design criteria depend on the type of structure and inspection capability, a typical configuration involves a 0.05-in (1.3-mm) quarter-circular corner crack located along the bore of a fastener hole (see Reference 4 for other cases). Smaller initial crack sizes can be assumed for design analyses provided that an inspection demonstration program verifies detection to 90% probability with 95% confidence (recall Section 4.3).

16.6 MATERIALS SELECTION

Although employing construction materials with superior yield, fatigue, fracture, and corrosion resistance is paramount for damage tolerant structures, this goal is complicated by competing demands for material properties. The lower wing skin of a transport aircraft is primarily loaded in tension, for example, and requires good fracture toughness and fatigue properties to resist cracking. In contrast, the upper wing skin is compression dominated, requiring superior yield strength and stiffness to resist buckling and inelastic deformation. Material properties needed for other components vary with location, depending on the particular failure mechanisms that occur in that member. In addition, one also desires that materials have low cost and density, are easy to manufacture and repair, are readily available, along with a host of other considerations. Since no material possesses all of the desired attributes, materials selection is a complex subject that requires many compromises. The reader is, for example, encouraged to study the excellent book by Ashby [21] for an in-depth discussion of the role of materials selection in mechanical design. In addition, Sections 5, 6, and 7 of Reference 22 provide detailed treatment of the fatigue and fracture properties of ferrous and nonferrous alloys, composites, ceramics, and glasses.

The materials selection dilemma was previously demonstrated in Chapter 6 (see Figures 6.8–6.10) by the inverse relationship between fracture toughness and yield strength. (Section 6.5.2 discusses early fracture problems that resulted when designs emphasized materials with high tensile strength at the expense of ductility.) As metallurgists have gained a better understanding for the various parameters that control the strength-versus-toughness issue [23, 24], they have made significant strides in developing alloys that give good combinations of strength and toughness (e.g., recall Figure 6.9). Aluminum alloy 2024-T3, for example, was the material of choice for aircraft fuselage skin applications for several decades, beginning with the DC-3. The more recent 2524-T3 offers improved toughness and fatigue crack growth properties without sacrificing yield strength, however, and led to its selection for fuselage skins in the Boeing 777 and Airbus A340–500/600 commercial transports [24].

It is difficult to offer general statements regarding trade-offs between corrosion resistance and other material properties. Reference 24, for example, compares stress corrosion cracking properties of several aluminum alloys with their tensile yield stress and plane strain fracture toughness but found no systematic correlation between any of these properties. Thus, one must consider test results for a particular alloy when designing to resist these failure mechanisms. Aluminum alloy 7055 offers significantly improved combinations of toughness, strength, and stress corrosion resistance in comparison to the 7075 and 7178 alloys, which gave unsatisfactory fracture and corrosion performance in early jet aircraft. Alloy 7055-T7751 has, for example, been adopted for the upper wing skins of the Boeing 777 and the A340–500/600 aircraft [24].

There is, however, an interesting competition between corrosion and fatigue failure mechanisms for clad and bare materials. Aluminum alloys are frequently clad with a thin layer of pure aluminum to protect against corrosion (the cladding corrodes sacrificially with respect to the parent alloy; recall Section 9.3.2). Although the cladding is quite effective in resisting corrosion, it is softer than the parent material and can quickly crack under cyclic loading, facilitating fatigue failure. The trade between corrosion and fatigue resistance is shown by the fatigue life curves in Figure 16.18 [23]. Here constant-amplitude stress–life data (i.e., S–N curves) are compared for aluminum alloys 2024-T3 and 7075-T6 under a variety of conditions: clad versus bare alloys and air versus seawater environments. When tested in laboratory air, where there is little corrosion, the S–N curves show superior fatigue resistance for both bare alloys, with 2024-T3 outperforming 7075-T6. When the tests are conducted in saltwater, however, the advantage of the aluminum cladding becomes apparent, with both clad materials being superior to their bare counterparts. The preference for clad 2024-T3 relative to clad 7075-T6 is not as pronounced in this case, however, as for the air results with the respective bare materials. Thus, if corrosion is an issue, the cladding has an obvious benefit. If, however, corrosion is not a concern, the more expensive cladding is actually detrimental to fatigue strength.

A final example of the complex pitfalls associated with materials selection is shown in Figure 16.19 [23], where load history is shown to have a pronounced effect

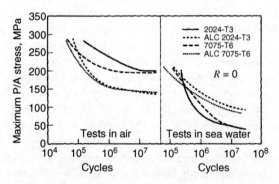

Figure 16.18. Cyclic stress–life curves for clad (ALC) and bare sheets of 2024-T3 and 7075-T6 aluminum alloys tested in air and in sea water. (From p. 808, Ref. 23.)

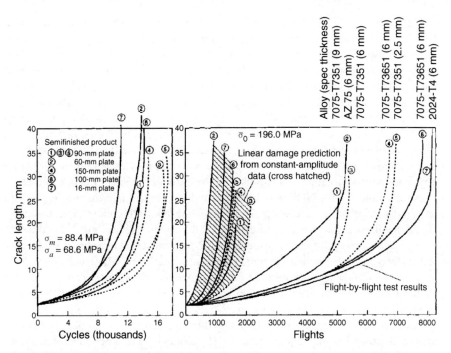

Figure 16.19. Comparison of fatigue crack growth curves for various aluminum alloys under constant- and variable-amplitude loading. (From p. 809, Ref. 23.)

on the results. Fatigue crack growth curves are compared here for several aluminum alloys and specimen thicknesses. Note that material 3 gave one of the longest fatigue crack growth lives for *constant-amplitude* loading whereas material 7 had the fastest crack growth. Material 7 had the longest *variable-amplitude* life, however, whereas material 3 gave one of the shortest lives for the flight-by-flight loading. This *reverse in ranking* associated with different fatigue loadings indicates that variable-amplitude crack growth depends on many competing factors (e.g., cyclic hardening, crack retardation, alloy strength–toughness combinations, microstructure interaction, environmental influences). Clearly, the designer must be very cautious about general conclusions from materials ranking fatigue tests and should base decisions on experiments that simulate the expected service conditions as closely as possible. (These results further emphasize the importance of accurately estimating the original design loads as discussed in Section 16.4.)

16.7   DESIGN FOR INSPECTION

The inspection task may be greatly simplified by considering NDE during the structural design stage. One may, for example, minimize the need for inspections by selecting damage tolerant materials and design concepts that lead to large critical crack sizes and/or crack arrestment mechanisms (e.g., the pressure vessel "flapping" shown in Figure 16.10). Or, one may employ the slow-crack-growth approach of Figure 16.17 so that inspections are not required during service (only an initial postmanufacturing inspection is needed).

Assuming, however, that repeated inspections are necessary, perhaps the most important design task is to facilitate access for NDE. Fracture critical components should not, for example, be buried deep inside the structure where significant disassembly is needed and/or the NDE interrogation signal (e.g., ultrasonic pulse) must pass through multiple layers of structure. If stack-ups are necessary, however, they should be planned so that key members are readily accessible (i.e., on top), avoiding multiple-layer inspections. Use of borescope ports can also be helpful for providing access to internal structure and was used in the design of much of the B-1 aircraft.

An example of how NDE was successfully incorporated into the original design and construction of an underwater pipeline is reported in Reference 25. In this case, the oil pipeline ran for several hundred miles at the bottom of the North Sea and would have been extremely difficult to inspect had not internal access been provided by special surface-based "launchers" mounted on floating platforms. These launchers allow inspection equipment to be inserted into the pipeline at various locations directly from the ocean surface. The extra construction cost incurred by this NDE access is expected to be recovered by inspection savings during the life of the pipeline.

Another NDE design consideration would be to minimize the number of components and assemblies that require inspection by eliminating fastener holes or avoiding complex fastener systems that are difficult to inspect. The use of taper-lok fasteners on the B-1 bomber, for example, turned out to significantly complicate subsequent NDE on this aircraft [20]. These tapered fasteners are drawn tightly into a tapered hole to provide beneficial residual stresses that improve the fatigue life of the joint. Although it was originally planned that the taper-loks would not require inspection during the life of the aircraft, more severe aircraft usage than anticipated required that they be removed for inspection. This process proved to be very difficult and time consuming and caused concern that removing the interference fit fasteners could lead to additional damage to the fastener holes. If inspections cannot be accomplished without fastener removal, it may be better to employ straight shank fasteners and to introduce the desired residual stresses by another technique (e.g., hole coldworking as described in Section 17.6.1).

One should also be cognizant of the limitations of various NDE methods when selecting materials (e.g., is the material magnetic, an electrical conductor). In summary, the key issue here is to involve the NDE team at early stages of the design and to consider design choices that will aid the inspection task following initial manufacturing and subsequent service.

16.8 ACCEPTANCE OF NEW TECHNOLOGY FOR STRUCTURAL DESIGN

Successful incorporation of new technology into structural applications is a critical issue both for the designer who must decide when it is safe to accept new concepts

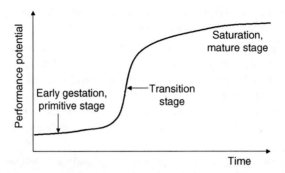

Figure 16.20. Technology maturation S-curve, showing the growth in performance potential of a new technology over time.

and to the researcher who needs to understand the criteria that will be used to evaluate long-term acceptance of his or her research. Technology transition is a difficult task, however, and having good ideas does not always guarantee product success. Indeed, as stated by R. N. Foster [26, page 24]: "Technology leaders tend to become technology losers. A few companies manage transitions to new technological fields effectively, but many others are unable even to begin the process, and most find it impossible to complete the move successfully." The United States, for example, has been criticized for missing opportunities that were later seized by Japan. Although not the first to introduce transistors, integrated circuits, computers, fiberoptics, lasers, video recorders, or color television, Japan became very successful in these technology-based industries, due in part to their strengths in manufacturing, design, and advancements in design aspects of the development cycle [27].

As shown in Figure 16.20, the life-cycle of many technologies can be described by an S-curve with three distinct phases [28]. The *primitive stage (revolutionary technologies)* occurs early in life when technology matures slowly but provides opportunities for major changes. Next is the *transition stage* (competitive technologies) characterized by a period of rapid growth and greatly improved understanding of the technology. Various alternative approaches begin to appear at this time as companies attempt to capitalize on the new technology. The final *mature stage (base technology)* is characterized by much slower growth as a few large firms may dominate the technology and it becomes "basic" to business.

There is a limited window of opportunity during the design process when new technology can be incorporated in a new product. Introducing technology too soon can result in schedule delays, cost overruns, inferior product, and "surprises" in service. Acting too late can mean missed opportunities and profits that may be captured by competitors. Lincoln [29] has identified five factors that are essential to transition new structures and materials technology from the laboratory to full-scale aircraft development:

- *Stabilized Material and/or Material Properties*   Here the material composition and processing must be fixed (no more laboratory "tweaking"). There must be specific processing specifications, material qualification and acceptance standards, and manufacturing instructions.

- *Producibility*   Since scale-up can be a lengthy process, the ability to produce the final desired product form in adequate quantity for full-scale production must have been demonstrated. Forming parameters and heat treatments must be specified, along with appropriate inspection procedures for the entire manufacturing process. All manufacturing methods must be compatible with the existing shop environment and all applicable U.S. Environmental Protection Agency (EPA) standards.

- *Characterized Mechanical Properties*   Mechanical properties must be well characterized by the preliminary design review and be established to sufficient accuracy to predict structural weight by the time of full-scale development. This characterization includes measurement of yield, ultimate, and bearing strengths; elastic modulus; elongation; fracture toughness; fatigue crack growth; stress corrosion cracking; and dimensional stability (e.g., creep, coefficient of thermal expansion).

- *Prediction of Structural Performance*   The ability to understand and accurately predict all failure modes associated with the use of a new structural technology is essential. Unanticipated failures that occur during full-scale tests, or even worse during service, can have disastrous consequences on the success of a new product. Although many conventional materials have "graceful failure modes" that are well understood, new and exotic materials that are designed to resist these common failures often have different limitations that are not easily analyzed or even recognized (i.e., "unknown unknowns"). Implementation of composite materials, for example, has been delayed by the research needed to understand their unique failure modes (e.g., interlaminar stresses, impact damage, fiber breakage, matrix failure, moisture absorption).

- *Supportability*   Finally, the use of new technology must be supportable in the field. Personnel must be trained to make appropriate inspections and repairs (both during manufacturing and in service) and must have access to well-equipped maintenance centers with all of the necessary tools and procedures for working with the new structure.

Although Lincoln's five criteria provide a specific approach to evaluate the technical maturity of structures and material technology for successful aerospace design, similar metrics are needed for research on new NDE methods. The following list is offered in the spirit of Lincoln's criteria as a means for determining when an *NDE method* is suitable for practical applications:

- *Address Real Inspection Need*   The proposed NDE method must detect realistic damage in actual structures and materials. It must exceed present capability with a significant, demonstrated improvement in detection capability and/or reliability compared to existing methods or provide some complementary advantages. Since replacing existing equipment could be an expensive undertaking, often involving retraining and qualification of personnel, there should also be significant cost–benefit advantages to existing methods.

- *Demonstrated for Field-Level Inspections*   The new NDE method must be applicable under service conditions by field personnel and provide timely and affordable results.

- *All NDE Procedures Established/Documented*   This criterion requires that all training methods and qualification standards be developed and published in an approved form suitable for field use by inspection personnel.

- *Reliability and Accuracy of Method Established*   Probability-of-detection curves should be established with realistic types of structural components and anomalies. Blind (or preferably double-blind) tests should have been conducted with recognized calibration standards, and the robustness of the method and procedures should be well demonstrated and documented.

- *Confidence in Method Established*   The new NDE method should be accepted by NDE technicians, local management, the end user of inspected components, and the NDE industry at large. There should be no unknown unknowns resulting from its use, and the approved operating procedures should protect against potential abuse of the method (i.e., it should be "idiot proof").

- *Safety Concerns*   All potential hazards regarding use of the NDE method and equipment need to be recognized and understood and appropriate safety procedures, including training and oversight, be developed and established.

16.9 CONCLUDING REMARKS

The goal of this chapter has been to discuss various design issues that impact structural integrity. The key requirement is for the designer to diligently anticipate *all* potential failure modes, including the possibility for preexistent manufacturing or service-induced damage. Only after carefully analyzing all possible failure scenarios can the designer provide the means to prevent failure from occurring during the life of the structure. As indicated, when considering a new structure, one has a variety of geometric, materials selection, and inspection tools that can work together to prevent catastrophic fracture. Maintaining structural integrity is, however, a lifelong process that extends beyond initial design and manufacturing to include in-service inspections, maintenance, and repair. Issues associated with the integrity of old structures are discussed in the next chapter.

PROBLEMS

16.1 Briefly discuss the importance to structural design of having an accurate understanding of the forces that must be resisted in service. What design decisions depend on the load history?

16.2 Briefly discuss the load history that you would anticipate for the following types of structures:

(a) Helicopter rotor blade

(b) Floor beams in a large office building

(c) Transport aircraft fuselage

(d) Transcontinental gas pipeline

16.3 Assume a wide panel that contains a through-thickness center crack of length $2a$ is made from a material whose fatigue crack growth properties are given by the following power law:

$$\frac{da}{dN} = 5 \times 10^{-10}\, \Delta K^4$$

Here the fatigue crack growth rate da/dN is measured in inches per cycle and ΔK is given in ksi-in$^{1/2}$. Assume that the panel is subjected to a remote $R = 0$ cyclic stress $\Delta\sigma = 10$ ksi. In the context of Figure 16.13, prepare a plot of the influence of initial detectable crack size a_0 on the required inspection period. Assume in this case that fracture occurs when the crack growth rate $da/dN = 0.01$ in/cycle.

16.4 Assume that two wide panels have the same thickness and are loaded in parallel. Both panels are subjected to an $R = 0$ cyclic stress of 10 ksi and have the same fatigue crack growth properties as problem 16.3. Consider one panel to be the primary member and the other to be the secondary member in the context of the multi-load-path, easily detectable *secondary*-crack scenario discussed in Figure 16.14. Assume that the primary panel has an initial crack length $2a = 2$ in and the secondary panel contains an initial crack $2a = 0.2$ in. If the secondary-member crack can be readily detected when $2a = 1$ in, determine the safe inspection period needed to prevent fracture of both members. Assume that fracture of a given panel occurs when the fatigue crack growth rate $da/dN = 0.01$ in/cycle and that when the primary member fails, its load is carried by the secondary member.

16.5 Repeat problem 16.4 in the context of the multi-load-path, easily detectable *primary*-member failure scenario shown in Figure 16.15. In this case, the safe inspection period begins when the *primary* member fractures (i.e., when $da/dN = 0.01$ in/cycle).

16.6 Repeat problem 16.4 in the context of the multi-load-path, easily detectable *subcritical* primary-member crack scenario shown in Figure 16.16. In this case, assume that the safe inspection period begins when the *primary*-member crack length $2a = 3.0$ in.

16.7 Assume that two 4-in-wide by 1-in-thick bars are made from 2024-T3 aluminum. Each of the long bar has a centrally located 1-in-diameter open hole. Assume that the holes are inspected with an NDE method capable of finding all quarter-circular corner cracks that are larger than 0.05 in. The bars are loaded in parallel in a manner such that the total load is divided evenly with each bar carrying a constant-amplitude remote tensile stress that cycles between 5 and 15 ksi. Assume one bar is the primary and the other is the secondary member in a multi-load-path design. Determine the safe inspection period needed to prevent fracture of both members in the context of the multi-load-path, easily detectable secondary-crack scenario discussed in Figure 16.14. In this case assume that the corner crack in the hole of the secondary member is easily detectable when it penetrates the back face of the bar to become a through-thickness crack. Use the AFGROW computer program described in the Appendix to answer this question. (Use the 2024-T3 material properties given in the AFGROW material data base and assume that all load is transferred to the secondary member after primary failure.)

16.8   Repeat problem 16.7 in the context of the multi-load-path, easily detectable primary-member failure scenario shown in Figure 16.17. In this case the safe inspection period begins when the primary member fractures.

16.9   Repeat problem 16.7 in the context of the multi-load-path, easily detectable subcritical primary-crack scenario shown in Figure 16.16. In this case, assume that the safe inspection period begins when the corner crack in the primary member penetrates the back face to become a through-thickness crack.

16.10   A wide tension panel is made from a thin sheet that is fastened to two symmetric stringers, as shown in Figure 16.8. The sheet material has a threshold stress intensity factor $\Delta K_{\text{th}} = 2$ ksi-in$^{1/2}$, a fracture toughness $K_c = 40$ ksi-in$^{1/2}$, and $R = 0$ fatigue crack growth properties that are given by $da/dN = 10^{-9} \times (\Delta K)^4$ (units of da/dN are inches per cycle and ΔK are ksi-in$^{1/2}$). The stringers are located 14 in apart and the panel contains a central crack of length $2a$. The stress intensity factor for this configuration has been determined to be as follows (here K is normalized with the remotely applied stress σ).

$$a \leq 2 \text{ in:} \qquad \frac{K}{\sigma} = (\pi a)^{1/2} \text{ in}^{1/2} \qquad \text{(note } K \text{ increases with } a)$$

$$2 < a \leq 3 \text{ in:} \qquad \frac{K}{\sigma} = 2.5 \text{ in}^{1/2} \qquad \text{(note } K \text{ constant with } a)$$

$$3 \leq a \leq 7 \text{ in:} \qquad \frac{K}{\sigma} = \frac{24}{\pi a} \text{ in}^{1/2} \qquad \text{(note } K \text{ decreases with } a)$$

$$a \leq 7 \text{ in:} \qquad \frac{K}{\sigma} \quad \text{increases}$$

(a) Sketch the dimensionless stress intensity factor solution K/σ as a function of crack length a.

(b) What is the biggest remote *static* stress σ that can be applied to the panel if the crack is not to grow beyond the stringer (i.e., $a = 7$ in)?

(c) What is the biggest $R = 0$ cyclic stress $\Delta\sigma$ that can be applied to the stiffened panel if the fatigue crack is not to grow beyond the stiffener (i.e., $a = 7$ in)?

(d) If this stiffened panel is subjected to a remote stress that cycles between 0 and 5 ksi, how many cycles will it take for the half-crack length a to grow from a size $a = 2$ in to $a = 7$ in?

16.11 Two 20-in-long bars are pinned at A, B, and C, as shown in Figure P16.11. A horizontal force of 100,000 lb is applied at B. Bars AB and BC are 20 in long and 2 in wide and are both are assumed to contain a 0.1-in-long edge crack along one side.

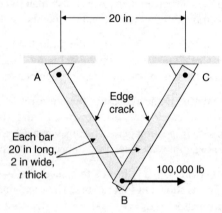

Figure P16.11

(a) Determine the weight of the *lightest* pair of bars that will resist complete failure of the two-bar system. Assume that each bar is initially straight and flat and is made from one of the steel, aluminum, or titanium materials given in Figure 6.8. (The bars do not need to be made from the same material or have the same thickness t.) For purposes of this problem, assume that the density of all steel, aluminum, and titanium alloys are 0.28, 0.1, and 0.16 lb/in³, respectively, and that the elastic modulus E for all steel, aluminum, and titanium alloys are 29×10^6, 10.5×10^6, and 16×10^6 psi respectively. Also assume that fracture is controlled by the plane strain K_{Ic} fracture toughness values given in Figure 6.8 (i.e., ignore potential thickness effects on fracture toughness in this case).

(b) Repeat your calculation if a cheaper (i.e., less capable) NDE method is used to inspect the bars so that the initial edge crack length is 0.2 in.

(c) Repeat your calculation if a "perfect" inspection is performed and both bars are free of initial cracks.

16.12 Significant research with composite materials has been underway since the 1960s, and although composites show considerable promise for aircraft structural applications, they still have not achieved widespread use in aircraft. Briefly discuss the benefits of composite materials and why you believe that adoption of composite technology for aircraft has progressed relatively slow. (*Hint:* Consider criteria for transitioning new technology for structural applications.)

16.13 One of the current trends in industry is to create integrated design teams where the designer works closely with manufacturers, suppliers, maintenance personnel, end users, and so on.

(a) Briefly discuss how interactions between the designer and maintenance supervisor could influence an initial structural design. Give some specific examples of design changes that could result from this collaboration.

(b) Briefly discuss how interactions between the designer and NDE supervisor could influence an initial structural design. Give some specific examples of design changes that could result from this collaboration.

(c) Briefly discuss how interactions between the designer and materials producer could influence an initial structural design. Give some examples of design changes that could result from this collaboration.

AFGROW Problems

The following problems employ the AFGROW software described in the Appendix.

16.14 Assume that a 4-in-wide by 0.5-in-thick panel has a 0.5-in-diameter hole that contains a small *quarter-circular* corner crack located at the intersection of the hole bore with one of the free surfaces. If the panel is made from 2024-T3-clad aluminum [L-T (long-transverse) direction] and is subjected to 15 ksi remote cyclic stress ($R = 0$), prepare a plot in the context of Figure 16.13 that shows the influence of the initial corner crack size on the required inspection period. Let the initial corner crack radius vary between 0.01 and 0.5 in. Use the NASGRO database option of AFGROW to select the material properties.

16.15 Assume that two 3.0-in-wide by 0.3-in-thick panels are made from 7050-T7651 plate (T-L orientation) aluminum (use the NASGRO database option in AFGROW to obtain material properties). Both panels are loaded in parallel with 20 ksi remote cyclic stress ($R = 0.3$), and both are assumed to contain an initial *semicircular* surface crack. Consider one panel to be the primary member and the other to be the secondary member in the context of the multi-load-path, easily detectable *secondary*-crack scenario discussed in Figure 16.14. Assume that the semicircular surface crack in the primary member has an initial 0.1-in radius whereas that in the secondary member has an initial 0.02-in radius. If the secondary crack can be readily detected when its total surface length is 0.5 in, determine the safe inspection period needed to prevent fracture of both members.

16.16 Repeat problem 16.15 in the context of the multi-load-path, easily detectable *primary* failure scenario shown in Figure 16.15. In this case, the safe inspection period begins when the *primary* member fractures.

16.17 Repeat problem 16.15 in the context of the multi-load-path, easily detectable *subcritical* primary-member crack scenario shown in Figure 16.16. In this case, assume that the safe inspection period begins when the *primary*-member crack first penetrates completely through the panel thickness (i.e., when the surface crack depth and plate thickness are 0.3 in).

16.18 Assume that a 2-in-wide by 1-in-thick bar contains a 0.5-in-diameter hole. The bar is made from titanium alloy Ti-6–4 alpha–beta forging and is subjected to a remotely applied constant amplitude $R = 0.25$ cyclic stress of 25 ksi. Assume the open hole contains an initial quarter-circular corner crack (radius of 0.05 in) located where the hole bore intersects with one of the free surfaces. Determine the safe inspection period in the context of Figure 16.17 if the corner crack can be in-service detected when its surface length is 0.1 in. What would be the life of the structure if it was based on the initial 0.05-in manufacturing crack? Use the Harter T-model database option in AFGROW to obtain the material properties.

REFERENCES

1. T. Swift, "Damage Tolerance Certification of Commercial Aircraft," *ASM Handbook,* Vol. 19: *Fatigue and Fracture,* ASM International, Materials Park, Ohio, 1996, pp. 566–576.

2. M. P. Kaplan and J. W. Lincoln, "The U.S. Air Force Approach to Aircraft Damage Tolerant Design" in *ASM Handbook,* Vol. 19: *Fatigue and Fracture,* ASM International, Materials Park, Ohio, 1996, pp. 577–588.

3. C. F. Tiffany and J. N. Masters, "Applied Fracture Mechanics," *Fracture Toughness Testing and Its Applications,* ASTM Special Technical Publication 381, American Society for Testing and Materials, West Conshohocken, Pennsylvania,1965, pp. 249–277.

4. J. L. Rudd, "Air Force Damage Tolerance Design Philosophy," in *Damage Tolerance of Metallic Structures: Analysis Methods and Applications,* ASTM STP 842, American Society for Testing and Materials, West Conshohocken, Pennsylvania, 1984, pp. 134–141.

5. *Airplane Damage Tolerance Requirements,* MIL-A-83444, Air Force Aeronautical Systems Division, Wright-Patterson AFB, Ohio, July 1974.

6. Anonymous, "Part 25—Airworthiness Standards: Transport Category Airplanes," Federal Airworthiness Requirements, *Federal Register,* March 31, 1998.

7. *Standard NDE Guidelines and Requirements for Fracture Control Programs,* MSFC-STD-1249, George C. Marshall Space Flight Center, NASA, Washington, D.C., September 11, 1985.

8. T. L. Lomax, *Structural Loads Analysis for Commercial Transport Aircraft: Theory and Practice,* AIAA Education Series, American Institute of Aeronautics and Astronautics, Washington, D.C., 1995.

9. J. M. Potter and R. T. Watanabe (Eds.), *Development of Fatigue Loading Spectra,* ASTM Special Technical Publication 1006, American Society for Testing and Materials, West Conshohocken, Pennsylvania, 1989.

10. F. J. Giessler, S. J. Duell, and R. F. Cook, *Handbook of Guidelines for the Development of Design Usage and Environmental Sequences for USAF Aircraft,* AFWAL TR-80-3156, Flight Dynamics Laboratory, Wright-Patterson AFB, Ohio, February 1981.

11. D. J. Peery and J. J. Azar, *Aircraft Structures,* McGraw-Hill, New York, 1982.

12. G. E. Lambert and D. F. Bryan, *The Influence of Fleet Variability on Crack Growth Tracking Procedures for Transport/Bomber Aircraft,* Technical Report AFFFDL-TR-78-158, Flight Dynamics Laboratory, Wright-Patterson AFB, Ohio, November 1978.

13. C. C. Poe, Jr., "The Effect of Riveted and Uniformly Spaced Stringers on the Stress Intensity Factor of a Cracked Sheet," in *Proceedings of the Air Force Conference on Fatigue and Fracture of Aircraft Structures and Materials,* AFFDL-TR-70-144, Air Force Flight Dynamics Laboratory, Wright-Patterson AFB, Ohio, 1970, pp. 207–215.

14. T. Swift, "Development of the Fail-Safe Design Features of the DC-10," in *Damage Tolerance in Aircraft Structures,* ASTM STP 486, American Society for Testing and Materials, West Conshohocken, Pennsylvania, 1970, pp. 164–214.

15. T. Swift, "Fracture Analysis of Stiffened Structure," in *Damage Tolerance of Metallic Structures: Analysis Methods and Application,* ASTM STP 482, American Society for Testing and Materials, West Conshohocken, Pennsylvania, 1984, pp. 69–107.

16. T. Swift, "Widespread Fatigue Damage Monitoring Issues and Concerns," paper presented at The Fifth International Conference on Structural Airworthiness of New and Aging Aircraft, Hamburg, Germany, June 16–18, 1993.

17. T. Swift, "Damage Tolerance Capability," in *Fatigue of Aircraft Materials,* Proceedings of the Specialists Conference Dedicated to the 65th Birthday of J. Schijve, Delft University Press, Delft, 1992, pp. 351–387.

18. T. Swift, "Unarrested Fast Fracture," in *Durability of Metal Aircraft Structures,* Proceedings of the International Workshop on Structural Integrity of Aging Airplanes, March 31–April 2, 1992, Atlanta, Georgia, Atlanta Technology Publications, Atlanta, Georgia, 1992, pp. 419–442.

19. D. Broek, *Elementary Engineering Fracture Mechanics,* 3rd rev. ed., Martinus Nijhoff, The Hague, The Netherlands, 1986.

20. J. W. Lincoln, USAF, Wright-Patterson AFB, Ohio, personal communication, January 2001.

21. M. F. Ashby, *Materials Selection in Mechanical Design,* 2nd ed., Butterworth-Heinemann, Oxford, U.K., 1999.

22. S. Lampman (Tech. Ed.), *ASM Handbook,* Vol. 19: *Fatigue and Fracture*, ASM International, Materials Park, Ohio, 1996.

23. R. J. Bucci, G. Nordmark, and E. A. Starke, Jr., "Selecting Aluminum Alloys to Resist Failure by Fracture Mechanisms," in *ASM Handbook,* Vol. 19: *Fatigue and Fracture,* ASM International, Materials Park, Ohio, 1996, pp. 771–812.

24. R. J. Bucci, C. J. Warren, and E. A. Starke, Jr., "Need for New Materials in Aging Aircraft Structures," *Journal of Aircraft,* Vol. 37, No. 1, pp. 122–129, January/February 2000.

25. T. Coughlin, "The Holy Grails of NDE," *NTIAC Newsletter,* Nondestructive Testing Information Analysis Center, Vol. 24, No. 4, page 1, December 2001.

26. R. N. Foster, "A Call For Vision in Managing Technology," *Business Week,* May 24, 1982, pp. 24–33.

27. R. E. Gomory and R. W. Schmitt, "Science and Product," *Science*, Vol. 240, pp. 1131–1132, 1203–1204, May 27, 1988.

28. W. D. Compton, *Engineering Management: Creating and Managing World-Class Operations*, Prentice-Hall International, London, 1997.

29. J. W. Lincoln, "Structural Technology Transition to New Aircraft," *Proceedings of the Fourth Symposium of the International Committee on Aeronautical Fatigue,* Ottawa, Ontario, Canada, 1987, Engineering Materials Advisory Services, West Midlands, U.K., 1987.

CHAPTER 17

STRUCTURAL INTEGRITY OF AGING SYSTEMS

17.1 OVERVIEW

This chapter discusses special safety issues associated with continued operation of older structures. Maintaining the integrity of aging systems is a particularly important challenge in an era of fiscal constraint and/or when long lead times are required to obtain replacement structures. The USAF, for example, has projected continued operation of large numbers of its KC-135 tanker aircraft and T-38 jet trainers beyond the year 2020 [1]. Since the KC-135 fleet was produced between 1954 and 1965 and the T-38, which first flew in 1959, ceased production in 1972, these aircraft may see well over 50 years of active service, significantly beyond their original design lives. Similar examples can be cited for other military and commercial aircraft, space craft (i.e., the space shuttle), as well as bridges, nuclear reactors, and so on. The fact that these structures have reached "old age" with potential for further productive use is testimony to the soundness of their original design and construction and continuous maintenance upkeep on the structure. Indeed, premature retirement could be an unnecessary waste of these valuable assets, *provided they can continue to be employed in a manner that protects public safety.*

Determining the "effective age" of such structures can be problematic, however, as "calendar time" does not provide a complete indication of structural health. A component's actual life depends on the severity of fatigue, corrosion, and other time dependent damage that develops in service. Moreover, the formation of fatigue cracks and/or corrosion depends upon the nature of prior usage as well as on the effectiveness of structural maintenance, repairs, and modifications. In this context, Lincoln [2] has defined an aging structure as one that requires operational restrictions or changes in maintenance procedures due to use beyond its original design life, corrosion, widespread fatigue damage, repairs, or maintenance. Some specific structural concerns that develop with age are described in the following section. Subsequent sections then discuss various procedures for extending the lives of older structures.

17.2 SUMMARY OF POTENTIAL THREATS TO AGING SYSTEMS

While there are many economic and public policy matters involved with the problem of aging structures, key technical issues include the potential for service-induced damage caused by fatigue, corrosion, repairs, and/or accidental trauma. In particular, one must determine how the structure's original damage tolerance could be compromised by the symptoms of age and then employ an inspection program to detect those symptoms well before they lead to premature "death." Although beyond the scope of this text, it should be noted that many other factors, such as degraded electrical wiring and electronic devices, increased operational and maintenance costs, or general obsolescence, may also limit the final structural life.

17.3 WIDESPREAD FATIGUE DAMAGE

Widespread fatigue damage (WFD) is the simultaneous occurrence of fatigue cracking at many different locations and is characteristic of structures that have seen extensive cyclic loads. Widespread fatigue damage can occur as multiple-site damage (MSD), where cracks develop at similar details in a single component (e.g., a row of fastener holes), or as multiple-element damage (MED), where adjacent components experience simultaneous cracking. Figure 17.1, for example, shows a large lead crack that has formed along a row of fastener holes on one side of a stringer. Although one objective of the stringer is to arrest cracks before they completely sever the panel (recall Section 16.5.1), it is possible for small MSD on the opposite side to link up with the lead crack and extend across the stringer. Thus, WFD can compromise the discrete-source residual strength provided by stiffeners and other damage tolerance structural features.

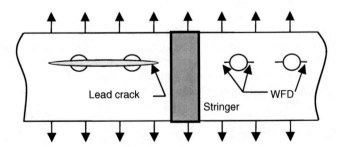

Figure 17.1. Schematic representation of a large lead crack and small WFD cracks in a stiffened panel.

The classic example of the WFD problem is the 1988 Aloha Airlines B-737 incident shown previously in Figure 1.6. Recall that small fatigue cracks developed along a row of fastener holes, coalesced, and jumped across adjacent stiffeners, failing a large section of fuselage skin [3]. This incident clearly demonstrated the deleterious effects of WFD and led to much concern about this problem. Swift [4], for example, has referred to WFD as a "dreaded disease in aging airframes" and reports that Professor James W. Mar has compared it to AIDS in humans, where "the immune system (in this case fail safe or damage tolerance capability) starts to deteriorate."

Piascik and Willard [5] benchmarked the WFD that could be expected in an aging aircraft fuselage by a detailed examination of lap joint panels taken from the full-scale fatigue test of a commercial transport fuselage that had seen 60,000 pressurization cycles. Their destructive examination of 419 fastener holes found a total of 340 cracks at 239 of the holes (i.e., 57% were cracked). Most cracks were found in the countersink region and were attributed to high local stresses and to fretting between various components of the joint. The fatigue crack growth rate for small crack sizes [0.002–0.4 in (0.05–10 mm)] was determined by fractographic analysis and was predicted reasonably well by LEFM analyses of the lap joint. Luzar and Hug [6] also conducted a destructive analysis of lower wing skin sections from two retired B707 aircraft that had seen 78,416 and 57,382 flight hours. They found cracks in 2631 fastener holes joining the lower skin and stringers, with approximately two-thirds of the cracks located in the stringers and one-third in the skin. Although the cracks were of different shapes and locations, Huzar and Hug converted them to the same equivalent corner crack hole geometry (by stress intensity factor matching) and provided statistical crack size distributions for several wing locations (recall EIFS distributions discussed in Section 9.5). The mean crack sizes for these various distributions varied from 0.005 to 0.015 in (0.13–0.38 mm).

Thus, it is quite possible for aging structures to have large numbers of small fatigue cracks, and apprehension about WFD led to a March 31, 1998 revision of FAA airworthiness regulations directed toward continued operation of older aircraft. FAR 25.571 [7], for example, now states:

> This action amends the fatigue requirements for damage-tolerant structure on transport category airplanes to require a demonstration using sufficient full-scale fatigue test evidence that widespread multiple-site damage will not occur within the design service goal of the airplane; and inspection thresholds for certain types of structure based on crack growth from likely initial defects. This change is needed to ensure the continued airworthiness of structures designed to the current damage tolerance requirements, and to ensure that should serious fatigue damage occur within the design service goal of the airplane, the remaining structure can withstand loads that are likely to occur, without failure, until the damage is detected and repaired.

17.3.1 Residual Strength

The particular concerns about WFD deal with determining the size and number of small cracks that compromise lead crack residual strength and the time when those

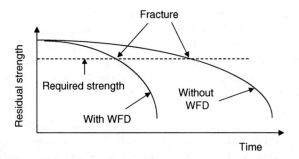

Figure 17.2. Reduction in lead crack residual strength associated with WFD.

cracks occur [4]. The potential reduction in residual strength is shown schematically in Figure 17.2, where the lead crack residual strength is compared with and without the occurrence of WFD. As discussed previously in Section 1.3.7, the strength of structural components gradually degrades with time as damage develops through fatigue and/or corrosion processes. The lead crack (i.e., discrete-source damage) residual strength can, moreover, be significantly reduced below the pristine structure value if WFD is present. Heinimann and Grandt [8], for example, in Figure 17.3

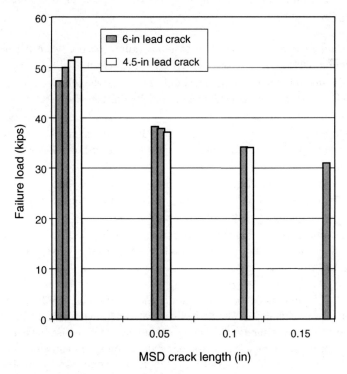

Figure 17.3. Reduction in lead crack residual strength caused by WFD in 15-in-wide stiffened 2024-T3 aluminum panels [8].

summarize tests with 10 stiffened 2024-T3 aluminum panels that contained large lead cracks [6 or 4.5 in (15 or 11 cm) long] along with much smaller MSD cracks [0.05 in ≤ average MSD length ≤ 0.15 in (1.3 mm ≤ average MSD length ≤ 3.8 mm]. Both the lead and MSD cracks were located along a row of open holes. Note that the *large-crack* residual strength is significantly reduced by *much smaller* WFD on the opposite side of the stiffener. Thus, small cracks can still be quite damaging, even when large cracks are already present.

Prediction of the reduction in residual strength caused by WFD requires an appropriate fracture criterion and the ability to analyze the interaction between adjacent cracks. Analyzing the link-up and fracture of small MSD cracks can be problematic in the case of an aircraft fuselage due to the large-scale yielding that can develop in the ductile materials and thin skin thickness [e.g., 0.040–0.10 in (1.0 to 2.5 mm)] associated with that structure. Although rigorous fracture analysis of MSD can be quite difficult, it may be possible to make "engineering estimates" of this failure mode. Cherry et al. [9], for example, summarize residual strength tests and predictions with unstiffened 2024-T3 aluminum panels that contained 2.4- or 3.2-in- (6.1- or 8.1-cm-) long lead cracks and small MSD cracks [0.005–0.015 in (0.13 to 0.38 mm)] located along a row of open holes. The panels were tested in two configurations: 9.0 in (23 cm) panel width with 0.09 in (2.3 mm) specimen thickness and 15 in (38 cm) width with 0.04 in (1 mm) thickness. Several simple MSD failure criteria were examined, including the following:

- *The net ligament loss criterion* assumes failure occurs when the net section ligament yields (recall Fedderson criterion in Section 6.7.1).
- *The K-apparent criterion* assumes failure occurs when the maximum stress intensity factor equals a constant value (*K* apparent) determined for the particular sheet thickness in question.
- *The ligament yield criterion* proposed by Swift [4] assumes failure occurs when the crack tip plastic zones of the lead and adjacent MSD cracks touch (see Figure 17.4).
- *The average displacement criterion* proposed by Jeong and Brewer [10] states that failure is based on superposition of two models for the displacements occurring on the ligament between adjacent cracks.
- *The average stress criterion,* also proposed by Jeong and Brewer [10], assumes failure occurs when the average stress between two crack tips equals the material ultimate strength.

Figure 17.5*a* compares the actual failure loads with those predicted by the failure criteria described above. The Swift ligament yield model gave the most accurate results for this study, although the net ligament loss criterion also gave good predictions (except for showing a systematic dependence on specimen width, as would be expected by the Fedderson criterion). Results from a similar study involving 15-in- (38-cm-) wide *stiffened* panels with and without MSD are given in Figure 17.5*b* [8]. Here comparison of predicted and measured link-up loads (i.e., the force when MSD

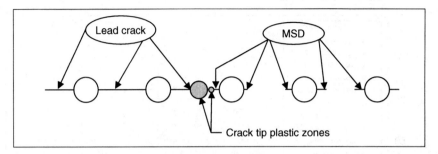

Figure 17.4. Schematic representation of Swift ligament yield model that assumes failure occurs when the plastic zones of lead and MSD cracks touch [4].

cracks begin to coalesce) shows that, again, the ligament yield model predicted crack coalescence quite well, whereas the K-apparent approach consistently overestimated the force to cause the first coalescence of cracks. The K-apparent method did a better job, however, of predicting the second MSD link-up. Although the ligament yield model performed well in these cases, it should be emphasized that this simple criterion may not be suitable for all WFD applications. References 11 and 12 describe other research to develop more sophisticated fracture criteria (such as the crack tip opening angle approach) for the MSD problem.

17.3.2   Fatigue Crack Growth

In addition to predicting the fracture stress of components with WFD, it is essential to determine when such cracking can develop in the structure [4], since continued operation with WFD is extremely dangerous. Although the onset of WFD in a particular structure depends on many factors, including usage, quality of construction and material, and environment, and cannot be predicted precisely, probabilistic methods can be used to establish bounds on when WFD compromises residual strength. As described below, one such approach is to determine the statistical variability in fatigue life at the onset of WFD by employing Monte Carlo simulations that entail repeated applications of a deterministic calculation with statistical variations of input parameters.

The multi-degree-of freedom fatigue crack growth model described previously in Section 7.10.2, for example, was successfully applied to the MSD problem in References 13–17. Recall that the predictive scheme is based on a deterministic computation for the growth, coalescence, and final fracture resulting from various combinations of small cracks and large lead cracks that occur along a row of holes in a flat sheet. The multi-degree-of-freedom model allows individual cracks to grow independently, until interacting and linking up with adjacent cracks or until growing into an uncracked hole (i.e., a "crack stopper"). The algorithm also employed a notch fatigue analysis (recall Section 9.2.4) to predict crack nucleation at pristine holes or to continue growing cracks that were momentarily stopped at the holes.

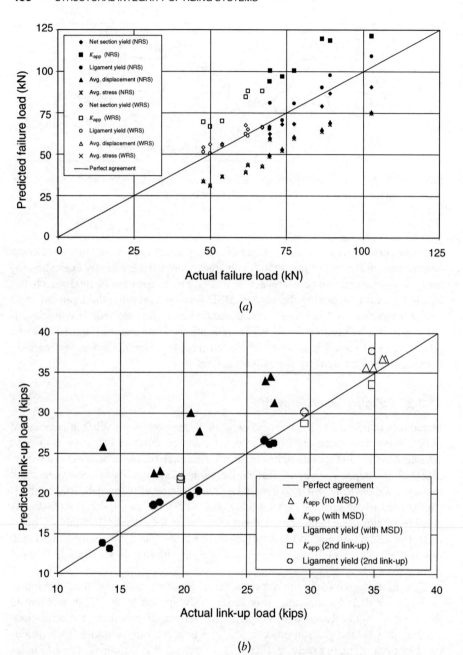

Figure 17.5. (*a*) Comparison of measured and predicted failure loads for 9-in-wide (designated by NRS) and 15-in-wide (designated by WRS) unstiffened 2024-T3 aluminum panels containing 2.4- or 3.2-in-long lead cracks along a row of 0.125-in-diameter open holes with 0.005–0.015-in MSD cracks [9]. (*b*) Comparison of actual and predicted link-up loads by two different link-up criteria for 15-in- (38-cm-) wide, stiffened, 2024-T3 aluminum panels containing lead and in most cases MSD cracks [8].

Numerical predictions for lead and WFD fatigue crack growth [13–15] were compared previously with test results for both stiffened and unstiffened specimens in Figure 7.28. Since the WFD model did an excellent job of predicting fatigue life for those open-hole tests, the approach was applied to lap joint specimens, where loads are transferred across rivets installed in precracked holes [16, 17]. Constant-amplitude fatigue tests were conducted with single lap specimens made from 0.09-in- (2.29-mm-) thick 2024-T3 bare aluminum. The average precrack lengths were typically on the order of 0.013–0.044 in (0.34–1.12 mm), and they emanated radially from both sides of the countersunk rivet holes. Grandt and Wang [16] summarize in Figure 17.6 the measured and predicted fatigue crack growth lives for five lap joint tests, indicating that it is possible to obtain reasonable life predictions provided that one has appropriate stress intensity factor solutions for the lap joint problem. (Reference 18 gives more details of the numerical and experimental program.)

Although the basic WFD analysis was deterministic in nature, fatigue usually demonstrates significant "scatter" due to many sources (e.g., manufacturing quality, material variability, applied loading). In order to estimate the variation that could be encountered in the reduction of residual strength with time, Monte Carlo calculations were used to determine the scatter in life to fracture resulting from uncertainty in input parameters such as initial crack size, material properties, or fastener fit. In this procedure, various inputs to the deterministic life analysis are randomly chosen from statistical distributions of the given parameters, and fatigue life is calculated. Repeated calculations with other randomly selected inputs allow one to determine the statistical variation in life associated with uncertainty of a given parameter.

Grandt and Wang [16] employed Monte Carlo simulations to characterize the variation in lead crack residual strength associated with MSD growth in a lap joint. Those

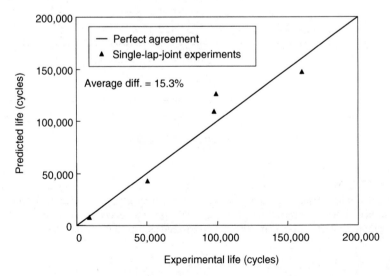

Figure 17.6. Summary of predicted and measured fatigue crack growth lives for lap joint fatigue tests [16].

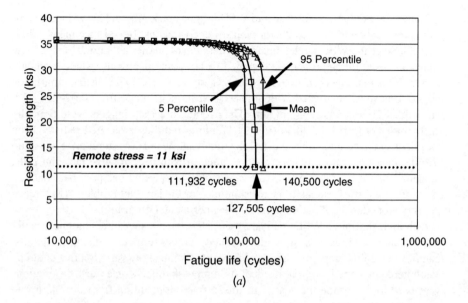

(a)

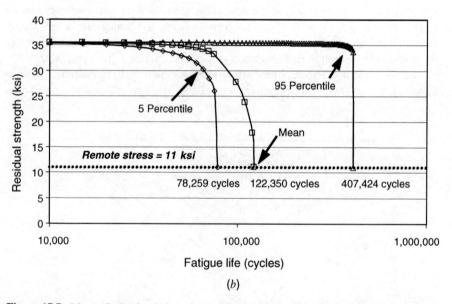

(b)

Figure 17.7. Monte Carlo simulations for the distribution in residual strength of a riveted lap joint. (*a*) The hole expansion level is fixed at 2.06% and randomly selected initial WFD cracks are located along the top rivet row. (*b*) The hole expansion level and the initial WFD cracks are randomly selected from experimentally measured statistical distributions [16].

calculations considered a 31-in- (79-cm-) wide joint with three rows of 30 rivets that was subjected to a constant-amplitude ($R = 0.05$) peak cyclic stress of 11 ksi (76 MPa). The statistical input variables include initial flaw size distributions and variability in interference fit associated with rivet installation. One row of fasteners was assumed to contain WFD cracks randomly selected from the EIFS distribution obtained by Luzar and Hug for the retired B-707 aircraft [6]. The variation in fastener interference (and its subsequent effect on rivet stress fields) was based on measurements from specimens that contained large numbers of rivets installed with a conventional rivet gun [18]. Monte Carlo simulations (1000 runs) were performed to determine the influence of both variation in rivet installation (i.e., fastener interference) and initial WFD on fatigue life.

The variability in residual strength degradation as a function of elapsed cycles when the WFD cracks grow is shown in Figure 17.7. Here fracture was assumed when the residual strength drops to the 11 ksi (76 MPa) peak cyclic stress. The mean, 95th and 5th percentile variations in residual strength are shown. Figure 17.7*a* is for a case when the rivet interference was assumed to be constant and the initial WFD at the top row of rivet holes was randomly selected from the EIFS distribution. Figure 17.7*b* presents a case where both the initial WFD cracks and the fastener interference level varied randomly. Although these are hypothetical calculations intended to demonstrate the effect on residual strength caused by variations in the growth of WFD, they do indicate how Monte Carlo simulations can be used to determine an estimate of the probability of fracture. (Several other cases are discussed in References 16 and 18.) A key point here is that the rapid drop in residual strength due to WFD is sensitive to many variables. Thus, once WFD is present, it would be most difficult, if not impossible, to rely on an inspection program to permit continued operation in the presence of WFD.

Although the above example uses equivalent initial flaw size distributions obtained from tear-down inspections, the EIFS data could also be based on models that estimate WFD caused by material inhomogeneities [19], precorroded material state [20], fretting, or the results of nondestructive inspection capability (i.e., probability-of-detection curves). Similar Monte Carlo calculations reported in Reference 21 also incorporated statistical variations in fatigue crack growth properties and EIFS. One of the main goals for these types of probabilistic studies is to determine the key parameters that induce the most variability in life and, thus, help designers and fleet managers focus on the most critical issues needed to prevent the formation of WFD.

17.4 CORROSION

Corrosion is another key type of service-induced damage that limits the life of older structures. As discussed in Section 9.3, corrosion occurs in many forms and can develop whether the structure is in use or not. The extent of corrosion depends on local environmental conditions and maintenance and can vary widely within a given fleet. Moreover, many older structures were constructed from alloys that, in retrospect, had poor corrosion resistance and often employed coatings and other maintenance materials that have been subsequently banned by stricter environmental protection regulations. Thus, as corrosion prevention systems deteriorate with age and use, they often need to be replaced with newer and more expensive chemical processes, adding to the growth in maintenance costs. Corrosion maintenance has, for example, been reported to be the single most expensive maintenance item for the USAF [1], totaling $1 billion for 2001, having grown from $720 million in 1990 and $795 million in 1997.

Corrosion poses several problems for aging structures, not the least of which are detection and classification of damage. While many forms of corrosion (e.g., pitting or exfoliation) are readily apparent on surfaces that can be examined visibly, intergranular corrosion or stress corrosion cracking that occurs away from exposed surfaces is more difficult to detect. Moreover, corrosion in hidden areas (e.g., between lap joints, beneath installed fasteners, or in multilayer stack-ups) is much harder to locate without costly disassembly and/or sophisticated NDE methods.

Even when found, however, corrosion classification remains problematic, often employing general terms such as "mild," "moderate," or "severe." Some USAF studies [2], for example, have characterized mild corrosion as that which penetrated < 0.001 in (25 μm) below the surface, moderate between 0.001 and 0.01 in (25–250 μm) penetration, and severe as that which extends over 0.01 in (250 μm) in depth. Reference 1 defines three increasing degrees of corrosion damage to aircraft structure. *Level I corrosion* is local or widespread impairment occurring between inspections that can be reworked or blended out according to limits specified by the manufacturer. Level I corrosion can also result from a unique event or be based on past history that shows it to be a form of light corrosion although cumulative blending might exceed allowable limits. *Level II corrosion* requires a single rework that exceeds permissible limits or wide spread corrosion that requires a single blend-out approaching allowable limits. *Level III corrosion* is the most severe form of damage found during inspection and is deemed to be an urgent airworthiness concern demanding immediate action.

This terminology is somewhat arbitrary, however, since the actual corrosion impact on the remaining structure can seldom be so easily quantified. Clearly, general thickness loss associated with corrosion increases local stresses and would hasten fatigue or other failure mechanisms. Pitting corrosion also causes stress concentrations that serve as sites for fatigue crack nucleation, whereas other local damage could be associated with increased stresses caused by bulging or pillowing in corroded lap joints or, perhaps, with faster fatigue crack growth rates in corroded areas [22].

An approach to characterize the effect of corrosion on fatigue life in terms of two geometric changes (thickness loss and surface roughness due to pitting) is proposed by Scheuring and Grandt [23]. Here reduction in thickness would lead to an increase in applied stress whereas surface pitting would be represented by an equivalent initial crack size. These two parameters could then be used with LEFM calculations to determine the remaining fatigue life of the corroded structure (assuming no further corrosion occurs). Examples of nondestructive eddy current measurements of these two geometric changes due to corrosion were shown previously in Figure 13.17. (See Problems 17.17–17.20 at the end of this chapter for additional discussion of this proposed two-parameter approach for characterizing corrosion damage.)

17.5   OTHER THREATS TO OLDER STRUCTURES

Although fatigue and corrosion, along with fretting, are perhaps the major causes for materials degradation in older structures, other threats to structural integrity are also possible. Creep can occur from elevated temperatures or embrittlement may result from radiation exposure (a particular issue for nuclear reactors). One important concern for many older components deals with the fact that actual stress histories may be significantly different from that assumed for the original design analyses. Missions may have changed, for example, as it is quite common for service loading to be much more severe than that assumed for design purposes. There may also have been accidents that caused overstressed areas or there may be unreported structural modifications and repairs that alter stress distributions and, thus, the accumulated fatigue or stress corrosion damage. Indeed, failure to track field-level repairs and modifications and anomalous loading could lead to a significant underestimation of the actual fatigue damage.

Another issue for older structures is the potential for basic mechanical properties to degrade over time with exposure to corrosion and service loading [1]. This concern does not seem serious for aging aircraft, however, based on studies that have compared mechanical properties obtained from specimens taken from retired aircraft with pristine materials. Reference 24, for example, showed that tensile stress–strain curves, fatigue–life data (S–N curves), and fatigue crack growth (da/dN–ΔK) properties were largely unchanged in several "aged" aluminum aircraft alloys (2024-T3, 7075-T6, and 7178-T6) when the effect of thickness loss is accounted for. Although S–N curves for corroded specimens did, of course, fall below smooth polished specimens, that difference was attributed to stress concentrations associated with local pitting and largely disappeared when the corroded specimens were also polished to a smooth finish.

A final threat to continued operation of older structures is gradual loss of corporate memory regarding their original design, construction, operation, and maintenance history. Indeed, as aging systems outlive the technical careers of those responsible for overseeing structural integrity, it is easy for critical records and operational data to become misplaced or lost. New employees may not possess the skills or training to perform critical inspection and maintenance tasks, and new management may not even appreciate the necessity for those actions. Thus, rigorous recordkeeping and continued oversight is needed to ensure strict adherence to the inspection and maintenance plans required to maintain structural integrity.

17.6   LIFE EXTENSION CONCEPTS

There may come a point when one must consider a life improvement program to either overcome deficiencies that arise during planned service or extend operation beyond the original life goals. The first step is to perform a new damage tolerance analysis that takes into account prior incidents of fatigue or corrosion, the actual repair and maintenance history, and the desired future usage (i.e., projected load histories). If

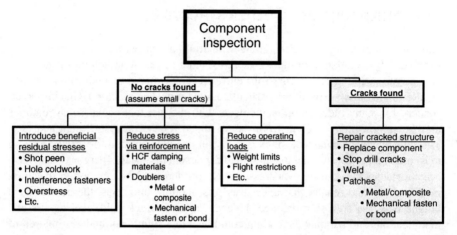

Figure 17.8. Summary of potential actions to extend the fatigue life of an aged structure [25].

the structure were originally damage tolerant, this new study would confirm whether or not sufficient protection was provided to fracture critical areas and if existing maintenance plans are adequate. If the structure was not designed to damage tolerant principles, this analysis is essential to identify potential areas of future vulnerability. The new analyses should employ the latest calculation methods and software, reflect the most recent assessment of structural condition, account for potential modifications or repairs, and consider projected usage. Since the updated analysis will, in most cases, be conducted with more sophisticated software and incorporate more accurate input data than available for the original design calculations, it is possible that some overly conservative assumptions and large safety factors can be relaxed, and there may, in fact, be more life remaining than originally calculated.

Figures 17.8–17.10 overview a number of actions that can be taken to renovate older structures that could have developed fatigue cracking, corrosion, or other forms of degradation during prior service [25].† Note that all of the processes begin with a rigorous NDE to locate any existing service-induced trauma. It is, in fact, essential to determine the actual structural condition before serious thought can be given to extending the service life beyond the initial design goals.

Considering Figure 17.8, for example, a fatigue inspection leads to two possible results: Cracks are found or not found. If fatigue cracks are detected, the structure must be repaired at that time. This refurbishment might involve replacing the entire component with a new member or repairing the cracked area by welding, patching, or stop drilling. (Recall that stop drilling involves drilling a hole through the crack tip to eliminate the singular crack tip stress field.) Stop drilling is not to be confused with another effective method to repair small cracks at fastener holes, which is to remove the crack by redrilling the hole to oversize and reinserting a larger diameter fastener.

---

†Gunderson attributed the original organization of the concepts shown in Figures 17.8–17.10 to C. F. Tiffany.

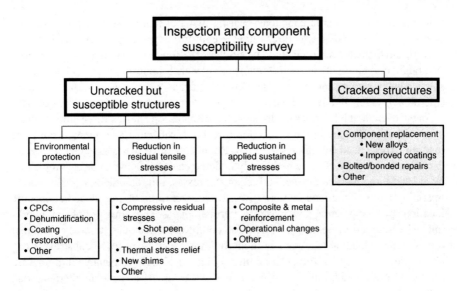

Figure 17.9. Summary of potential actions to extend the stress corrosion cracking life of an aged structure [25].

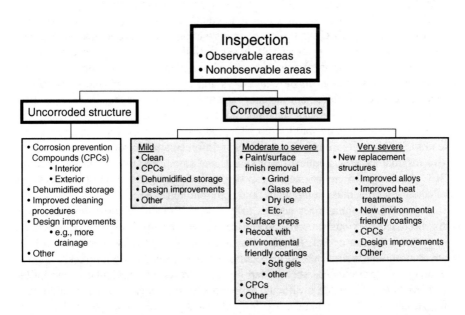

Figure 17.10. Summary of potential actions to extend the corrosion life of an aged structure [25].

Patching the cracked area adds load-bearing material that is mechanically fastened or bonded to the parent structure. Either metal or composite patches could be employed, and the patch material could be similar or different from the existing structure. Although attaching patches with bolts is simpler than bonding, bolts add additional stress concentrations and weight [1]. Bonded patches have been the subject of considerable research and offer an effective method to repair damaged structure [26–28].

In the event that NDE does not locate fatigue cracks, one must still assume the presence of cracks that are at the inspection threshold, particularly in view of the prior loading experienced by older structures. Reduction of local stresses can, however, decrease the severity of these potentially undetected cracks and can be accomplished by at least three methods: beneficial residual stresses, reinforcement, or reduction in operating loads. Beneficial compressive stresses can often be introduced at specific locations by shot peening, low plasticity burnishing, laser peening, or hole coldworking. As discussed in Section 17.6.1, these compressive residual stresses counteract remote tensile loading and can significantly improve fatigue life. Local stresses can also be decreased by reinforcement with doublers or, if high-frequency vibrations are an issue, through addition of damping materials. Doublers provide an alternate load path to shed load in critical areas and, again, may be fabricated from either composite or metal materials and be mechanically fastened or bonded to the parent structure. Finally, the fatigue life may be extended by a general reduction in structural loading through restrictions on future usage imposed by weight limits or other means. One must be cautious, however, when modifying future missions to ensure that sequencing effects do not result in unanticipated consequences (i.e., elimination of potential crack retardation associated with overloads, as discussed in Section 7.4).

Figure 17.9 summarizes life extension actions following NDE focused on stress corrosion cracking (SCC). In this case, the inspection should be conducted along with a general assessment of prior and future susceptibility to SCC. In the event NDE locates cracks, the component must be replaced or repaired as before. Since significant improvements have been made with alloys and coatings that are more resistant to SCC, component replacement in older structures should consider employing these newer materials. If no cracks are found but the structure is still deemed susceptible to stress corrosion, one should, if possible, reduce the sustained tensile stresses that lead to SCC by operational changes, local reinforcements, introduction of compressive residual stresses, application of shims, and so on. The other contributor to the SCC problem—the environment—can be dealt with by restoration of coatings, use of corrosion prevention compounds, and dehumidified storage.

Figure 17.10 summarizes actions to be taken following a corrosion inspection. As discussed previously, corrosion can occur over long periods of time whether a structure is in use or not. In areas that are readily observable, corrosion will be apparent though normal visual inspection. Hidden areas that do not receive proper drainage or tight crevices such as lap joints, however, are particularly susceptible to corrosion and are more difficult to inspect. If such areas can be disassembled for visual inspection, general corrosion will usually be readily apparent. The more difficult challenge is to locate corrosion in areas that cannot be easily disassembled. Here ultrasonics, thermal imaging, and/or other more specialized inspection methods are required. The pillowing phenomenon, characterized by surface bulging caused by expansion of corrosion products trapped in lap joints, can be a significant aid to identifying corrosion-prone areas and may often be located by enhanced visual methods such as D-sight.

Even if the inspection does not reveal corrosion, one must continue to focus on preventative maintenance. Corrosion protective compounds should continue to be applied on a regular basis along with periodic cleaning. Drainage should be checked and, if possible, improved. Long-term storage should employ desiccants and other procedures to avoid corrosive environments. (Reducing relative humidity to 30–40%, for example, significantly decreases corrosion in stored aircraft [1].) Indeed, the best long-term solution to corrosion is through prevention rather than by repair.

Dealing with corroded structure depends on the extent of damage that has occurred. Severe corrosion requires component replacement, and, again, if possible, newer corrosion resistant alloys and coatings should be employed. One should also attack the root cause to avoid recurring problems with the new component (e.g., provide adequate drainage, ensure that seals are in place, follow preventative maintenance). Grinding out the affected areas, replacing surface finishes, and continuing use of CPCs (corrosion prevention compounds) may treat moderate to severe corrosion. When grinding out corroded material, however, one must be careful not to increase stresses that could lead to fatigue problems, and for this reason, surface removal is usually limited to 10% of the component thickness. General cleaning and continued adherence to the corrosion prevention maintenance plan may be all that is necessary to treat mild corrosion.

17.6.1 Beneficial Residual Stresses

One of the most effective ways to extend the service life of components is to introduce compressive residual stresses in areas subjected to large tensile loads. The compressive stresses counteract applied tension and can significantly reduce formation and growth of fatigue or stress corrosion cracks. As discussed below, these beneficial stresses can be established by several procedures, including shot peening, low plasticity burnishing, laser shock peening, or hole coldworking, and can extend the lives of both pristine and damaged structures.

The benefits of surface peening have long been known, beginning with blacksmiths who recognized that repeated striking on the surface of metal components with a rounded hammer could increase service life. The more formalized process of shot peening is shown schematically in Figure 17.11. Here a component is bombarded with small spherical pieces of metal, glass, or ceramic shot that cause overlapping surface indentations. The plastic deformation resulting from the shot impact leaves a thin layer of residual stress that is compressive at the surface but is then balanced by subsurface tensile stresses. The compressive stresses are particularly beneficial at the surface, which is prone to abuse and is the usual site for fatigue cracking. The magnitude of the compressive region depends on the material and extent of peening but generally exceeds half of the material yield strength. Although the beneficial compression can penetrate as much as 0.035 in (1 mm) below the surface [29] the depth

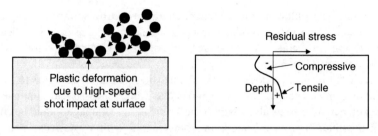

Figure 17.11. Introduction of compressive residual stresses to extend component life by shot peening component surface.

of penetration is typically much smaller for many applications. Since the subsurface residual tensile field is distributed over a larger area and has a smaller magnitude than the compressive stresses, its potential negative influence is usually more than compensated by the compressive benefits at the surface, resulting in an overall improvement in component life.

Low plasticity burnishing and laser shock peening are similar to shot peening except that pressure is applied to the component surface by pushing with a mandrel or by repeated pulses from a high-power laser. Although laser shock peening is more expensive and time consuming than conventional shot peening, deeper compressive stress regions (and, thus, longer lives) can be achieved, making the process especially cost effective for expensive components such as turbine blades and disks. In all of these cases, the depth of penetration results from incompressible plasticity that depends on component thickness and/or overall restraint. Subsequent fatigue analyses must consider both the magnitude of the residual stress at the free surface and its gradient into the depth of the component (including the influence on stress ratio).

Hole expansion is another practical procedure for introducing controlled residual stresses around fastener holes. As shown schematically in Figure 17.12, hole cold-working is typically accomplished by pulling an oversized mandrel through an open hole that is protected by a lubricated sleeve (split or seamless). Hole expansion (on the order of 2–5%) by the mandrel causes plastic deformation and results in residual

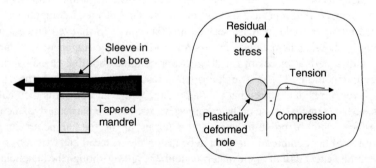

Figure 17.12. Introduction of compressive hoop stresses at the bore of a hole by mandrel hole coldworking leaves.

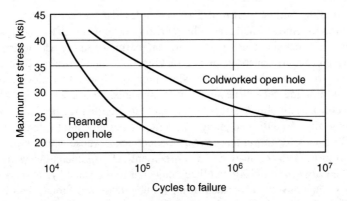

Figure 17.13. Stress–life (S–N) curves for 2024-T851 aluminum specimens containing cold-worked and nonworked holes. (From [30], courtesy of Fatigue Technology, Inc.)

hoop stresses when the mandrel is removed. The hoop stresses are compressive in the area near the edge of the hole and are balanced by a region of tensile stresses remote from the hole. As before, the beneficial effects of the large compressive hoop stresses usually far outweigh the tensile residual stresses, which are smaller in magnitude and located away from the large stress concentration at the hole bore.

Hole expansion can significantly extend the fatigue lives of both pristine and damaged holes and can be used for new construction or to remedy service cracking problems. Figure 17.13 compares S–N curves for coldworked and nonworked open holes in 2024-T851 aluminum specimens and shows the significant life improvement obtained by coldworking [30]. A dramatic example of the process's benefits for aged structure is given in Figure 17.14 [31]. Originally motivated by fatigue problems on the C-5 aircraft, this experimental study was conducted to determine if coldworking

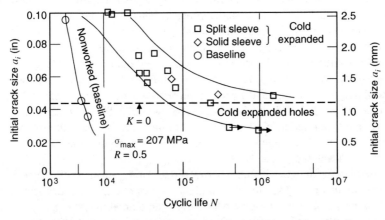

Figure 17.14. Increase in the fatigue life of precracked 0.25-in- (6.4-mm-) diameter holes in 0.25-in- (6.4-mm-) thick 7075-T6 aluminum sheet caused by mandrel hole coldworking [31, 32].

could extend the fatigue lives of *precracked* holes. The concern was that a planned repair action to remove small cracks by redrilling fastener holes to an oversize could miss some undetected cracks, so an additional margin of safety was desired.

The test program consisted of precracking 0.25-in- (6.4-mm-) diameter holes in 0.25-in- (6.4-mm-) thick 7075-T6 aluminum specimens to various initial crack lengths. A baseline set of specimens was subjected to a peak 30 ksi (207 MPa) $R = 0.5$ constant-amplitude stress history and fatigue life measured as a function of initial crack size. As indicated in Figure 17.14, the baseline fatigue lives were all less than 7500 cycles for the initially cracked specimens, whose starting crack lengths ranged from 0.03 to 0.1 in (0.8–2.5 mm). Identical specimens that were mandrel coldworked *after precracking* and then subjected to the same cyclic load, however, gave much longer lives. Two-orders-of-magnitude life improvement was obtained for the smaller initial crack sizes, with run-outs (i.e., no failure) of 10^6 cycles achieved for the smallest cracks [0.03 in (0.8 mm) length].

Stress intensity factor solutions developed for the coldworking process in Reference 32 indicate that life improvement is due to a significant reduction in K for cracks emanating from the coldworked holes. (The linear superposition procedure discussed in Chapter 8 and demonstrated in Example 8.4 was employed for that analysis.) The run-out crack sizes shown in Figure 17.14 were subsequently analyzed [32], and it was found that the residual compressive stresses kept the crack closed during remote tensile loading for small crack sizes, so that effectively $\Delta K = 0$, preventing further crack growth. This crack size is designated by the dotted line in Figure 17.14 and provides a LEFM explanation for the run-out specimens that agrees reasonably well with the measured run-out crack sizes.

17.6.2 Materials Replacement Options and Issues

A key repair issue for older structures is whether or not to substitute newer alloys for components that have developed fatigue and corrosion during service. In many cases, replacement with the original material may be the cheapest option, particularly in view of the fact that small-quantity procurements might preclude the expensive testing and recertification required to qualify new alloys for the given application [1]. As discussed by Bucci et al. [33], however, there are good reasons to consider material substitution for older systems. Many aircraft built before 1970, for example, employed materials selected on the basis of strength considerations but which often exhibited poor fracture and corrosion properties. These earlier materials include aluminum alloys 2024-T3, 7075-T6, 7178-T6, and 7079-T6, and indeed, their shortcomings led to service cracking problems that motivated the new damage tolerance analysis concepts and emphasis on nondestructive inspection methods that are the subject of this volume.

These material deficiencies also spurred research on many new alloys and processing methods that now give superior combinations of strength, fracture, fatigue, and corrosion properties (recall Figure 6.9). New aluminum alloys, for example, include 2524, 7050, 7055, 7155 and the overaged corrosion resistant T7X tempers for the 7000 series alloys. It is also now possible to produce thick plates with less damaging porosity than in the past and to forge components with improved grain flow directions and reduced residual stresses that minimize the possibility for SCC [33]. While many of these new materials and processes have been successfully incorporated into newer designs, their use as replacements in older structures has been much slower to occur. Procedures to streamline acceptance of these newer technologies for retrofits of aging systems could make them cost-efficient solutions for those applications as well [1].

17.7 INSPECTION ISSUES FOR OLDER STRUCTURES

Clearly, NDE plays a critical role when considering use of structures beyond their original design lives. As mentioned previously, the actual extent of fatigue, corrosion, or other forms of service-induced damage resulting from prior usage must be determined and evaluated before the structure can be safely returned to service and/or modified for additional use. Indeed, one important conclusion from this inspection might be that there is such widespread damage that there is no remaining life that can be safely exploited and retirement is the only correct decision.

There are both positive and negative factors that influence the inspection of aging systems. On the plus side, potential "hot spots" may have been identified through prior experience, so that many critical locations are known and in some cases have resulted in development of specialized NDE solutions. Moreover, there may be a cadre of qualified inspectors who are familiar with the intricacies of the particular structure and inspection methods in question. The longer a system is in service, however, the more difficult it is to maintain this corporate knowledge as these key individuals retire or move on to other positions.

Despite these potential advantages, older systems pose many difficult challenges for the inspector. Years of accumulated dirt, corrosion, or multiple layers of paint and other coatings may hinder access to critical areas or complicate disassembly of components for inspection. Undocumented repairs and ad hoc modifications are also possible, so that the current structure often does not conform to the original documentation of inspection procedures. It is also conceivable that results from prior inspections have been misplaced or lost or that existing NDE equipment is obsolete or inoperative, requiring the inspection team to "start from ground zero." Moreover, since many inspections of older structures will be conducted in the field, there may not be access to all of the sophisticated NDE methods that were available in the original manufacturing facility.

There have been several studies regarding NDE requirements for aging structures. Reference 1 (page 64), for example, offers the following general recommendation. "Emphasis should be placed on NDE technique design and development aimed at improved detection reliability and defect characterization, cost-effective validation and qualification procedures, transferability to a range of applications, and interdisciplinary coordination with other elements of the aging aircraft strategy."

Some specific NDE research topics recommended [1] for aging aircraft include the means to make rapid inspections for fatigue cracks under installed fasteners in order to prevent the onset of WFD [i.e., quickly find crack lengths on the order of 0.04–0.1 in (1–2.5 mm)] at multiple locations. A particular need is to detect and characterize cracks in multilayer assemblies. Techniques to discover and quantify hidden corrosion without component disassembly as well as the detection of stress corrosion cracking in thick sections are also needed for older structures.

17.8   CONCLUDING REMARKS

As discussed in Chapter 16, designers of new structures have the advantage of employing new technology and the freedom to develop creative approaches to guarantee structural integrity. When considering continued operation of aging systems, however, engineers often have less flexibility and are faced with several special safety issues associated with component age. Chief among these are the need to ensure that damage tolerance does not succumb to widespread fatigue damage, corrosion, changes in usage, or structural modifications and repairs that have occurred during years of prior service. The problem is further complicated by potential changes in ownership or loss of corporate memory that could result in failure to properly implement the inspection and maintenance actions required to guarantee structural integrity or even to understand and appreciate the need for attention to those details.

Through continuous vigilance to all aspects of structural integrity, it may be possible to prolong safe operation of some structures indefinitely. There will eventually come a time, however, when the costs associated with the rigorous NDE, maintenance, and repair requirements become prohibitive and the structure must be retired on an economic basis. At this point the structural integrity engineer may take satisfaction in knowing that the maximum use was obtained from that particular asset.

PROBLEMS

17.1   Define an aging structure in the context of structural integrity.

17.2   Briefly describe the special safety issues associated with prolonged operation of aging structures. Give examples of both technical concerns (e.g., failure modes) and administrative challenges (e.g., management, economic, human factors) associated with maintaining the integrity of aging systems.

17.3   Conduct an Internet search for articles about aging structures and prepare a brief case history on one particular application. What was the original design life of the structure in question? What special safety issues arise in continued operation of this structure? What actions are being proposed or taken to extend the service life?

17.4   Discuss the role NDE should play in combating the widespread fatigue damage problem. In the context of the inspection interval scenarios discussed in Section 16.5.2, what should the inspection period be in a structure that contains widespread fatigue damage?

17.5 Discuss how widespread fatigue that develops late in life could defeat crack arrestment features that were originally designed into the structure.

17.6 Discuss the issue of materials replacement in the repair and maintenance of aging structures.

17.7 Discuss the various ways in which the cyclic stresses in an existing structure can be modified in order to extend fatigue life.

17.8 Discuss various techniques that can be used to extend the fatigue life of an older structure.

17.9 Discuss the various ways in which the stress corrosion life of an existing structure can be extended.

17.10 Since corrosion often takes many years to develop, discuss whether it would be more cost effective to invest regular financial resources in preventative maintenance when the structure is new and there is little evidence of corrosion or to save those funds to spend on inspection and repair after corrosion actually begins to occur. Consider the fact that funds from delayed expenditures could be invested in interest-bearing accounts.

17.11 The purpose of this problem is to compare the decision to replace an automobile or electronic device with that of replacing or maintaining an aging structure.

(a) Discuss the issues one would consider when deciding whether to purchase a new automobile or to continue to operate an older one.

(b) Discuss the issues one would consider when replacing a personal computer or an electronic entertainment system.

(c) Compare the considerations in parts a and b with those associated with deciding whether or not to continue operation of a fleet of aging aircraft or to keep an old highway bridge in service.

17.12 Which of the nondestructive inspection methods discussed in Part III would be most effective for detecting widespread fatigue damage? Explain your answer.

17.13 Which of the nondestructive inspection methods discussed in Part III would be most effective for detecting hidden corrosion damage? Explain your answer.

17.14 Discuss how shot peening the surface of a component to improve its fatigue life would influence subsequent inspections?

17.15 Discuss the criteria you would use to evaluate the suitability of a new nondestructive inspection method that has been proposed for a particular aging structure application.

17.16   Discuss how a new detailed damage tolerance evaluation conducted on an aging structure could be considered as a life extension technique. How might the new analysis differ from the original analyses that were conducted when the structure was originally designed?

Problems 17.17–17.20 deal with a proposed technique [23] to quantify corrosion damage in terms of two geometric parameters: thickness loss and surface roughness. It is suggested that general thickness loss resulting from corrosion leads to an increase in stress (due to a reduction in load-bearing area), whereas surface roughness (e.g., pitting) could be expressed in terms of equivalent initial flaw size (EIFS, recall Section 9.5). These two parameters would then be used to prepare constant-life diagrams, as shown in Figure P17.17. Here thickness loss (represented by a given increase in applied stress) and surface roughness (characterized by an EIFS) are used as inputs to a fatigue crack growth analysis to compute fatigue life for a given corroded structural configuration and stress history. The premise is that combinations of thickness loss and EIFS that give the same fatigue life represent equivalent states of corrosion from the standpoint of resisting fatigue for a particular loading. Points that fall below the constant-life curve correspond to acceptable combinations of thickness loss and surface roughness (i.e., they will not lead to failure within the desired fatigue life), whereas points above the curve represent an unacceptable state of corrosion damage for the given application.

17.17   The goal of this problem is to determine a theoretical constant-life curve for different combinations of thickness loss and EIFS, as shown in Figure P17.17. To this end, assume the structure of interest is a *wide*, 0.10-in-thick 2024-T3 aluminum sheet (clad, L-T direction) subjected to a remotely applied constant amplitude $R = 0$ cyclic tensile stress of 25 ksi. Assume that corrosion results in a uniform decrease in thickness combined with surface roughness that can be described in terms of an initial *semicircular surface crack* (EIFS). If 10,000 *additional* cycles of constant-amplitude loading are needed for the corroded sheet, determine combinations of thickness loss and semicircular surface crack depth (EIFS) that give the same 10,000-cycle life. For purposes of analysis, assume that the applied stress is directly related to thickness loss. A 0.01-in decrease in thickness, for example, reduces the load-bearing corroded thickness to 0.09 in and increases the applied stress to 27.78 ksi. The goal then is to find the semicircular crack depth (EIFS) that yields a 10,000-cycle life for the 27.78 ksi cyclic stress. Find at least four combinations of thickness loss and EIFS depth and plot them in the constant-life diagram format shown. Use the AFGROW software described in the Appendix to solve this problem (employ the NASGO material option in AFGROW to determine the appropriate material properties).

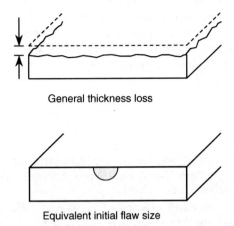

General thickness loss

Equivalent initial flaw size

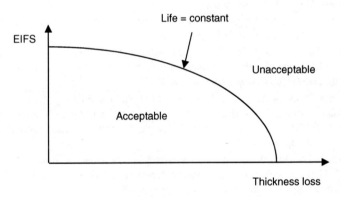

Figure P17.17

17.18 Repeat problem 17.17 assuming that the corroded component must resist 20,000 cycles of loading rather than the original 10,000 cycles specified.

17.19 Repeat problem 17.17 assuming that the original uncorroded component must carry 20 ksi constant-amplitude ($R = 0$) cyclic stress rather than the original 25 ksi cyclic stress. Assume that 10,000 load cycles are required for the corroded member as originally specified in problem 17.17. Compare the results from problems 17.17–17.19, and discuss how the two-parameter characterization of the "acceptable" corrosion state depends on component load and the desired remaining life. How might the result also depend on component geometry?

17.20 Briefly discuss the various NDE methods that might be used to implement the two-parameter characterization of corrosion discussed in problems 17.17–17.19. What inspection techniques might be employed to measure component thickness in order to determine general thickness loss by corrosion? What NDE methods might be used to characterize surface roughness in order to determine the EIFS measure of corrosion damage? How would one correlate the measured surface roughness with EIFS? What assumptions would be implicit in that correlation?

REFERENCES

1. Committee on Aging of U.S. Aircraft, National Materials Advisory Board, National Research Council, *Aging of U.S. Air Force Aircraft,* Final Report, Publication NMAB-488-2, National Academy Press, Washington, D. C., 1997.

2. J. W. Lincoln, "Aging Aircraft—USAF Experience and Actions," Fifteenth Plantema Memorial Lecture, in R. Cook and P. Poole (Eds.), *ICAF 97, Fatigue in New and Aging Aircraft,* Vol. I, Proceedings of the 19th Symposium of the International Committee on Aeronautical Fatigue, Edinburgh, June 18–20, 1997, Materials Advisory Services, West Midlands, U.K., 1997, pp. 3–38.

3. R. G. O'Lone, "Safety of Aging Aircraft Undergoes Reassessment," *Aviation Week & Space Technology,* May 16, 1988, pp. 16–18.

4. T. Swift, "Widespread Fatigue Damage Monitoring Issues and Concerns," paper presented at the Fifth International Conference on Structural Airworthiness of New and Aging Aircraft, Hamburg, Germany, June 16–18, 1993.

5. R. S. Piascik and S. A. Willard, "The Growth of Multi-Site Fatigue Damage in Fuselage Lap Joints," in C. E. Harris (Ed.), *The Second Joint NASA/FAA/DoD Conference on Aging Aircraft,* NASA/CP-199-208982/Part 2, Langley Research Center, Hampton, Virginia, January 1999, pp. 397–407.

6. J. Luzar and A. Hug, "Lower Wing Disassembly and Inspection Results of Two High Time USAF B707 Aircraft," paper presented at the 1996 USAF Structural Integrity Program Conference, San Antonio, Texas, December 3–5, 1996.

7. Anonymous, *Part 25—Airworthiness Standards: Transport Category Airplanes,* Federal Airworthiness Requirements, *Federal Register,* March 31, 1998.

8. M. B. Heinimann and A. F. Grandt, Jr., "Analysis of Stiffened Panels with Multiple Site Damage," in *Proceedings of 1996 USAF Structural Integrity Program Conference,* WL-TR-97-4055, Wright Laboratory, Wright-Patterson AFB, Ohio, June 1997, pp. 655–682.

9. M. C. Cherry, S. Mall, B. Heinimann, and A. F. Grandt, Jr., "Residual Strength of Unstiffened Aluminum Panels with Multiple Site Damage," *Engineering Fracture Mechanics,* Vol. 57, No. 6, pp. 701–713, 1997.

10. D. Y. Jeong and J. C. Brewer, "On the Linkup of Multiple Cracks," *Engineering Fracture Mechanics,* Vol. 51, No. 2, pp. 233–238, 1995.

11. D. S. Dawicke and J. C. Newman, Jr., "Evaluation of Various Fracture Parameters for Predictions of Residual Strength in Sheets with Multi-Site Damage," in *Proceedings of the First Joint DoD/FAA/NASA Conference on Aging Aircraft,* Vol. II, Ogden, Utah, July 1997, pp. 1037–1326.

12. J. E. Ingram, Y. S. Kwon, K. J. Duffie, and W. D. Irby, "Residual Strength Analysis of Skin Splices with Multiple Site Damage," in C. E. Harris (Ed.), *The Second Joint*

NASA/FAA/DoD Conference on Aging Aircraft, NASA/CP-199-208982/Part 2, January 1999, Langley Research Center, Hampton, Virginia, pp. 427–436.

13. E. J. Moukawsher, A. F. Grandt, Jr., and M. A. Neussl, "Fatigue Life of Panels with Multiple Site Damage," *Journal of Aircraft,* Vol. 33, No. 5, pp. 1003–1013, September/October, 1996.

14. D. G. Sexton, "A Comparison of the Fatigue Damage Resistance and Residual Strength of 2024-T3 and 2524-T3 Panels Containing Multiple Site Damage," M.S. Thesis, School of Aeronautics and Astronautics, Purdue University, West Lafayette, Indiana, August 1997.

15. M. B. Heinimann and A. F. Grandt, Jr., "Fatigue Analysis of Stiffened Panels with Multiple Site Damage," in *Proceedings of the First Joint DoD/FAA/NASA Conference on Aging Aircraft,* Vol. II, Ogden, Utah, July 8–10, 1997, pp. 1263–1305.

16. A. F. Grandt, Jr. and H. L. Wang, "Analysis of Widespread Fatigue Damage in Lap Joints," *SAE 1999 Transactions, Journal of Aerospace,* Sec. 1, Vol. 108, pp. 262–269, 1999.

17. H. L. Wang and A. F. Grandt, Jr., "Fatigue Analysis of Multiple Site Damage in Lap Joint Specimens," in *Fatigue and Fracture Mechanics,* Vol. 30, ASTM STP 1360, American Society for Testing and Materials, West Conshohocken, Pennsylvania, 2000, pp. 214–226.

18. H. L. Wang, "Evaluation of Multiple-Site Damage in Lap Joint Specimens," Ph.D. Thesis, Purdue University, West Lafayette, Indiana, 1998.

19. A. F. Grandt, Jr. and A. J. Hinkle, "Predicting the Influence of Initial Material Quality on Fatigue Life," in *Proceedings of 17th International Committee on Aeronautical Fatigue Symposium on Durability and Structural Reliability of Airframes,"* Volume I, ICAF 93, Stockholm, Sweden, June 9–11, Engineering Materials Advisory Services, Warley, U.K., 1993, pp. 305–319.

20. A. F. Grandt, Jr., T. N. Farris, and B. M. Hillberry, "Analysis of the Formation and Propagation of Widespread Fatigue Damage," in *ICAF 97, Fatigue in New and Aging Aircraft,* Vol. I, *Proceedings of the 19th Symposium of the International Committee on Aeronautical Fatigue,* Edinburgh, June 18–20, 1997, Engineering Materials Advisory Services, West Midlands, U.K., pp. 115–133.

21. S. M. Rohrbaugh, B. M. Hillberry, B. P. McCabe, and A. F. Grandt, Jr., "A Probabilistic Fatigue Analysis of Multiple Site Damage," in C. F. Harris (Ed.), *Proceedings of FAA/NASA International Symposium on Advanced Structural Integrity Methods for Airframe Durability and Damage Tolerance,* NASA Conference Publication 3274, Part 2, September 1994, pp. 635–652.

22. R. Kinzie, "USAF Cost of Corrosion Maintenance," presented at *the Sixth Joint FAA/DoD/ NASA Conference on Aging Aircraft,* September 16–19, 2002, San Francisco, California.

23. J. N. Scheuring and A. F. Grandt, Jr., "Quantification of Corrosion Damage Utilizing a Fracture Mechanics Based Methodology," in *Proceedings of The Third Joint FAA/DoD/ NASA Conference on Aging Aircraft,* Albuquerque, New Mexico, September 20–23, 1999.

24. J. N. Scheuring and A. F. Grandt, Jr., "Mechanical Properties of Aircraft Materials Subjected to Long Periods of Service Usage," *Journal of Engineering Materials and Technology, Transaction of the ASME,* Vol. 119, pp. 380–386, October 1997.

25. A. Gunderson, personal communication, circa 1999.

26. S. F. Adams, "Expansion of the WR-ALC Fatigue-Arrest Composite Repair Capability," in C. E. Harris (Ed.), *The Second Joint NASA/FAA/DoD Conference on Aging Aircraft,* NASA/CP-199-208982/Part 1, January 1999, Langley Research Center, Hampton, Virginia, pp 201–223.

27. K. Y. Blohowiak, K. A. Krienke, J. H. Osborne, J. J. Mazza, G. B. Gaskin, J. R. Arnold, W. S. DePiero, and J. Brescia, "Bonded Repair Techniques Using Sol-Gel Surface Preparations," in C. E. Harris (Ed.), *The Second Joint NASA/FAA/DoD Conference on Aging Aircraft*, NASA/CP-199-208982/Part 1, January 1999, Langley Research Center, Hampton, Virginia, pp. 313–320.

28. A. A. Baker, S. C. Galea, and I. G. Powlesland, "A Smart Patch Approach for Bonded Composite Repair/Reinforcement of Primary Airframe Structures," in C. E. Harris (Ed.), *The Second Joint NASA/FAA/DoD Conference on Aging Aircraft*, NASA/CP-199-208982/Part 1, January 1999, Langley Research Center, Hampton, Virginia, pp. 328–338.

29. J. Champaigne, *Shot Peening Overview*, Electronics, Mishawaka, Indiana, January 2001.

30. L. F. Reid, Fatigue Technology, personal communication, October 2002.

31. G. J. Petrak and R. P. Stewart, "Retardation of Cracks Emanating from Fastener Holes," *Engineering Fracture Mechanics*, Vol. 6, No. 2, pp. 275–282, 1976.

32. W. H. Cathey and A. F. Grandt, Jr., "Fracture Mechanics Consideration of Residual Stresses Introduced by Coldworking Fastener Holes," *Journal of Engineering Materials and Technology*, Vol. 102, pp. 85–91, January 1980.

33. R. J. Bucci, C. J. Warren, and E. A. Starke, Jr., "Need for New Materials in Aging Aircraft Structures," *Journal of Aircraft*, Vol. 37, No. 1, pp. 122–129, January/February 2000.

CHAPTER 18

CONCLUDING REMARKS

18.1 OVERVIEW

The goal of this volume is to provide the technical foundation to design and operate structures for extended periods of safe operation. As shown in Figure 18.1, structural integrity may be compared to a three-legged stool, with the "legs" being rigorous NDE, slow subcritical crack growth, and high residual strength. These three factors work together to ensure that structural damage will not lead to catastrophic failure during the operational life of the component.

Nondestructive evaluation is needed both at the onset of service to ensure that manufacturing errors do not go undetected and later in life to locate service-induced damage. Initially one seeks material flaws and construction anomalies to prevent production of defective components. Later in life one searches for service-induced damage (e.g., fatigue, corrosion, fretting, foreign object damage, creep) that could degrade safety. In-depth discussions of the major inspection procedures for locating these various forms of trauma are provided in Part III. Since no inspection is perfect, some cracks escape detection, and as discussed in Chapter 4, the "inspection limit" can be characterized in terms of probability of detection to determine a statistical upper bound for the largest damage that can exist following NDE.

Focus now turns on determining the growth and subsequent fracture associated with this undetected damage. It is known that cracks will grow by fatigue, creep, and/or environmental attack. Chapter 3 introduces LEFM procedures for predicting crack growth life, and these concepts are further developed in Part II to consider practical issues associated with complex structural configurations and operating environments. The fatigue crack growth life may be extended, for example, by selecting materials and geometric arrangements that are resistant to crack growth, by keeping

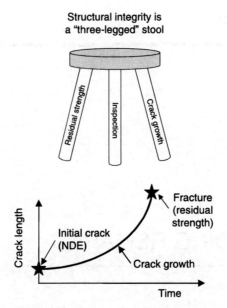

Figure 18.1. Structural integrity compared to a three-legged stool.

cyclic stresses small, by introducing compressive residual stresses through shot peening or coldworking, and by protecting against environmental influences. Fatigue crack growth life is especially sensitive to the initial crack size, placing additional burden on NDE to make sure that the initial undetected crack is as small as possible.

It must be emphasized that all damage will eventually grow over time depending on the service loads and environment. Although slow, stable crack extension occurs in most structures and is not unexpected, there will be a time when cracks grow to a size and/or occur in sufficient numbers that the member can no longer carry its intended load. At this point the structure loses its residual strength, resulting in catastrophic fracture. This condition must be anticipated in advance and the structure be retired or repaired before such failure can occur. Residual strength may be increased by selecting "tough" materials or by providing crack arrestment mechanisms such as redundant load paths or other impediments to rapid crack growth. Again, it is incumbent on NDE to detect damage before it reaches catastrophic size and to identify local failures that have led to redundant components carrying additional load.

18.2 SPECIALIZED APPLICATIONS

Much of the discussion in this text assumes the presence of dominant cracks that can be detected before growing to failure in a manner that can be predicted by LEFM techniques. In this case, one determines the largest crack size that can be missed by NDE and then establishes the period of allowable service before fracture could occur. The member is either retired at that time (with an appropriate safety factor) or subjected to additional inspections before returning to service. While this crack growth/inspection-based approach has found success for many types of high-performance structures, there are some special applications that may require alternate types of analyses. A few such cases are briefly summarized below.

Rotorcraft and turbine engines are two applications where large operating stresses and severe service environments may result in extremely small critical crack sizes that cannot be readily discovered by conventional NDE. Structural use of composite materials represents another class of problems where failure does not occur from the normal mode I crack growth that is the bane of most structural metals. In this case, alternate failure modes besides fatigue and fracture must be prevented during the life of the component, requiring specialized analyses and testing.

18.2.1 Rotorcraft Structures

Federal Aviation Administration regulations [1] allow fatigue analyses for rotorcraft structure to be conducted by (a) flaw tolerant safe-life evaluation or (b) fail-safe (residual strength after flaw growth) evaluation or, if the first two procedures "cannot be achieved within the limitations of geometry, inspectability, or good design practice," allow (c) safe-life evaluation that demonstrates "that the structure is able to withstand repeated loads of variable magnitude without detectable cracks" for specified time intervals. Whereas some in the damage tolerant community would prefer the rigorous crack growth analyses provided by the fail-safe approach (b), proponents of the flaw tolerant safe-life method (a) believe that the inspection requirements needed to implement the fail-safe approach could lead to unrealistically short inspection intervals and/or small-crack-size detection limits [2, 3].

These two fatigue analysis methods differ in the initial damage assumption. The fail-safe (residual strength after flaw growth) approach assumes preexistent *cracks* that will *grow* during service and which must be detected before causing failure. Although the "flaw tolerant" safe-life evaluation also assumes preexistent "flaws," they are *not* treated as *cracks* and must not increase in size during the life of the structure. These initial flaws could be nicks and dings or material anomalies resulting from manufacturing, maintenance, corrosion, or accidental damage (but *not* fatigue cracks) and may be divided into two categories: barely detectable and clearly detectable [2]. The barely detectable flaws represent an upper bound to the largest *undetectable* flaw and must not "initiate" a propagating crack within the retirement time of the component. The clearly detectable flaws are those that can be readily located but which will *not* initiate a propagating crack within the given inspection interval [2]. The maximum stress levels associated with these flaw tolerant components would be established with stress–life fatigue tests (i.e., $S–N$ curves) with flawed (but not "cracked") components. In this case, the fatigue life is defined by the *onset* of detectable flaw growth (i.e., crack "initiation" at the preexistent flaw) and would be subjected to statistically based scatter factors.

18.2.2 Turbine Engines

Damage tolerance concepts have been studied for turbine engines since the 1970s [4] and were formally adopted by the USAF for design and management of LCF components in the 1980s along with a comprehensive Engine Structural Integrity Program (ENSIP). A recently updated handbook [5] contains a detailed description of the ENSIP damage tolerant design philosophy and turbine engine inspection requirements. This comprehensive document also contains a summary of ENSIP "lessons learned," along with a detailed description of shortcomings of prior designs that resulted in some engines with poor structural integrity.

Turbine engine applications remain problematic, however, as the extremely high stress levels and operating temperatures have led to development of highly specialized materials that are difficult to analyze and often result in extremely small critical crack sizes. Most success with damage tolerance criteria has been with low-cycle-fatigue components, where, as shown in Figure 18.2 [6], it is possible to couple nondestructive inspection with crack growth calculations to establish reasonable inspection intervals that protect engine integrity from growing damage. Although inspection requirements can be quite demanding for these applications, the resulting detectable crack growth life is a large fraction of total fatigue crack growth life and can often be protected by a rigorous NDE program. [Reference 5 gives typical NDE limits for engine applications, which can be on the order of 0.01 in (0.25 mm).]

The damage tolerance approach has not been as successful in the high-cycle-fatigue arena, however, where the detectable crack growth portion of fatigue life is a much smaller fraction of total life than for LCF (see Figure 18.2). In HCF, for example, small vibratory stresses such as those imposed on a turbine blade can occur at such large frequencies (on the order of 100–10,000 Hz) that it is possible to accumulate many millions of cycles before a crack slowly grows to a detectable size and then suddenly extends to failure in an extremely short time. For example, ENSIP states that engine components should have a HCF life of 10^9 cycles [5]. In contrast, low cycle fatigue lives accumulated over similar periods of operation are

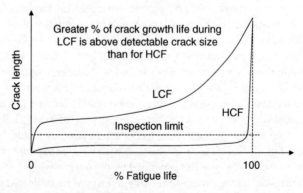

Figure 18.2. Conceptual difference between detectable crack growth portion of LCF and HCF in turbine engine components. (After Reference 6.)

controlled mainly by the start/stop cycle of engine operation (which may be an hour or two in length, resulting in cyclic frequencies of 0.01–1 Hz) and are measured in terms of thousands of cycles. Thus, an LCF inspection interval may take months of service to accumulate and would not be an unreasonable operating burden, whereas a much longer HCF inspection interval, measured in terms of cycles, would expire much more quickly in real time and could be impracticable to implement.

A rash of HCF engine failures during the 1990s focused major research activities to this problem and has led to significant advances in HCF life analysis methods. There are, for example, three primary sources of damage that have been identified as contributing to HCF failures: LCF crack formation, fretting, and foreign object damage [6]. In the first instance, one focuses on traditional fatigue crack formation resulting from large plastic strains during the relatively few LCF cycles, whereas in the second case, attention is directed to fretting action at attachment points, such as the turbine disk/blade dovetail joints, where high contact stresses can lead to crack formation that can then grow by high-frequency far-field stresses [7, 8]. Foreign object damage resulting from ingestion of small particles can also cause notches in turbine blades with equivalent surface crack lengths of 0.03 in (0.8 mm) or fatigue concentration factors K_f on the order of 3.0 [5]. In all of these cases, the existing damage must not suddenly extend by the much smaller stress amplitudes encountered by high-frequency vibrations. Threshold ΔK and endurance limit fatigue concepts are important factors in determining the maximum damage size that will not propagate under subsequent HCF loading [9].

18.2.3 Composite Materials

Composite materials offer many advantages for structural applications, including high strength-to-weight ratios, good resistance to fatigue damage, and the ability to be tailored for specific applications. These materials are obtained by embedding high-strength fibers (e.g., glass, graphite, boron, Kevlar, ceramics) in a relatively ductile matrix (e.g., thermoset or thermoplastic polymers, aluminum, titanium). Structural components are then built up by bonding thin unidirectional lamina into thicker laminates that usually involve cross orientations of the fiber directions. The number of plies and their stacking sequence are key factors that determine the bulk properties of the final composite member.

Thus, composite materials are by their very nature highly anisotropic and subject to different failure modes than those encountered by typical structural metals. They are, for example, highly susceptible to impact damage, delaminations, fiber breakage, matrix failure, and moisture absorption and often have poor compressive strength when damaged. Indeed, development of practical methods to analyze and predict the various failure modes for composite materials, along with their relatively large costs, has slowed their wide use in structural applications. Composite materials must, however, still satisfy damage tolerance and fatigue criteria when used in aircraft structure. FAR Part 23.573, for example, requires the structure to carry "ultimate load with damage up to the threshold of detectability," and "the damage growth, between initial detectability and the value selected for residual strength demonstrations factored to obtain inspection intervals, must allow development of an inspection program suitable for application by operation and maintenance personnel" [10]. Although these requirements are similar to those of metal structures, the unique nature of composite materials requires special analysis and testing methods that are beyond the scope of the present text.

18.3 PREVENTING FAILURE OF THE FAILURE PLAN

The key step in avoiding structural failures is to have a *fracture prevention plan* that anticipates all potential modes of failures and then employs structural design, inspection, and maintenance concepts that will preclude catastrophic failure. The damage tolerance principles discussed in this text provide an orderly framework to ensure that fracture resistant structures are designed and periodically inspected to ensure that preexistent or service-induced damage does not lead to failure. In spite of these formal fracture control plans, accidents, although rare, still happen. When such failures do occur, it is essential to determine where the plan broke down in its inability to preserve the integrity of the structure. Such malfunctions can occur from two sources: failure to properly implement the fracture control plan or failure caused by some issue not anticipated by the original plan.

Perhaps the most common source of *failure of the failure prevention plan* is inadequate execution of the prescribed procedures, either by negligence or by ignorance. The May 25, 1979, crash of a DC-10 airliner that killed 273 people in Chicago, for example, resulted when an unapproved procedure for removing the engine–pylon assembly caused a 10-in-long crack in the pylon bulkhead that led to the engine and pylon separating from the wing on take-off [11]. Although this improper maintenance procedure was common practice, existing oversight procedures failed to prevent its use and led to the worst U.S. aviation disaster at the time. In a similar vein, the largest civil engineering structural failure in U.S. history occurred when a suspended walkway at the Kansas City Hyatt Regency Hotel collapsed on July 17, 1981, killing 114 people and causing 200 injuries [12]. Again, this catastrophe resulted from failure to follow existing building codes and a last-minute change in an attachment detail that was improperly approved without analysis. Clearly, failure to follow prescribed procedures can have dire consequences on structural integrity.

The second type of "failure plan failure" results from a deficiency in the plan itself and is perhaps even more insidious. The 1965 rocket motor case and 1969 F-111 aircraft failures discussed in Chapter 1 (see Figures 1.2 and 1.4) are two such examples where existing design codes in the 1950s and 1960s emphasized use of high-strength materials at the expense of ductility. The problem in those cases was that the need to protect structures from initial manufacturing damage was not fully appreciated or understood by designers and their regulating authorities (i.e., the damage tolerant design principles that are the subject of this book were not yet developed).

A more recent example of a new failure mode not adequately covered by existing design criteria is the infamous 1988 Aloha Airlines incident also discussed in Chapter 1 (see Figure 1.6). In that case, widespread fatigue damage that developed in the older aircraft defeated the crack arrestment mechanisms built into the structure (recall

Section 17.3) and led to loss of a large segment of fuselage skin. The possibility for repeated incidents of this type was deemed so serious that FAR 25.571 was updated in 1998 to account for this potential degradation of structural integrity unique to aging aircraft.

Another related example from the civil engineering arena occurred on December 13, 2000, with the sudden fracture of two out of three steel girders that supported a 200-ft (60-m) section of a major interstate highway bridge in Milwaukee, Wisconsin. These fractures occurred without the presence of fatigue cracking and left the bridge in near collapse, requiring it to be dismantled [13]. This *brittle fracture* of normally ductile material was attributed in part to a unique bracing configuration that led to a *highly constrained* joint that prevented out-of-plane deformations and reduced ductility (recall the influence of constraint on fracture toughness discussed in Section 6.5). Since this failure mode had not been observed before, subsequent analyses focused on determining how common this type of bridge construction was and whether it should be prevented in the future. Thus, both this incident and the Aloha Airlines failure demonstrate the need to anticipate "out-of-the-box" failure scenarios not covered by existing design criteria. This challenge is especially important when considering the adoption of new technology (e.g., advanced materials, manufacturing methods, or inspection techniques) for structural design.

18.4   FINAL EXHORTATION

As the examples in the preceding section indicate, guaranteeing structural integrity requires continuous attention to detail and constant inquiry into the possibility for new and unforeseen failure modes. One must have a design and maintenance plan that anticipates all potential failure mechanisms along with the discipline to see that the plan is executed in complete detail throughout the operational life of the structure. This challenge can be especially difficult for older structures that have changed operators and/or are maintained by new personnel not familiar with the original failure prevention procedures.

Ensuring structural integrity is a vital responsibility, however, that must be approached with extreme caution and the utmost diligence. Technical competence backed by rigorous management attention and oversight is essential to success. Indeed, the engineer charged with designing and maintaining high-performance structures accepts a public trust that must be honored at all costs. There is no room for shortcuts or casual efforts when evaluating structural integrity. It is hoped that students of this volume understand and accept these challenges and will continuously develop and improve their technical and management skills as they strive to protect the public from structural failures.

REFERENCES

1. Anonymous, "Fatigue Evaluation of Structures," in *Federal Aviation Administration Airworthiness Standards: Transport Category Rotorcraft*, Paragraphs FAR 29.571, 2002.

2. D. E. Tritsch and D. O. Adams, "Practical Application of Damage Tolerance and Flaw Tolerance to the Design and Management of Rotor Structures," paper presented at the American Helicopter Society 56th Annual Forum, Virginia Beach, Virginia, May 2–4, 2000.

3. J. D. Cronkhite, C. Rousseau, C. Harrison, D. Tritsch, and W. Weiss, "Research on Practical Damage Tolerance Methods for Rotorcraft Structures," paper presented at the American Helicopter Society 56th Annual Forum, Virginia Beach, Virginia, May 2–4, 2000.

4. W. D. Cowie, "Turbine Engine Structural Integrity Program (ENSIP)," *Journal of Aircraft*, Vol. 12, No. 4, pp. 366–369, April 1975.

5. Anonymous, *Engine Structural Integrity Program (ENSIP)*, MIL-HDBK-1783B, February 15, 2002.

6. T. Nicholas, "Critical Issues in High Cycle Fatigue," *International Journal of Fatigue*, Vol. 21, pp. S221–S231, 1999.

7. M. P. Szolwinski and T. N. Farris, "Mechanics of Fretting Fatigue Crack Formation," *Wear*, Vol. 198, pp. 93–107, 1996.

8. M. P. Szolwinski, J. F. Matlik, and T. N. Farris, "Effects of HCF Loading on Fretting Fatigue Crack Nucleation," *International Journal of Fatigue*, Vol. 21, pp. 671–677, 1999.

9. P. J. Golden, B. Bartha, A. F. Grandt, Jr., and T. Nicholas, "Fatigue Crack Growth Threshold of Fretting Induced Cracks in Ti-6Al-4V," in *Proceedings of Sixth National Turbine Engine HCF Conference*, Jacksonville, FL, March 5–8, 2001.

10. Anonymous, "Damage Tolerance and Fatigue Evaluation of Structure," in *Federal Aviation Administration Airworthiness Standards: Part 23, Normal Utility, Acrobatic, and Commuter Category Airplanes*, Paragraph FAR 23.573, 2002.

11. "Airlines Pay Fines for DC-10 Errors," Associated Press, *The Champaign-Urbana News-Gazette*, November 17, 1979, p. C-1.

12. H. Petroski, *To Engineer Is Human: The Role of Failure in Successful Design*, St. Martin's Press, New York, 1985.

13. J. W. Fisher, E. J. Kaufmann, W. Wright, Z. Xi, H. Tjiang, B. Sivakumar, and W. Edberg, *Hoan Bridge Forensic Investigation Failure Analysis Final Report*, for the Wisconsin Department of Transportation and Federal Highway Administration, June 2001.

APPENDIX: AFGROW USERS GUIDE

JAMES A. HARTER

A.1 INTRODUCTION

This appendix is intended to serve as a basic users guide for the crack growth life prediction program, AFGROW, developed by the U.S. Air Force Research Laboratory, Air Vehicles Directorate. This public domain software (Version 4.0008.12.11) may be obtained from the John Wiley & Sons web site:

www.wiley.com/go/grandt

The latest version may also be obtained from the official AFGROW web site: http://fibec.flight.wpafb.af.mil/fibec/afgrow.html.

AFGROW is a very comprehensive crack growth program based on linear elastic fracture mechanics (LEFM) principles. The most common operations and functions of AFGROW are covered in this guide. Additional information is also available through the on-line help that can be accessed by pressing the F1 key while AFGROW is running.

A.2 AFGROW INSTALLATION

A.2.1 Installation Process

AFGROW uses the Install Shield program to generate the installation program required to copy and register the required program files to an individual PC. If the single-file method is used, the dialog in Figure A.1 appears.

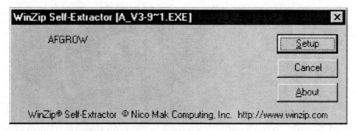

Figure A.1. WinZip self-extractor dialog.

The installation procedure starts when the user selects the setup button in the dialog box. Several dialog windows will appear during the installation process and users should simply follow the instructions provided until the installation process is complete. The default directory path is C:\Program Files\AFGROW. This is the recommended path, but it may be changed as desired. Users should never attempt to install multiple copies and/or multiple versions of AFGROW on a single computer.

A.2.2 Uninstalling AFGROW for Windows

AFGROW is a fairly complex code that includes several files, libraries, and registrations that need to be removed properly before installing a new version (or simply to clear AFGROW out of a computer). The proper way to remove AFGROW is to use the Add/Remove Programs dialog in the Windows control panel. This procedure should also be followed before upgrading to a newer version of AFGROW. Any files that have been saved by a user for any version of AFGROW will not be removed as a result of this process

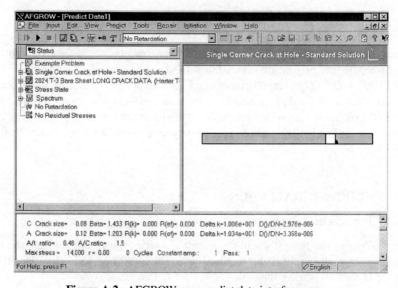

Figure A.2. AFGROW user predict data interface screen.

A.3 INTERFACE FEATURES

The AFGROW user interface is divided into three frames, as shown in Figure A.2. *Note:* The frames are resized by clicking on a frame boundary and dragging it to the desired position.

A.3.1 Main Frame

The main frame appears on the upper left-hand side of the window and is used as the workhorse frame of AFGROW. The main frame has several functions that may be accessed by the user, as indicated in Figure A.3. Information on the use of the main-frame functions is available on the on-line help (view menu) in AFGROW.

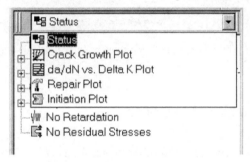

Figure A.3. Main frame status window.

A.3.2 Animation Frame

The upper right-hand frame is referred to as the animation frame since it shows a drawing of the crack plane (*AFGROW assumes planar crack growth*) and presents an animation of crack growth during the prediction process. The specimen view may be zoomed by dragging an area with the mouse (or using the view, zoom menu option) enlarged or diminished by simply resizing the animation frame (Figure A.4).

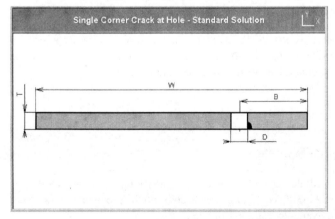

Figure A.4. Single corner crack at hole animation frame.

A.3.3 Output Frame

The lower frame is referred to as the output frame, since it is the default location for the results of the life analyses (Figure A.5). The output data consist of crack length, beta values, stress ratio, stress intensity, crack growth rate, and spectrum data.

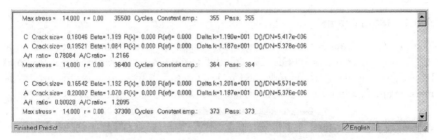

Figure A.5. Output frame showing typical data display.

A.4 AFGROW MENU SELECTIONS

The menu bar provides access to all of the features of AFGROW (Figure A.6). Many of the features available through the AFGROW menu bar are common to most Windows programs. Information on the most frequently used features that are unique to AFGROW is presented in the following sections.

Figure A.6. Menu bar showing access to AFGROW features.

A.4.1 Input Title

The title option is provided as a documentation tool. You can enter up to 80 characters in the title line to describe the problem being modeled. Furthermore, an additional 1000 characters may be stored in the comments area.

A.4.2 Input Material

Input material toolbar icon

The Material dialog provides a means for specifying the crack growth material properties used by AFGROW (Figure A.7). Details on the use of any of the material property options are available in the on-line help.

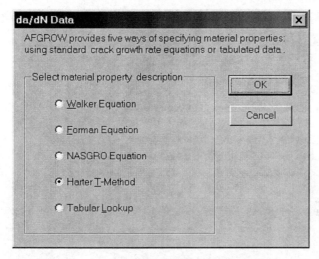

Figure A.7. Material da/dN data dialog window.

A.5 INPUT GEOMETRY AND DIMENSIONS

Input geometry and dimensions toolbar icon

Most crack growth life prediction programs available today are capable of calculating the life for a number of structural geometries with single (or symmetric) cracks. AFGROW has taken a further step that allows users to predict the lives of structures with two independent through-thickness cracks. For the purpose of differentiating these capabilities, single (or symmetric) crack models are being called "classic" models in AFGROW. The two-crack models use a separate, unique user interface.

A.5.1 Classic Geometry Models (Figure A.8)

There are two types of "classic" stress intensity factor solutions available in AF-GROW:

- standard stress intensity solutions and
- weight function stress intensity solutions.

In addition to these solutions, users can input their own solutions through the user input beta option. However, to use this option, the user must first select either the one- or two-dimensional user-defined geometry from the Standard Solutions dialog. The user can also choose one of the standard solutions and apply a beta correction based on the ratio of the actual stress distribution to the standard stress distribution.

The standard crack geometries in AFGROW consist of several models for which closed form or tabular stress intensity factor solutions are available. Solutions for several geometries are built into the code and are referred to as application-defined

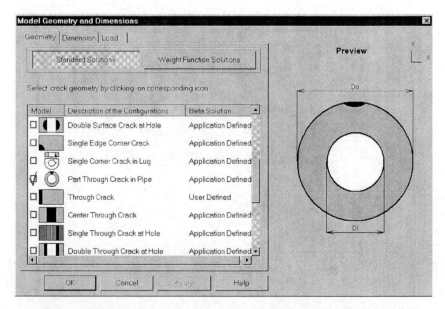

Figure A.8. Model geometry and dimensions interface showing geometry tab.

solutions. Both through-the-thickness (one-dimensional) and part-through (two-dimensional) crack models may be analyzed. AFGROW also allows user-defined stress intensity solutions to be input in the form of beta factors at various crack lengths. The beta factor is defined as $\beta = K/\sigma\sqrt{\pi x}$, where x is the appropriate crack length.

The crack length dimension in the thickness direction is designated as the A-dimension, and the crack length in the width direction is the C-dimension. An angle, ϕ, is used in these equations to determine the stress intensity value for the two-dimensional crack growth dimensions (A and C). Here ϕ is the *parametric angle* defined in Figure A.9. The angle is measured from a line in the c direction beginning at the crack origin. Care was taken in AFGROW to use the angle for each crack dimension that tends to match the published finite element results near the free edges. The default angles used for each crack dimension are documented in the on-line help.

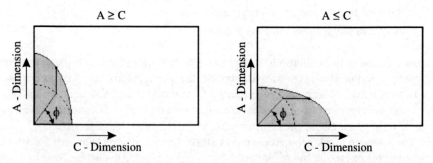

Figure A.9. Definition of crack dimensions a and c and parametric angle ϕ.

A.5.2 Model Dimensions

The options selected in the Dimension dialog allow one to define the particular crack configuration of interest. In the case of part-through flaws, the user may elect for AFGROW to maintain a constant crack shape (check "keep A/C = constant" option) or to allow the A and C crack dimensions to grow naturally. The preview window shows the input dimensions defined by the user when the Apply button is clicked. (See Figure A.10.)

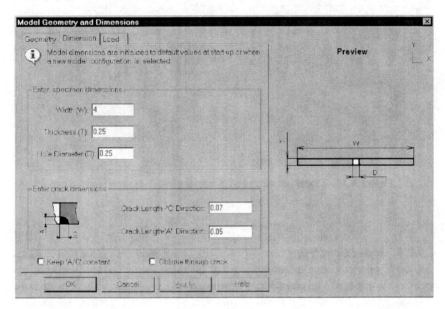

Figure A.10. Model geometry and dimension interface showing dimension tab.

A.5.3 Model Load (Figure A.11)

Since some crack models have multiple-load-case solutions, AFGROW allows the user to combine these solutions using the superposition method. To use this option, the ratio of the tension, bending, or bearing stress to the reference stress must be input for each load case to be considered. AFGROW shows the definition of each type of stress in the Load tab of the Model dialog. The reference stress is simply the product of the spectrum multiplication factor (SMF) and the current spectrum maximum or minimum value (see following section). Since AFGROW uses a single-channel spectrum, the inherent assumption is that each load case is in phase and the load case stress–reference stress ratio is constant. Therefore, the ratio may be determined for any applied reference stress. This approach allows a user to perform parametric studies for any number of stress levels by simply changing the value of SMF in the spectrum dialog. It is, however, up to the user to be aware of the definitions of the reference stress and the load case stress to use this capability correctly. Every attempt is made to identify the definition of the load case stresses.

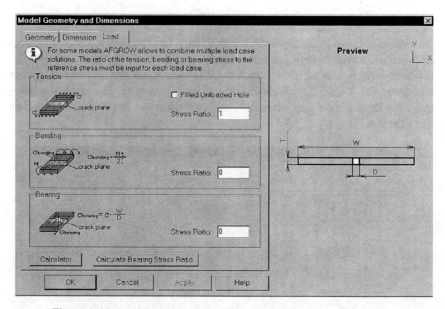

Figure A.11. Model geometry and dimension interface showing load tab.

A.5.3.1 Input Spectrum

Input spectrum toolbar icon

AFGROW operates on the assumption that all input stress (or load) spectra have been "cycle counted" to ensure that maximum and minimum values in each line of the spectrum file actually define a complete cycle (recall Section 9.2.6). Crack growth occurs as applied stress (or load) values change as a function of time. A cycle begins at a certain stress level and is completed when the stress returns to the starting value. AFGROW provides a cycle-counting program, which is accessible through the AFGROW tools menu (additional information is available through the help function provided with that program). Although AFGROW provides a cycle-counting algorithm, the user should be aware that other acceptable counting methods are also available in the literature. AFGROW allows the user to choose from several input options for spectrum data, as indicated below. (See Figure A.12.)

Spectrum Multiplication Factor (SMF) The SMF is multiplied by each maximum and minimum value in the user input stress spectrum. Both normalized (maximum value 1) and nonnormalized spectra may be used. The SMF value simply makes it easy to increase and decrease the severity of an input spectrum.

Residual Strength Requirement (P_{xx}) Here P_{xx} is simply the maximum value of stress (or load for models using load instead of stress input) that the structure *must* be able to carry at *all* crack sizes. It is *not* multiplied by the SMF. The residual strength

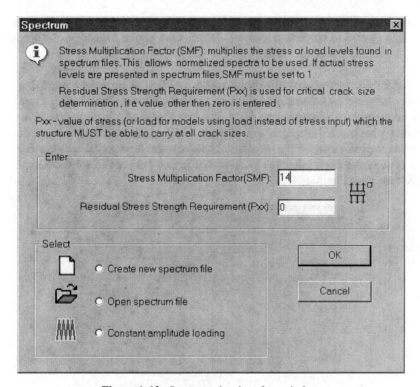

Figure A.12. Spectrum data interface window.

input is very useful for cases in which one does not know when the maximum stress (or load) will occur but wishes to check for failure at this condition at all times. If this value is set to zero (default), failure will occur based on the current applied stress (or load). If P_{xx} is not zero, failure occurs when the crack length reaches critical conditions for the P_{xx} load.

Create New Spectrum File   This option opens the Spectrum wizard, which guides the user through several steps to create a spectrum by manually entering the spectrum data. The number of lines of spectrum data is limited to 25 for each subspectra entered manually (see the on-line help). This is simply a practical limit since most users would not want to manually enter huge amounts of data in order to avoid the possibility for data entry errors. There is also an option in this wizard to read spectrum and/or subspectrum data from user-supplied text files. In this way, users can combine spectrum data as required to create a larger complete spectrum.

Open Spectrum File   This option opens the Windows standard Open File dialog. The Open File dialog will initially look in the AFGROW directory by default, but the spectrum files may be located in any directory. AFGROW will remember the directory where the last spectrum was located and will look in that directory the next time this option is used. The user may select any previously created AFGROW spectrum for use in a given life prediction analysis. Information on the format for the spectrum files is given in the on-line help.

A.5.3.2 Constant-Amplitude Loading   AFGROW provides a method to generate a simple constant-amplitude loading spectrum. The stress ratio (R) is the ratio of the minimum to maximum stress level. The maximum value in the spectrum is automatically normalized (set equal to 1.0), so the SMF value *must* be set to be the desired maximum peak stress for the spectrum. The block size is used to determine the number of constant-amplitude cycles considered in one pass of the constant-amplitude spectrum. (See Figure A.13.)

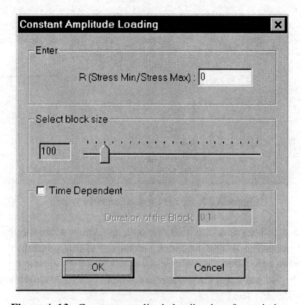

Figure A.13. Constant-amplitude loading interface window.

A.6   INPUT RETARDATION   Input retardation toolbar icon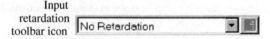

Crack growth retardation models are used to account for potential load spectrum sequence effects on fatigue crack growth. In general, sequences that consist of large stresses followed by lower stresses will result in longer lives than when the sequence is reversed by applying the lower stresses first (recall Section 7.5). There are currently four choices of load interaction (crack retardation) models in AFGROW. The models can be accessed through either the standard menu dialog or by using the pull-down menu located on the toolbar. (See Figure A.14.)

Note that each model has a user-adjustable parameter to tune the retardation model to fit actual test data. Ideally, these parameters should be material constants that are independent of other variables such as spectrum sequence or load level. Some

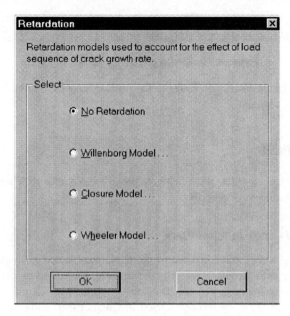

Figure A.14. Retardation menu dialog window.

models seem to work better than others in this regard, but there is need to reproduce experimental results for various types of retardation models. These load interaction models are provided at the user's discretion and responsibility. The details of the no-retardation, generalized Willenborg, closure, and Wheeler models are given in the on-line help.

A.7 PREDICT PREFERENCES

The preferences menu selection is one of the most important menu items in AFGROW. There are several optional settings which may be changed to suit the various require-ments of a given life prediction. The preferences are divided into five categories and are accessible through a tabbed dialog box, as shown in Figure A.15. The preferences dialog is accessible through the AFGROW menu *or* by right clicking anywhere in the output frame. The user sets the preference options with the buttons shown in Figure A.16. Use the Save button to save all parameter settings. These settings will be retained until changed by the user. The Default button will return the original

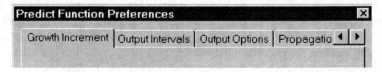

Figure A.15. Predict function preferences tabbed dialog box.

Figure A.16. Predict preferences option buttons.

AFGROW preference settings. Refer to the on-line help for a complete description of the available options.

A.8 EXAMPLES

Two short examples are described below to provide the reader with benchmark problems to aid in their use of AFGROW.

A.8.1 Edge Crack in Wide Plate

Consider a single edge crack in a semi-infinite plate subjected to constant-amplitude tension loading with fatigue crack growth properties described by the Paris law. An exact closed form solution for this problem is obtained in Chapter 3 and is given by Equation 3.23. That solution was used in Chapter 4 to solve Example 4.1 and will be repeated here with AFGROW. For purposes of this simulation, let the initial crack length be 0.1 in and the plate width W be 10,000 in (i.e., $a/W = 0.0001$ to approximate a semi-infinite plate width). Let the remotely applied $R = 0$ stress be 25 ksi, the fracture toughness be 50 ksi-in$^{1/2}$, and the material fatigue crack growth properties be given by $da/dN = 3.9 \times 10^{-10} (\Delta K)^{4.175}$.

To set up the problem in AFGROW, click the Classic Models icon on the tool bar to bring up the Model Geometry and Dimensions menu. Put a check in the Single Edge Through Crack box. Then click the Dimension tab and enter 10,000 for the width and 0.1 for crack length c direction. Note this tab also asks for a specimen thickness, but since this is a two-dimensional problem and loads will be input in terms of stress, any value can be input here. Now press the Load tab and make sure that the tension stress ratio 1 is selected. Recall here that the stress ratio is the combination of tension, bending, and bearing loads that will be superimposed to obtain the stress intensity factor solution. Since the loading in this case is all tension, the tension stress ratio is 1 and the bending and bearing ratios are 0. Now click OK and the menu closes.

Next, select the Material Data icon to input material properties. Although the Paris law is not one of the options listed, it may be effectively specified by appropriate choice of the Walker equation properties (press F1 to obtain a help description of the Walker equation). Select the Walker equation, click OK, and then enter values on the Walker Equation Data menu. Enter 3.9e-10 for C, 4.175 for n, and 1 for m. This latter choice for m will reduce the Walker equation to the Paris law. Now enter 50 for fracture toughness K_c and the plane strain fracture toughness K_{1c}. The other inputs on this screen are not used for this problem and may be left at their default values. Check OK to close the material window.

The final step is to specify the cyclic load by clicking the Spectrum icon. Enter 25 in the Stress Multiplication Factor box and check the constant-amplitude box (the

P_{xx} entry should be zero for this case). Check OK to bring up the Constant Amplitude Loading menu. Enter $R = 0$ (here R is the conventional stress ratio, which equals the minimum/maximum stress per cycle) and select a block size (the cyclic interval used to print crack length data). A value of 100 is reasonable for this case. Checking OK closes the spectrum input dialog. The problem is now ready to run by simply clicking the Run button (black triangle) on the tool bar. Checking the output frame for the solution data gives the final crack size $c = 1.0172$ that occurs after 2296 cycles, along with other details of the crack growth prediction. Note that this AFGROW solution compares very well with the "exact" solution given by Equation 3.23.

A.8.2 Corner Crack at Open Hole

For a slightly more complex example, consider a corner-cracked hole in a tension plate. Let the plate width be 4 in, the thickness be 0.5 in, the hole diameter be 0.5 in, and the initial crack be a 0.05-in quarter-circular crack located at the hole bore. Let the plate be made from 2024-T3 aluminum plate (L-T direction) and be subjected to a constant-amplitude $R = 0.3$ peak stress of 30 ksi.

As before, begin by going to the Classic Models icon to bring up the Model Geometry and Dimensions window. Check the Single Corner Crack at Hole box on the Geometry tab, and then from the Dimension tab, enter the 4.0 width W, 0.5 thickness T, and 0.5 hole diameter D. The initial quarter-circular corner crack dimensions $a = c = 0.05$ are also entered here. Since we want the crack shape to develop naturally, do *not* check the "keep A/C constant" box. Again select tension stress ratio 1 from the Load tab and close the Models menu by clicking OK.

Now specify material properties by using values given in the AFGROW library. Click the Materials Data icon, and in this case select the NASGRO option and press OK to open the Material—NASGRO Equation window. Now click on the Read button in the lower right-hand corner and select the Open Material Data Base option to bring up the Material-NASGRO Equation window. In succession select the 1000–9000 series aluminum, 2000 series, 2024-T3 Al, and finally the Clad; plt & sht; L-T options. Click OK at the top to close the Material Data Base browser and return to the Material-NASGRO Equation screen. Clicking OK here completes input of the material property data. For further information on the NASGRO materials database, click the F1 help button when the NASGRO window is open. (Note that the Harter-T Method option in the da/dN Data window also contains an alternate library of material input data.)

The applied stress history is again specified by going to the spectrum window, selecting 30 for the stress multiplication factor, selecting the constant-amplitude loading box, and pressing OK to bring up the Constant Amplitude Loading screen. Set the stress ratio to 0.3, pick the 100-cycle block size as before, and click OK to close the spectrum window. Finally, starting the AFGROW run by clicking the black triangle run button on the tool bar leads to the fatigue crack growth data shown. Note in the output frame that fracture occurs after 2597 cycles with a final through-the-thickness crack length $c = 0.66896$. Examining the output data more closely, one can determine the growth of the a and c crack dimensions as a function of elapsed cycles. In this case the corner-crack dimension a penetrated the wall thickness to transition into a through crack at 2190 cycles when $a = 0.47532$ and $c = 0.38934$ ($a/T = 0.95064$ and $a/c = 1.2208$).

A.9 CONCLUDING REMARKS

The goal of this appendix is to give a brief overview of the AFGROW fatigue crack growth prediction software. Although not intended to be a detailed instruction manual, it is hoped that this description will allow one to obtain and install this public domain software on a personal computer. The reader is encouraged to explore, with the aid of the on-line help provided by pressing the F1 key, the various capabilities of this powerful damage tolerance analysis tool. Several homework problems in Chapters 7, 16, and 17 have been formulated to use the AFGROW software.

INDEX

Lightning Source UK Ltd.
Milton Keynes UK
UKOW03n1840031013

218436UK00001B/44/P